ANIMAL
MODELS OF
OBESITY

ANIMAL MODELS OF OBESITY

Edited by

MICHAEL F.W. FESTING,
B.Sc., M.Sc., Ph.D., M.I.S., M.I.Biol.

*Medical Research Council Laboratory Animals Centre,
Carshalton, Surrey, U.K.*

MRC Laboratory Animals Centre Symposium Number 1

First published 1979 by
THE MACMILLAN PRESS LTD
London and Basingstoke
Associated companies in Delhi Dublin
Hong Kong Johannesburg Lagos Melbourne
New York Singapore and Tokyo

Typeset by Reproduction Drawings Ltd, Sutton, Surrey

British Library Cataloguing in Publication Data

Animal models of obesity
 1. Obesity   2. Laboratory animals—Diseases
 I. Festing, Michael Francis Wogan   II. Laboratory
 Animals Centre
 616. 3'98'07   RC628

 ISBN 978-1-349-04203-6   ISBN 978-1-349-04201-2 (eBook)
 DOI 10.1007/978-1-349-04201-2

Symposium Contributors

Margaret Ashwell, Division of Clinical Investigation, Clinical Research Centre, Watford Road, Harrow, Middlesex HA1 3UJ

Anne Beloff Chain, Department of Biochemistry, Imperial College of Science and Technology, Imperial Institute Road, London SW7

M. A. Cawthorne, Beecham Pharmaceuticals, Research Division, Nutritional Research Centre, Walton Oaks, Dorking Road, Tadworth, Surrey KT20 7NT

Amanda Donaldson, Department of Physiology, St. George's Hospital Medical School, Cranmer Terrace, London SW17 0RE

J. A. Edwardson, Department of Physiology, St. George's Hospital Medical School, Cranmer Terrace, London SW17 0RE

M. F. W. Festing, Medical Research Council Laboratory Animals Centre, Woodmansterne Road, Carshalton, Surrey SM5 4EF

J. S. Garrow, Division of Clinical Investigation, Clinical Research Centre, Watford Road, Harrow, Middlesex HA1 3UJ

D. A. Hems, Department of Biochemistry, St. George's Hospital Medical School, Blackshaw Road, London SW17 0QT

W. P. T. James, Assistant Director, MRC Dunn Nutrition Unit, Dunn Nutrition Laboratory, Milton Road, Cambridge CB4 1XJ

D. P. Lovell, Medical Research Council Laboratory Animals Centre, Woodmansterne Road, Carshalton, Surrey SM5 4EF

J. C. McCarthy, Dean, Faculty of Agriculture, University College Dublin, Glasnevin, Dublin 9, Republic of Ireland

C. J. Meade, Transplantation Biology Section, Clinical Research Centre, Watford Road, Harrow, Middlesex HA1 3UJ (Present address: Lilly Research Centre Ltd, Earlwood Manor, Windlesham, Surrey, UK)

D. S. Miller, Department of Nutrition, Queen Elizabeth College, University of London, Campden Hill Road, London W8

Nancy J. Rothwell, Department of Physiology, Queen Elizabeth College, University of London, Campden Hill Road, London W8

Jinan Sheena, Transplantation Biology Section, Clinical Research Centre, Watford Road, Harrow, Middlesex HA1 3UJ

M. J. Stock, Department of Physiology, Queen Elizabeth College, University of London, Campden Hill Road, London W8

P. Trayhurn, MRC Dunn Nutrition Unit, Dunn Nutrition Laboratory, Milton Road, Cambridge CB4 1XJ

D. A. York, Department of Physiology and Biochemistry, The University of Southampton, Medical and Biological Sciences Building, Basset Crescent East, Southampton SO9 3TU

Contents

Acknowledgements

This volume is based on the proceedings of a symposium organised by the Medical Research Council Laboratory Animals Centre, and held at the Zoological Society, Regent's Park, on 16th and 17th February 1978. I should like to thank John Bleby, D. A. York, M. A. Cawthorne and J. Rivers for chairing the four sessions. Thanks are also due to Drs J. A. Turton and M. A. Cawthorne for playing an important part in planning the scientific content of the meeting, and Colin V. Clark for his hard work in organising the practical aspects of the programme.

Introduction

Obesity is rarely recorded as the cause of death in humans, yet it is associated with a significant reduction in average lifespan, which may be quantified statistically, and it is also associated with an unquantifiable reduction in the quality of life of those who are substantially overweight. Moreover, it is clearly becoming an increasingly serious problem in developed countries.

According to data given by Brackenridge (1977), a 30 year old person 180 cm (5ft 11in) high weighs on average 79 kg (174 lb) and has a life expectancy of about 44 years. Mild obesity (103 kg, 227 lb) is associated with a reduction in life expectation of about 2 years, whereas severe obesity (143-151 kg, 315-333 lb) leads to a reduction in life expectation of 9-12 years.

Much of the extra mortality is associated with diabetes, cirrhosis of the liver and diseases of the gall bladder, and it is clear that obesity may be a symptom of underlying metabolic abnormalities rather than the cause. However, the mortality from accidents, cardiovascular disease, chronic nephritis, cerebral haemorrhage and even cancer are also increased, and there must be a strong suspicion that at least some of the extra mortality is due to the obesity, as such.

The reduction in the quality of life is impossible to quantify. In fact, there is a consistently low mortality ratio from suicide, which might imply that obese people tend to be happy. On the other hand, in Western culture at least, there is a strong association between slimness and beauty. The success of low-calorie foods and health farms, and the very large number of people who attempt to diet regularly, suggest that many people would rather be thinner than they are at present, and this is probably largely associated with a desire to look and feel better rather than solely a desire to prolong their life.

Unfortunately, in spite of years of research, very little is known about the way in which the human body controls the deposition of fat. No satisfactory method has yet been discovered which will allow an individual to control his weight and at the same time eat whatever he wants, even though a few lucky individuals never seem to have any difficulty in controlling their weight. As society becomes more affluent, the problem becomes more acute. Research on obese humans will probably answer some of the present queries. However, research on humans has obvious limitations. The fact that many laboratory animals also become obese, either spontaneously on a standard diet or as a result of dietary or other manipulation, opens up new areas for fundamental research. Although these animal models cannot be expected to be exact models of obesity in humans, they may still be of great value in studying the biochemical, physiological and pathological conditions necessary to the deposition of excess fat. In time such models should lead to a better understanding of the regulation of obesity, and thereby give some guidance as to the best way to treat obesity in humans.

Carshalton, Surrey, 1979 M.F.W.F.

REFERENCE

Brackenridge, R. D. C. (1977). *Medical Selection of Life Risks,* Undershaft Press, London

1
Normal variation
in body fat
and its inheritance

J. C. McCarthy
(Faculty of Agriculture, University College, Dublin, Ireland)

SUMMARY

The major generalisations about the inheritance of fatness in 'normal' mice are:

(1) Mice selected for high growth rate become very fat as adults.

(2) There are heritable differences between strains of mice in their propensity for fattening at the same stage of development.

The possible contribution of changes in metabolic and behavioural traits to these two types of heritable variation is discussed. It would seem that differences of the first type result from an excessive appetite and a relatively lower mainten-ance cost. The causation of differences among strains in their ability to partition energy between fat and lean tissues at the same stage of development is less well understood but such differences can be generated by different types of selection for growth rate.

INTRODUCTION

In an outbreeding population of animals the variation in any quantitative trait such as the weight or percentage of fat can be partitioned into components attributable to genetic and environmental factors. Quantitative geneticists are particularly inter-ested in the 'additive' genetic portion, which is the variance among individuals in a population due to differences in 'breeding values' (for details, see Falconer, 1960a, 1967). The ratio of the additive genetic variance to the total phenotypic variance is the 'heritability' of a trait. There are several published estimates of this genetic parameter for measures of fatness in farm animals. They range from about 0.4 to 0.5, indicating that about a half of the total variance in measures of fatness is due to genetic differences between animals. There are no comparable estimates of the heritability of fatness in laboratory animals, but, as we shall see, there is no reason to suspect that the underlying genetics of fat deposition is much different in these various species.

1

J. C. McCarthy

The overwhelming majority of quantitative studies of fat deposition in laboratory animals have been conducted with lines of mice which have been selected for increased or decreased weight at a fixed age or for a fast or slow rate of growth between two ages. (Such lines are referred to here simply as 'High' and 'Low', respectively.) The changes that occur in traits other than those selected for directly (body weight or weight gain) are called 'correlated responses'. It is these correlated responses in measures of fatness that occupy most of this review, which is confined almost exclusively to evidence from mouse work − in contrast to two recent reviews, which also dealt with this subject in other species of laboratory and farm animals (McCarthy, 1977; Roberts, 1978).

The typical pattern of fat deposition in a mouse is shown in figure 1.1. The amount of fat in the body increases as it grows and fat forms an increasing proportion of the total weight of the body. Most fat is deposited in identifiable adipose depots which grow at different rates as the animal ages (figure 1.2). There is a high correlation between the total weight of these depots and the weight of fat which can be extracted chemically from the whole body at lower and medium body weights in particular (figure 1.3). The correlation breaks down at very high body weights for reasons discussed by Allen (1978).

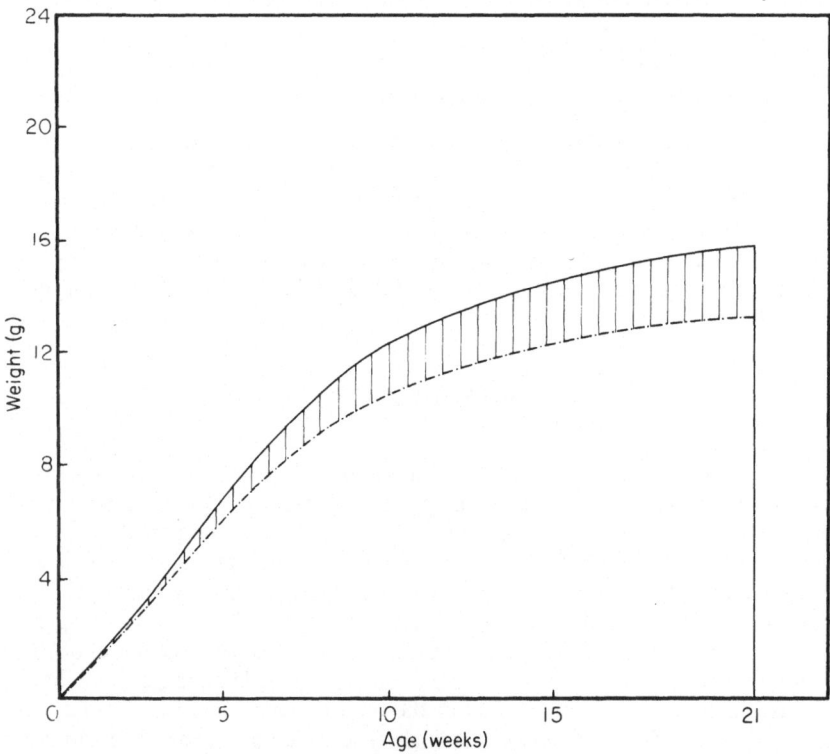

Figure 1.1   The typical pattern of changes in the weight of fat (shaded area) and the weight of other tissues (unshaded area) in the carcass of the growing mouse. Adapted from Hayes and McCarthy (1976).

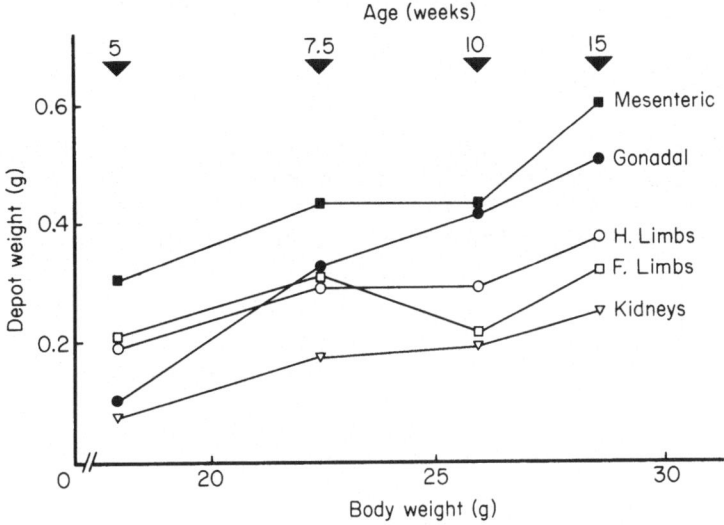

Figure 1.2 The typical pattern of changes in the weight of different adipose depots in the growing mouse. Adapted from Allen (1978).

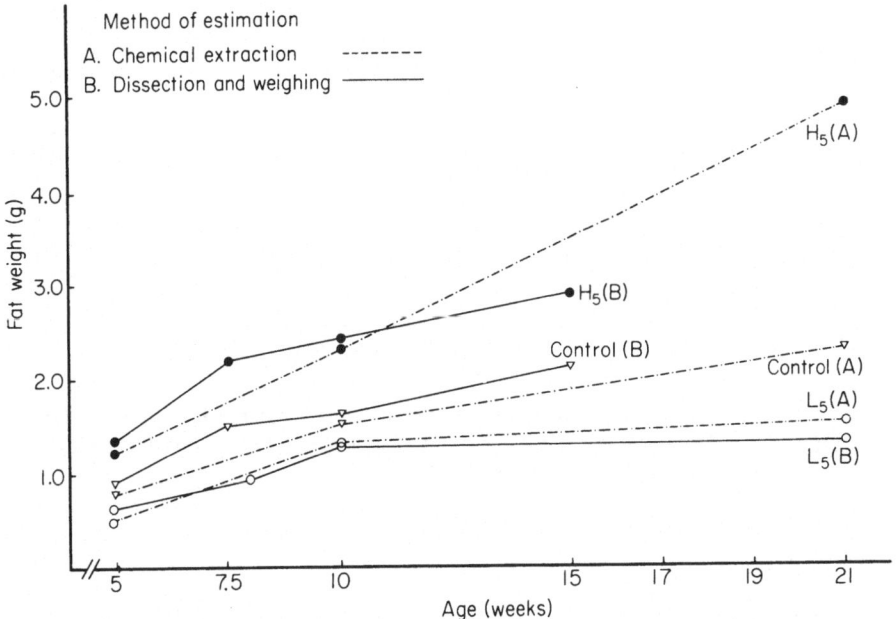

Figure 1.3 The pattern of changes at successive ages in estimates of the total weight of fat in lines of mice selected for High and Low weight at 5 weeks of age (H_5 and L_5) and for an unselected line (control) obtained by chemical extraction (Hayes, 1974) and dissection of adipose depots (Allen, 1978).

J. C. McCarthy

RESULTS OF SELECTION FOR BODY WEIGHT

Growth rate

Selection for high and low weight in an outbred population is extremely effect-ive in changing the mean body weight and the responses to selection continue for many generations. The resulting lines selected for High and Low body weight at 5 or at 10 weeks of age by the author are shown in figure 1.4.

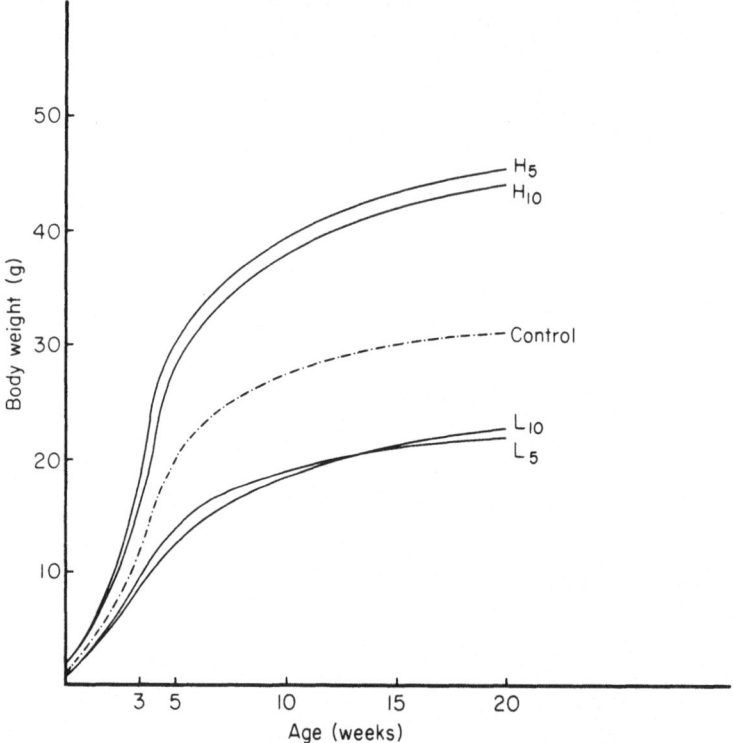

Figure 1.4 The average growth curves of lines of mice selected for High and Low weight at 5 or 10 weeks of age (H_5, L_5 and H_{10}, L_{10}) and of an unselected line (control). From Hayes and McCarthy (1976).

Fat deposition

The variation among lines in the pattern of fat deposition as animals get older is well illustrated by the results of Hayes and McCarthy (1976), who compared their High and Low lines at 5, 10 and 21 weeks of age (figure 1.5). While the lines did not differ in fatness at 5 weeks of age, it is clear that the High lines gradually became fatter and the Low ones leaner as they increased in age.

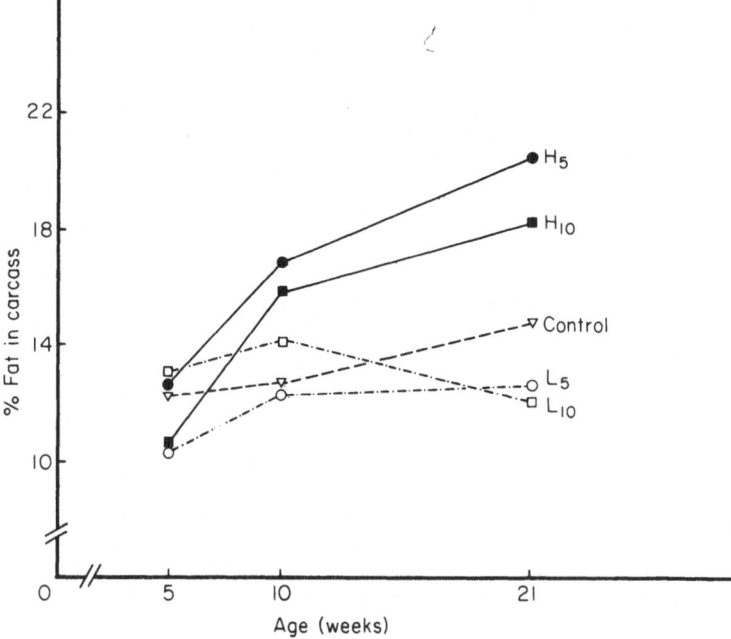

Figure 1.5   The pattern of changes at successive ages in the percentage fat in the carcass in High and Low lines described by Hayes and McCarthy (1976).

The effect of selection for body weight on fatness has been similar in many other experiments. High lines of mice were fatter than unselected ones at the age of selection, or at later ages in experiments of Fowler (1958), Biondini *et al.* (1968), Bakker (1974), Sutherland *et al.* (1974), McPhee and Neill (1976) and Eisen and Bandy (1977). In one particular case, in which a High line was compared with an unselected one at a young age, no difference in fatness was found (Lang and Legates, 1969). However, it is possible that this line *may* have become fatter at later ages. There have been relatively few other experiments in which Low lines have been compared with unselected ones. Fowler (1958) and Bakker (1974) recorded small decreases in percentage fat in Low lines relative to unselected ones at a series of ages.

Food consumption

The patterns of food consumption for the High and Low lines studied by Hayes and McCarthy are shown in figure 1.6. It is obvious that the rate of food consumption changed in a highly predictable way on selection for body weight. High lines ate more and Low ones less. The level of intake settled down at about 6 weeks of age in most lines and stayed about this level henceforth. At this stage the rate of food consumption in the High lines was about 60 per cent

J. C. McCarthy

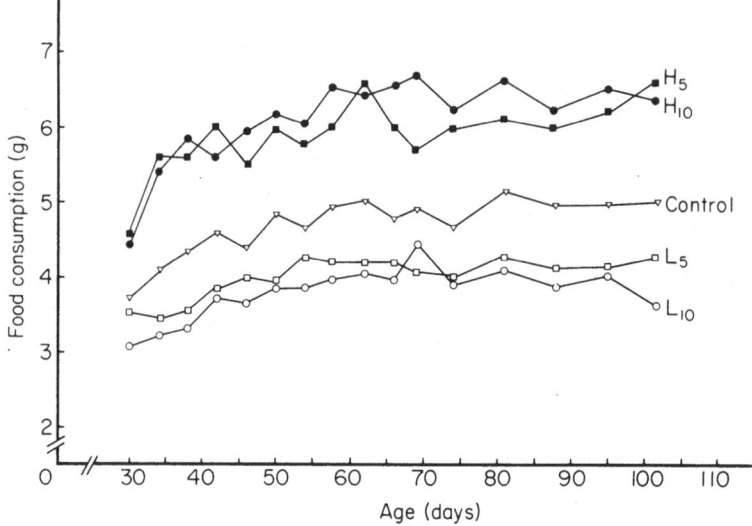

Figure 1.6   The pattern of average daily food consumption at successive ages in the lines described by Hayes and McCarthy (1976).

higher than in the Low ones. (The High lines were about twice as heavy as the Low ones at this age.) Other reports of greater rates of food consumption in High lines relative to their unselected controls have been given by Fowler (1962), Rahnefeld *et al.* (1965), Lang and Legates (1969), Timon and Eisen (1970), Sutherland *et al.* (1970), Stanier and Mount (1972) and Eisen and Bandy (1977).

Nothing has been published about the actual feeding behaviour of selected lines. A preliminary study of this aspect of food consumption in our High and Low lines (Petersen and McCarthy, unpublished) indicates that the main correlated change on selection for body weight is in meal size (table 1.1). The High line mice ate about 90 per cent more per day than Low mice. Both lines consumed the same number of meals on average. The time spent feeding per meal was about 80 per cent greater for High line mice. Thus, the rate of feeding was almost the same in both lines, i.e. about 2.4 mg/s, which is similar to the rate found on the particular diet used with other lines. The differences in meal size are produced by alteration of the number of feeding bouts contained within a meal so that the mean bout length is unchanged — about 5 s.

Table 1.1   Preliminary results of a comparison of the feeding behaviour of High and Low lines of mice (Petersen, unpublished data)

Line	Food consumed (g/day)	Meals per day	Duration of meals (s)
High	4.8	17	120
Low	2.5	17	67

Relative rate of development of fat

To test whether these effects on fatness at fixed ages were reflected in altered patterns of development of adipose tissue, several authors have examined the relationship between the weight of fat and the weight of the whole body or of the carcass. Hayes and McCarthy (1976) found that selection at 5 weeks of age did not alter the relationship but that selection at 10 weeks did (figures 1.7 and 1.8). In the latter case the High line was leaner and the Low line fatter than the control line at low weight and fat was deposited at an increased relative rate in the High line and a decreased relative rate in the Low one. This pattern of 'deferred' development of fat was also observed in other High body weight lines by Fowler (1958), Clarke (1969) and McPhee and Neill (1976). The contrasting pattern of 'enhanced' development of fat in the Low line was observed by Fowler (1958) in two Low body weight lines. These lines also had a lowered relative rate of fat deposition.

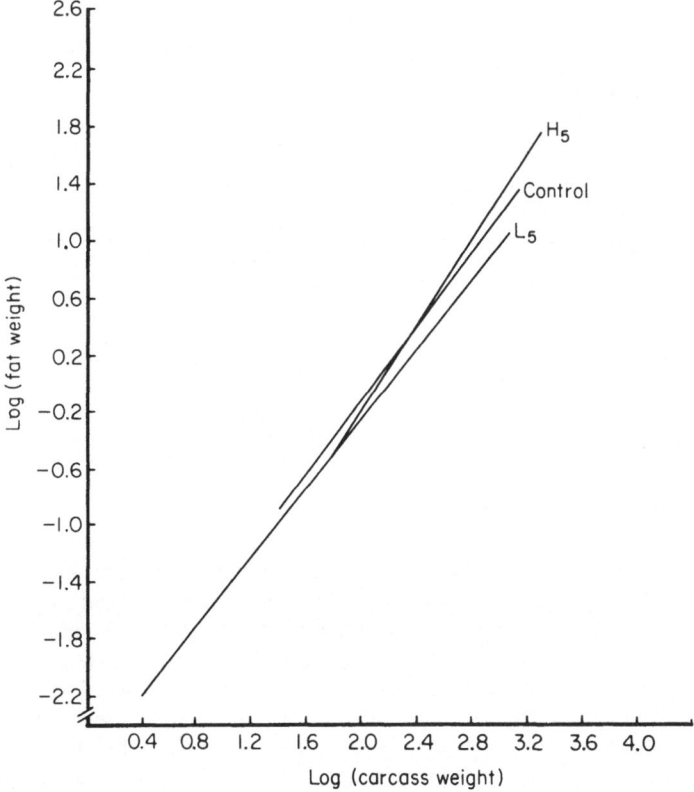

Figure 1.7   The relationship between log (fat weight) and log (carcass weight) in lines selected at 5 weeks of age for High (H_5) and Low (L_5) body weight. From Hayes and McCarthy (1976).

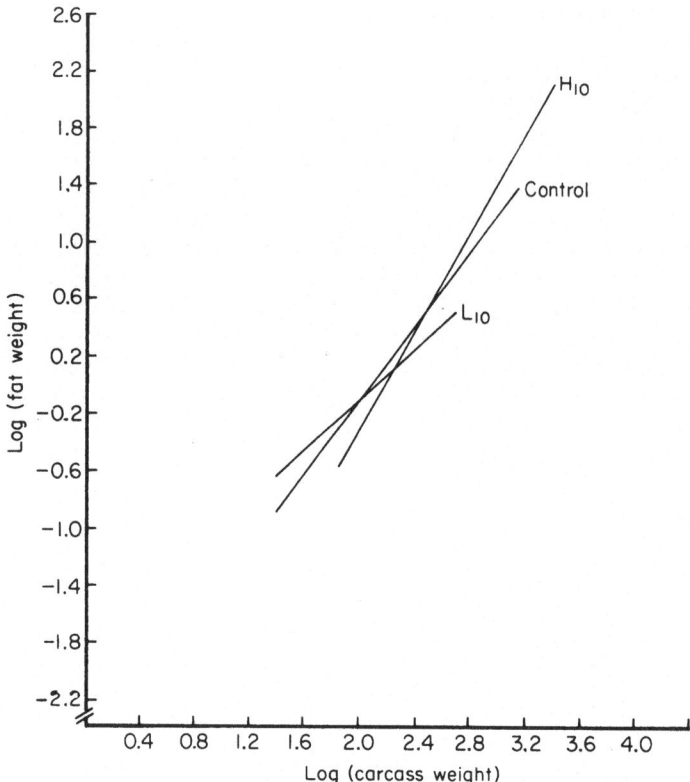

Figure 1.8 The relationship between log (fat weight) and log (carcass weight) in lines selected at 10 weeks of age for High (H_{10}) and Low (L_{10}) body weight. From Hayes and McCarthy (1976).

The distribution of adipose tissue

Allen (1978) weighed the adipose depots of the forelimb, hindlimb, kidney, mesentery and gonads at successive ages in the lines of mice referred to above in figures 1.1–1.8. He found that differences between lines already described for total fat were obvious for each depot at 5, $7\frac{1}{2}$, 10 and 15 weeks of age. He also noted differences in the distribution of fat among the depots at each age. For example, at older ages the proportion of the total fat in the gonadal depot was much greater in High than in Low lines. This correlated response was due to the fact that the lines were different in mean weight. The gonadal depot forms an increasing proportion of the total adipose tissue as a mouse increases in weight (see figure 1.2). But Allen also found that the distribution of fat among depots was different in mice of the same level of fatness from High and Low lines. In particular, the relative amounts of fat in the mesenteric and gonadal depots were quite different in the lines. Changes of the latter type were a developmental side-effect of selection for body weight. There is no other record of the possible genetic variation in fat distribution in laboratory animals.

THE NATURE OF GENETIC VARIATION IN FATNESS

Variation associated with differences in growth rate

The most obvious heritable differences in fatness occur between lines of mice selected for High and Low body weight. High lines, at later ages in particular, usually become very fat — values of up to 30 per cent of body weight have been recorded (e.g. Bakker, 1974). The simplest explanation of this is that the correlated change in the rate of food consumption in such lines leads to a pattern of food intake which at later ages supplies metabolisable energy which is not required for protein deposition and which is simply used for the deposition of fat. Let us consider the possible determination of the pattern of food intake in different lines.

Webster and his colleagues, working with the Zucker rat (*fa/fa*) (Pullar and Webster, 1974; Radcliffe and Webster, 1976), have suggested that long-term patterns of food consumption in *ad libitum* feeding regimes are regulated via an intimate link with the impetus for protein deposition. Their results showed that lean and fatty rats on *ad libitum* feeding regimes achieved identical rates of protein deposition even when diets of varied protein content which restricted growth were given. There is plenty of evidence of the obvious fact that the rate of protein deposition is greater in High body weight lines than in unselected or Low body weight ones (e.g. Fowler, 1958; Robinson and Bradford, 1969; Byrne *et al.*, 1973). The obvious differences in the slope of the relationship between food consumption and age in figure 1.6 up to 6 or 7 weeks of age can easily be interpreted as a reflection of differences in the rate of protein deposition.

Now let us consider the utilisation of energy in different lines. Much attention has been paid to the fact that animals from a High strain during growth, until about 5 or 6 weeks, are more efficient in converting food to weight gain than those from an unselected or Low one (for reviews, see Sutherland *et al.*, 1974; Roberts, 1978). The reason for this is not wholly clear, although it is usually assumed that because mice from a High line consume more food per unit body weight they have a higher proportion of their metabolisable energy available for growth relative to maintenance. While there are reports that High line mice eat more per unit body weight (e.g. Timon and Eisen, 1970), it is not a characteristic of the lines shown in figures 1.4 and 1.6, where it is obvious that the High lines, at later ages in particular, eat less per unit body weight than Low lines. Roberts (1974) found a similar contrast; High lines ate 25 per cent less per unit body weight in that case. Correlated changes in the activity of selected lines of mice also have to be considered. Although there is no adequate study published on the relative activity of selected lines, there is no doubt that mice from High lines are

less active than those from unselected or Low ones (Roberts, 1978). The low
activity of High lines must be one reason why they have sufficient energy to
deposit large amounts of fat at later ages, when Low lines are hardly growing at
all. There is also, of course, the possibility that the relative metabolic rate of High
lines is less — i.e. the energy expenditure per unit weight. Some evidence has been
provided by Fowler (1962) and Kownacki *et al.* (1975) that this is so at later ages.
It is probable that both the lower activity and the lower relative metabolic rate
provide High mice with a much greater balance of energy for fat synthesis at later
ages.

There is some other evidence which supports the notion implicit in the explanation put ,forward above that selection for increased body weight acts primarily
by increasing the frequency of genes which change the rate of protein deposition,
which, in turn, changes the rate of food consumption. The only line of mice which
was selected for rate of food consumption (Sutherland *et al.*, 1974) grew fast and
became very fat. The estimate of the genetic correlation between weight gain and
rate of food consumption obtained was high and thus implied that largely the same
genes affected growth rate and the rate of food consumption. But this does not
rule out the possibility that there are genes which can affect appetite independently
of growth rate. The results of experiments of a slightly different sort do not favour
this possibility, however. Falconer and Latyszewski (1952) selected two High
lines of mice for increased body weight at 6 weeks of age, one under *ad libitum*
feeding and the other as follows: from 3 to 6 weeks of age mice were given a
predetermined amount of food per day, 75 per cent of the consumption of unselected mice of the same age. There were significant responses in both lines despite the fact that the variance of body weight was reduced in the 'restricted' line
to about a third of that in the *ad libitum* one. The interesting point is that when
the line selected on restricted feeding was switched to *ad libitum* conditions, it
grew as fast as the other High line. Although food intake was not measured in
this experiment, the results suggest that heritable differences in growth rate, which
appear even when variability in rate of food consumption is damped, can result in
correlated changes in rate of food consumption in *ad libitum* conditions. This
suggests that the primary effect of genes affecting body weight is on the rate of
tissue deposition and that changes in the rate of food consumption are secondary.
The form of dietary restriction imposed by Falconer (1960b) in a repetition of
this experiment was different. One High line was selected for increased growth
rate from 3 to 6 weeks on *ad libitum* feeding of a normal diet, the other for the
same trait on an identical diet diluted with 50 per cent indigestible fibre which was
fed *ad libitum.* In this case responses to selection were significant in both regimes,
and when the line selected on the diluted diet was switched to the normal one, it
grew as fast and ate similar amounts of food. However, the form of food restriction
in this experiment was slightly peculiar in practice. Mice on the diluted diet sorted
through their food, not consuming the fibre; their energy intakes were depressed
below normal because of a quantitative rather than a qualitative restriction of diet.
There was no evidence that selection under this type of restriction, which might
be expected to favour 'gluttons' or 'scavengers', had any different effect from
selection on the normal diet. However, these experiments were not designed to test
the point at issue, i.e. the genetical control of correlated changes in appetite. The
preliminary study of our High and Low lines reported above is possibly of some

interest in this context, since all the signs are that selection for body weight altered meal-terminating mechanisms exclusively. There was no indication of an altered reaction to food among the lines — a change one might expect in lines selected directly for increased food consumption. The need for comparative studies of the effects of different schemes of selection for body weight — in different types of nutritional regimes and for various aspects of food consumption — is obvious.

Variation associated with differences in the 'partitioning' of energy

Hayes and McCarthy (1976) proposed a model of the effects of selection for body weight which also took account of the correlated changes described for the relative rate of development of adipose tissue. It is summarised in figure 1.9. They proposed that genetic variation in body weight was attributable to heritable differences in rate of gain associated with differences in rate of food consumption, and also with differences in the efficiency of conversion to weight gain of metabolisable energy. Differences in the efficiency of conversion of energy to weight gain were proposed to arise among animals because of the way they 'partitioned' their energy for the development of muscle and for the development of adipose tissue. It is generally accepted by nutritionists — e.g. Webster (1977) — that the development of a unit weight of muscle is considerably less demanding of energy than the development of a similar amount of adipose tissue, because there is about six times more water associated with protein in muscle than with fat in adipose tissue and the energetic costs of protein and fat synthesis are about the same. A consequence of the model is that selection at different ages would be expected to have different effects on fatness. For example, High lines selected at young ages would be expected to be fatter than High lines selected at later ages. (The reason why the lines selected later would be less fat is because at young ages with little fat deposited there could be little or no selection pressure for differences in the efficiency of conversion of energy to weight gain due to different amounts of fat and lean.) This expectation is confirmed by the results of Hull (1960), who selected for increased weight at 3, 4½ and 6 weeks of age. The line selected at 3 weeks was the fattest.

The model is consistent also with the effects of selection for body weight under restricted feeding described by Falconer and Latyszewski (1952) and Falconer (1960b). They selected for high growth rate or weight in *ad libitum* and restricted feeding regimes and then compared both sorts of lines in the *ad libitum* regime. Those selected under restriction then grew as fast and had slightly less fat in their carcasses. The reason the High lines, selected in restricted feeding regimes, were less fat, when tested in *ad libitum* conditions, than High lines selected in *ad libitum* conditions was that in the restricted regimes there was greater selection pressure for differences in body weight associated with efficiency and less for differences in weight associated with intake. Finally, the reason the Low lines described above, which were selected at later ages, had a high level of fat at low body weight was because they have been selected for low efficiency of conversion as well as for low intake.

In theory, if two lines grow at the same rate but one 'partitions' its food less efficiently by laying down relatively more fat, there should be a perceptible dif-

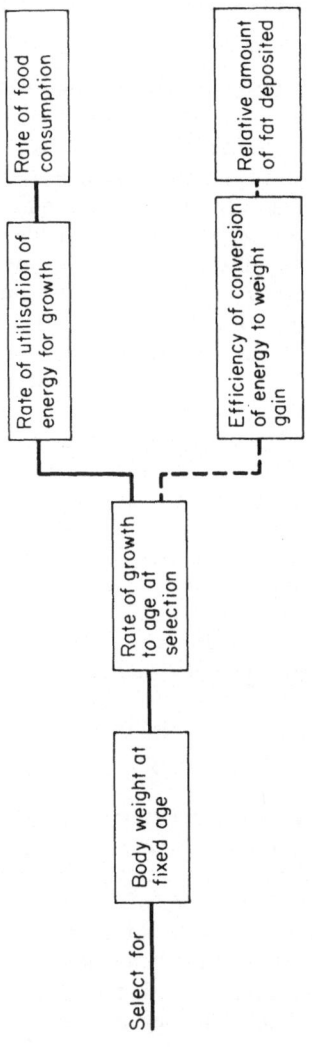

Figure 1.9 The causal chain of correlated factors influenced by selection for body weight is shown from left to right. The broken line indicates factors which show correlated responses on selection at *older* ages. From Hayes and McCarthy (1976).

ference in the amount of food eaten by the two lines. However, the only data with which to check this (those summarised in figure 1.6) do not agree with expectations.

Given that there are genetic differences in the ability to 'partition' energy between fat and other tissues, the occurrence of 'fat' and 'lean' strains of mice should be a fact. They do occur. Fowler (1958) described two High lines with 15 and 28 per cent fat, respectively, at 12 weeks of age. These lines had different relative rates of fat deposition but their growth rates were the same. Larsson (1967) found marked differences among inbred lines in their relative rates of fat deposition. How these developmental differences occurred in the first place, however, is as obscure as the genesis of well-known differences between breeds of farm animals in their predisposition to fatten at the same age or weight (Berg and Butterfield, 1976).

ACKNOWLEDGEMENT

The permission of Cambridge University Press to reproduce figures 1.4, 1.5, 1.7 and 1.8 from *Genetical Research* is gratefully acknowledged.

REFERENCES

Allen, P. (1978). Experimental studies of genetical variation in the pattern of fat deposition of mice and sheep. PhD Thesis (submitted to the National University of Ireland)

Bakker, H. (1974). Effects of selection for relative growth rate and body weight of mice on rate, composition and efficiency of growth. *Mededel. Landbouwhogeschool Wageningen*, 74-78

Berg, R. T. and Butterfield, R. M. (1976). *New Concepts in Cattle Growth*, Sydney University Press

Biondini, P. E., Sutherland, T. M. and Haverland, L. H. (1968). Body composition of mice selected for rapid growth rate. *J. Animal Sci.*, 27, 5-12

Byrne, I., Hooper, J. C. and McCarthy, J. C. (1973). Effects of selection for body size on the weight and cellular structure of seven mouse muscles. *Animal Prod.*, 17, 187-196

Clarke, J. N. (1969). Studies of the genetic control of growth in mice. PhD Thesis, University of Edinburgh

Eisen, E. J. and Bandy, T. (1977). Correlated responses in growth and body composition of replicated single-trait and index-selected lines of mice. *Theoret. Appl. Genet.*, 49, 133-144

Falconer, D. S. (1960a). *Introduction to Quantitative Genetics*, Oliver and Boyd, Edinburgh

Falconer, D. S. (1960b). Selection of mice for growth on high and low planes of nutrition. *Genet. Res.*, 1, 91-113

Falconer, D. S. (1967). Growth and fecundity in mice. In *Endocrine Genetics* (ed. S. G. Spickett and J. G. M. Shire), Cambridge University Press

Falconer, D. S. and Latyszewski, M. (1952). The environment in relation to selection for size in mice. *J. Genet.*, 51, 67-80

Fowler, R. E. (1958). The growth and carcass composition of strains of mice selected for large and small body size. *J. Agric. Sci.*, 51, 137-148

Fowler, R. E. (1962). The efficiency of food utilization, digestibility of foodstuffs and energy expenditure of mice selected for large or small body size. *Genet. Res.*, 3, 51-68.

Hayes, J. F. (1974). Personal communication

Hayes, J. F. and McCarthy, J. C. (1976). The effects of selection at different ages for high and low body weight on the pattern of fat deposition in mice. *Genet. Res.*, **27**, 389–403.

Hull, P. (1960). Genetic relations between carcass fat and body weight in mice. *J. Agric. Sci.*, **55**, 317–321

Kownacki, M., Keller, J. and Gebler, E. (1975). Selection of mice for high weight gains – its effect on the basal metabolic rate. *Genet. Polon.*, **16**, 359–363

Lang, B. J. and Legates, J. E. (1969). Rate, composition and efficiency of growth in mice selected for large and small body weight. *Theoret. Appl. Genet.*, **39**, 306–314

Larsson, S. (1967). Factors of importance for the etiology of obesity in mice. *Acta Physiol. Scand.*, Suppl. No. 294

McCarthy, J. C. (1977). Quantitative aspects of the genetics of growth. In *Growth and Poultry Meat Production* (ed K. N. Boorman and B. J. Wilson), British Poultry Science, Edinburgh

McPhee, C. P. and Neill, A. R. (1976). Changes in the body composition of mice selected for high and low eight week weight. *Theoret. Appl. Genet.*, **47**, 21–26

Pullar, J. D. and Webster, A. J. F. (1974). Heat loss and energy retention during growth in congenitally obese and lean rats. *Br. J. Nutr.*, **31**, 377–392

Radcliffe, J. D. and Webster, A. J. F. (1976). Regulation of food intake during growth in fatty and lean female Zucker rats given diets of different protein content. *Br. J. Nutr.*, **36**, 457–469

Rahnefeld, G. W., Comstock, R. E., Boylan, W. J. and Singh, M. (1965). Genetic correlation between growth rate and feed per unit gain in mice. *J. Animal Sci.*, **24**, 1061–1066

Roberts, R. C. (1974). Food consumption in mice selected for high and low body weight. *Genetics*, **74**, Supplement No. 2.2 (abstract)

Roberts, R. C. (1978). Side effects of selection in laboratory animals. *Livestock Prod. Sci.* (in press).

Robinson, D. W. and Bradford, G. E. (1969). Cellular response to selection for rapid growth in mice. *Growth*, **33**, 221–229

Stanier, M. W. and Mount, L. E. (1972). Growth rate, food intake and body composition before and after weaning in strains of mice selected for mature body weight. *Br. J. Nutr.*, **28**, 307–325

Sutherland, T. M., Biondini, P. E., Haverland, L. H., Pettus, D. and Owen, W. B. (1970). Selection for rate of gain, appetite and efficiency of food utilisation in mice. *J. Animal Sci.*, **31**, 1049–1057

Sutherland, T. M., Biondini, P. E. and Ward, G. M. (1974). Selection for growth rate, feed efficiency and body composition in mice. *Genetics*, **78**, 525–540

Timon, V. M. and Eisen, E. J. (1970). Comparisons of *ad libitum* and restricted feeding of mice selected and unselected for postweaning gain. I. Growth, feed consumption and feed efficiency. *Genetics*, **64**, 41–57

Timon, V. M., Eisen, E. J. and Leatherwood, J. M. (1970). Comparisons of *ad libitum* and restricted feeding of mice selected and unselected for postweaning gain. 2. Carcass composition and energetic efficiency. *Genetics*, **65**, 145–155

Webster, A. J. F. (1977). Selection for leanness and the energetic efficiency of growth in meat animals. *Proc. Nutr. Soc.*, **36**, 53–59

2

The inheritance of obesity in animal models of obesity

M. F. W. Festing (Medical Research Council, Laboratory Animals Centre, Carshalton, Surrey, UK)

SUMMARY

Obesity in laboratory animals may be inherited either as a single-gene Mendelian or as a polygenic character. A sharp distinction must be made between these two forms of inheritance. In the case of the Mendelian models, genetic theory suggests that there is a single genetic lesion leading to the production of a defective enzyme, which in turn leads to a multiplicity of metabolic abnormalities. In principle, it should be possible to detect the primary lesion, although in no case has this proved possible so far. In contrast, the polygenic models, which are usually found in inbred strains, probably arise from the presence of an extreme array of the genes which normally control the metabolic regulation of the amount of body fat. Thus, there is probably no single lesion. Moreover, the polygenic obesity cannot be so extreme as to induce total infertility, or the strain would die out. The polygenic models of obesity usually lie nearer the threshold of normality than the mutants, and are thus more likely to be curable by environmental manipulation. Similarly, fat but not obese 'normal' strains may easily be made obese by dietary and other environmental manipulation.

Comparative studies of different obese models seem to have been highly productive in showing both similarities and differences between models with entirely different modes of inheritance. The genetic tools are readily available to develop a much wider range of models by selection and backcrossing, should they be required. The technique of embryo fusion, which does not appear to have been used so far in investigating obesity, may have some value in the future.

ANIMALS AS MODELS OF HUMAN DISEASE

The development and use of certain disease conditions in laboratory animals as 'models' of similar conditions in man is currently a fashionable area of research.

In fact, it could be argued that laboratory animals are always used as models of the human (or some other) species which cannot be used directly for experimental purposes owing to ethical or economic factors.

There are a wide range of disease conditions which superficially appear to be similar in man and some of the common laboratory animals. These range from inborn errors of metabolism and metabolic diseases (Bulfield, 1977) to congenital abnormalities, hormonal deficiency or insensitivity, cancer of various types, and complex metabolic abnormalities leading to conditions such as obesity and diabetes. These disease models can be valuable in studying the possible aetiology, pathogenesis and treatment of the condition in humans, provided both the value and limitations of such models are fully understood.

VALUE OF ANIMAL MODELS

The value of studying animals rather than man is obvious to most research workers. Animals may be maintained in rigidly controlled conditions, fed on a standard diet and maintained under pathogen-free or germ-free conditions. They may be bred at will, and the genotype may be controlled through inbreeding and the propagation of mutant types. Animals may be placed on treatment regimens which impair their health, or even kill them, though most countries now have some laws designed to protect laboratory animals from undue abuse. Most laboratory animals are small, and are relatively economical to keep.

LIMITATIONS OF ANIMAL MODELS

The most obvious limitation of animal models is simply that there is no assurance that conditions which are superficially similar in animals and man are in fact similar at a more fundamental level. For example, the obese (ob) mutant in mice is used as a 'model' of obesity in humans, even though it is virtually certain that most human obesity is not associated with a single Mendelian recessive gene. Thus, such a mutant is in no way a model of the mode of inheritance of obesity in man. On the other hand, a mutant such as obese may well have metabolic similarities to obese humans, and it may respond to anorexic drugs, for example, in a similar manner to man. Thus, a model may be useful for a restricted set of investigations, even though the same model is recognised as being unrealistic for other types of investigation.

Another obvious limitation is the problem of scale. Garrow (1974) emphasised this point in considering energy balance in man by quoting Miller and Mumford (1966), who calculated that a 10 km walk would increase daily energy expenditure by about 3 per cent in the rat, 10 per cent in man and 22 per cent in the elephant, even though 10 km is presumably a relatively long walk for a rat, and a relatively short walk for an elephant. Thus, predicting an effect in man from an observation in a rodent may be extremely difficult even if it is known that in this respect man and the rodent behave in a biochemically similar manner.

In some cases useful models may bear very little obvious resemblance to the condition being studied. For example, it seems probable that much human obesity is inherited as a polygenic threshold character. The classical genetic analysis of such a character is that of Wright (1934), who studied atavistic polydactyly in guinea-pigs. Thus, it could be said that the inheritance of extra toes in guinea-pigs serves as a 'model' of the inheritance of obesity in humans.

INHERITED OBESITY IN LABORATORY ANIMALS

Previous reviews of obesity in laboratory animals have tended to classify the obese conditions according to their mode of inheritance. Thus, Bray and York (1971) recognised dominant inheritance, recessive inheritance, inbred obesity, hybrid obesity and miscellaneous conditions, and Stauffacher *et al.* (1971) classified the models into (1) single-gene mutants, (2) inbred strains and hybrids and (3) hereditary components strongly influenced by environmental factors. In this review the primary, and most important, distinction will be made between obesity inherited as a single Mendelian gene and obesity inherited as a polygenic character. There are a number of practical and theoretical differences between the two which need to be emphasised.

CHARACTERISTICS OF MENDELIAN MODELS OF OBESITY

In a number of cases obesity is inherited in a Mendelian manner (see table 2.1). In such cases the obesity arises as a result of a point mutation causing, presumably, the alteration or loss of a single polypeptide. Theoretically, the primary biochemical lesion should be detectable, though in practice there are so many metabolic abnormalities arising as a consequence of the primary lesion that it has proved impossible so far to detect the primary lesion in any of the mutants. Classical genetic methods may be used to map the mutant, and in some cases the linkage relationships of the obese mutants with other genetic markers are known. Linkage may be of some practical significance in the production of obese animals (see, for example, the paper by Lovell on the use of the mutant *misty* in the propagation of the *db* mutation).

Another practical advantage of mutants is that heterozygous or homozygous normal animals of the same strain serve as lean controls. In the case of recessive mutations, however, it should not automatically be assumed that homozygous normal (+/+) and heterozygous (x/+, where x represents any obese mutation) animals are exactly identical, even though both may be lean. At the biochemical level it is not unusual to find that heterozygotes may be different from both homozygotes. For example, Beloff-Chain *et al.* (1975) have shown that pituitary gland extracts from *ob*/+ and +/+ mice differ.

Another important characteristic of the Mendelian mutant models is that they may be combined with other mutants, or with any of the polygenic models of obesity, or they may be backcrossed to different inbred backgrounds. This could be useful as a method of studying the obese condition. For example, Wolff (1965) combined the pituitary dwarf mutant *dw* with the yellow obese mutant A^y in order to discover the effect of a deficiency of pituitary hormones on the obese condition, and Herberg and Coleman (1977) have emphasised that the expression of most of the obese mutants may be modified by the genetic background (this will be discussed in more detail later). In contrast, although F_1 hybrids could be produced by crossing some of the polygenically determined models of obesity, a true breeding colony combining the features of, say, the NZO and the KK strains of mice would be impossible to obtain.

Another feature of the Mendelian models of obesity is that the degree of abnormality may be very severe, and may lead to sterility. The stock is then maintained by heterozygous matings. In contrast, in the polygenic models such a degree of abnormality could not be tolerated, as the strain would be lost. Thus, in the polygenic models the degree of obesity is usually less and the onset of obesity is usually later than in Mendelian models.

CHARACTERISTICS OF THE POLYGENIC MODELS OF OBESITY

The polygenically determined models of obesity probably represent the extreme of normal variation in body fat content. The mode of inheritance is discussed in more detail below, but it can be said here that, in contrast with the Mendelian models, there is probably no single genetic lesion. Obese animals would arise as a result of having, for example, an extreme array of genes determining factors such as growth rate, metabolic rate, appetite, activity and social reactivity. The polygenic models have the disadvantage that there are no natural non-obese controls (the use of NZC as a control for NZO is, for example, of dubious validity). However, it could be argued that polygenic models are more realistic models of human obesity, which is probably predominantly of this type, than are the mutants.

Moreover, additional polygenic models of obesity could easily be engineered by selective breeding should more models be required, whereas mutants cannot be produced to order. A sharp distinction should therefore be made between the polygenic and the single-gene models of obesity. These two types will be considered in more detail below.

SPECIFIC SINGLE-GENE MODELS OF OBESITY

The single-gene models of obesity which have been reported so far are summarised in table 2.1, together with references describing their origin and characteristics, or with reference to a recent review article in which the mutants are described. Most of the mutants were described by Bray and York (1971), though some new ones have occurred since then.

Table 2.1 Single-gene models of obesity

Species	Locus	Allele(s)	Background	Reference(s)
Mouse	adult obesity and diabetes	Ad	–	Wallace (1975)
	agouti	A^y, A^{vy}, A^{iy}	YBR, VY, YS, C3H	
			C57BL/6	Dickie (1969)
	diabetes	$db, db^{2J}, db^{3J}, db^{ad}$	C57BL/Ks	Herberg and Coleman (1977)
			–	Falconer and Isaacson (1959)
	fat	fat	(C57BL/6)	Hummel and Coleman (1974)
	obese	ob	C57BL/6	Ingalls *et al.* (1950)
Rat	fatty	fa, fa^k	–, SHR (?)	Zucker and Zucker (1961) Koletsky (1973) Yen *et al.* (1977)

Adult obesity and diabetes (Ad)

This mutation was discovered in wild mice trapped in order to study the inheritance of warfarin resistance (Margaret Wallace, 1977, personal communication). It is inherited as an autosomal dominant, and is located on chromosome 7. Heterozygotes are intermediate between the normal (+/+) and homozygous mutant (Ad/Ad) animals, though they resemble the normals more closely than the mutants with respect to obesity. The mutation has not yet been described formally, though Wallace (1975) gave a brief description. The mutation is not yet placed on a defined genetic background, and no information is available on the effect of genetic background on the degree of obesity.

Mutants at the agouti locus

There are now 16 mutants known at the agouti locus on chromosome 2 of the mouse. These control the distribution of eumelanin (brown or black) and phaeomelanin (yellow) in the hair. Three of the mutants are associated with a large proportion of phaeomelanin, and, according to Dickie (1969), are also obese. The yellow mutant A^y was first described in 1883. It is dominant, and is lethal when homozygous. $A^y/-$ animals have a range of metabolic abnormalities leading to increased protein and fat deposition, a higher incidence of both spontaneous and induced tumours, an altered reaction to castration, and decreased serum insulin (Wolff, 1971). Viable yellow A^{vy} occurred spontaneously in C3H mice at The Jackson Laboratory. Homozygotes are fully viable, and the mutation is dominant.

An interesting feature of this mutation is the great phenotypic variability of genetically identical animals. Mice which are $A^{vy}/-$ on an inbred genetic background can vary from agouti in colour and slim in build through mottled and slightly fatter to bright yellow and obese (Wolff and Pitot, 1973). Intermediate yellow A^{iy} occurred spontaneously in C57BL/6J mice and is reported to be obese, though little work seems to have been reported on its characteristics so far. It is dominant and fully viable.

Mutations at the diabetes locus

The first mutation at this locus was reported by Falconer and Isaacson (1959), who named it *adipose (ad)*. Hummel *et al.* (1966) and Coleman and Hummel (1967) reported a mutation causing obesity and diabetes in C57BL/Ks mice and named it *diabetes (db)*. It was shown to be a recessive mutation located on chromosome 4. Tests of *ad* and *db* carried out by Falconer and Isaacson failed to show that *ad* and *db* were allelic. However, later tests by Lane (1973) showed that in fact *db* and *ad* are allelic. Failure of the initial tests was attributed to abnormal segregation ratios shown by a single male, on whom the tests were largely based. As there was already a substantial literature on *db*, Falconer suggested that the locus should continue to be known as the diabetes locus, and the symbol for adipose should be changed to db^{ad}. According to Herberg and Coleman (1977), two more mutations designated db^{2J} and db^{3J} have occurred more recently. None of the mutations at the diabetes locus are distinguishable from one another if they are on the same genetic background, but the expression of the mutations is dependent on genetic modifiers (Coleman and Hummel, 1975). In particular, on the C57BL/6J genetic background *db/db* animals are obese but not diabetic, and in fact are phenotypically indistinguishable from *obese (ob/ob)* animals. On the C57BL/Ks genetic background *db/db* mice develop diabetes. This will be considered in more detail later. A bibliography of some mutants at this locus has been published by Staats (1975).

The fat locus

A mutation named at *fat* was described briefly by Hummel and Coleman (1974). It occurred spontaneously in HRS/J, is recessive and is not allelic with *ob* or *db*. Homozygotes are recognisable by 4-5 weeks of age and reach 60 g by 6 months. It is hyperinsulinaemic, but diabetic symptoms have not yet been seen. Further studies will not be undertaken until the gene is placed on a C57BL/6J and C57BL/Ks genetic background.

The obese locus

The *obese (ob)* mutation occurred spontaneously in the V strain of mice maintained at The Jackson Laboratory, and was first reported by Ingalls *et al.* (1950). It is an autosomal recessive located on chromosome 6. It is now probably the second most widely studied mammalian mutant (after *nude*). The gene has been transferred to the C57BL/6J and the C57BL/Ks genetic backgrounds, though it is only normally available on the former. It is also maintained on a number of random bred and undefined backgrounds. As the gene interacts with the genetic background, it is important that in any critical investigation in which the characteristics of the *ob/ob* animals are described the genetic background should be specified. Failure to specify the genetic background may mean that the investigation is not repeatable.

The fatty locus in the rat

A mutation causing extreme obesity arose spontaneously in a cross involving Sherman and Merck M rats. This was first described by Zucker and Zucker (1961), who called it 'fatty' (*fa*), though it is sometimes known as the 'Zucker' rat. It is an autosomal recessive mutation, but its linkage relationships are known. Its characteristics have been reviewed recently by Bray (1977).

More recently, another obese mutation which occurred spontaneously in animals derived from a cross of the spontaneously hypertensive SHR and a Sprague-Dawley rat followed by selection for hypertension was described by Koletsky (1973). Homozygotes are obese, hypertensive and hyperlipaemic, have endocrine and metabolic disturbances and develop premature arteriosclerosis. Presumably many of these abnormalities arise from a combination of the mutant (which has been named 'corpulent', *cp*) and the genetic background strain with its hypertension and associated defects. Yen *et al.* (1977) have now shown that 'fatty' and 'corpulent' are allelic, and have proposed the gene symbol fa^k for the latter. It is not yet known whether the two are phenotypically distinguishable when placed on the same background.

THE POLYGENIC THRESHOLD MODEL

Both body fat and many of the variables, such as metabolic rate, spontaneous exercise and food consumption, which may influence fatness are under genetic control. The mode of inheritance is, however, complex, and environmental variables, such as the quality of the diet fed, can obviously have a strong influence. An individual is only classified as 'obese' if he exceeds some specified level of fatness, but there is no clear dividing line between the normal but fat animal and the pathologically obese animal. Under these circumstances, obesity may be regarded as having a polygenic threshold mode of inheritance (Falconer, 1960), illustrated in figure 2.1. The position of any one individual with regard to the threshold depends on the array of genes controlling fatness, acting through the underlying variables, that it happens to possess. The environment in which an individual is raised will also influence its fatness, and if the whole population is raised in the same environment, then the relative position of the threshold will itself be influenced by the quality of that environment. Thus, feeding a high-calorie diet would be expected to shift the threshold to the left, so that more animals would be classified as being pathologically obese. It is important to emphasise that,

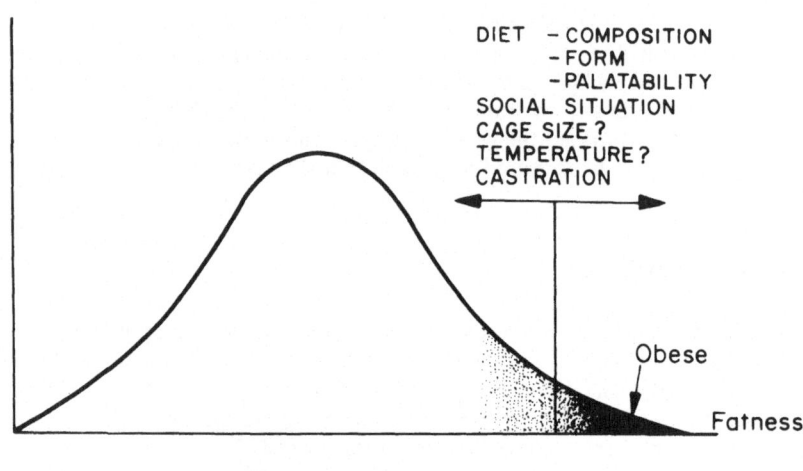

DIET – COMPOSITION
      – FORM
      – PALATABILITY
SOCIAL SITUATION
CAGE SIZE ?
TEMPERATURE ?
CASTRATION

Obese

Fatness

SPONTANEOUS EXERCISE
METABOLIC RATE
FOOD SATIATION THRESHOLD
SOCIAL REACTIVITY
TASTE DISCRIMINATION
FAT/OTHER PARTITION
(SINGLE GENES)

Figure 2.1 The polygenic threshold model of obesity. The position of an individual with respect to the threshold determining whether or not the animal is classified as obese depends on both genetic and environmental factors. There are a number of underlying factors, such as spontaneous exercise and metabolic rate, each of which is under the control of many genes. The position of the threshold relative to the whole population can be altered by altering environmental factors such as diet.

according to this model, obese animals are those which have an extreme array of genes present in normal populations which determine body fat. Thus, it is unlikely that any single genetic 'lesion' would be found in such cases. The number of genes involved is not known, and techniques for determining the number of genes controlling polygenic characters in mammals are poorly developed. Experience with other polygenic characters suggests that there may be a few 'major' genes (say fewer than ten) each of which has a relatively important effect, and many (possibly hundreds or even thousands) which have negligible effects individually, but taken together can have a strong influence.

Many of the polygenic models of obesity found in laboratory animals occur in inbred strains of mice, rats and hamsters. An inbred strain is produced as a result of 20 or more generations of brother × sister mating (Staats, 1976). Each strain may be regarded as a single genotype which results from the chance fixation of genes at all loci which are usually polymorphic. Thus, by chance an array of 'fatness' genes may be fixed which make the animals of that strain slim, intermediate or fat.

In the case of characters which have a high heritability, such as fatness (McCarthy, 1978), individual strains may fall well outside the range found in a normal segregating colony, as it is known that one result of inbreeding is to increase the total additive genetic variance, which is expressed as a difference between inbred strains, while at the same time the genetic variation within an inbred strain is reduced virtually to zero (Falconer, 1960). Thus, some inbred strains may be very obese, while others are very lean. There will still be some phenotypic variation within an inbred strain due to environmental and chance effects. Differences between inbred strains, on the other hand, will be due to genetics, and studies of several inbred strains for body composition should give something approaching a normal distribution (if an appropriate scale of measurement is used). For example,

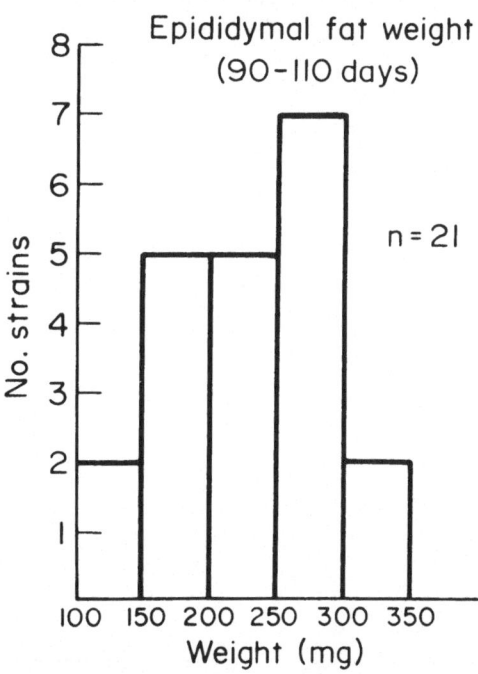

Figure 2.2 Weight of the epididymal fat pad in 21 inbred strains of non-obese mice on a standard laboratory mouse diet (unpublished data of M. F. W. Festing and M. E. Robbins).

figure 2.2 shows epididymal fat weight in 21 inbred strains of mice, none of which are classified as obese. This histogram is not too different from a normal distribution, in view of the rather small number of strains involved. It has been established in rats that the weight of individual fat stores is highly correlated with total body fat (Grewal *et al.*, 1973).

Variation in body fat among inbred strains is presumably dependent on variation in a number of the underlying variables indicated in figure 2.1. Festing (1977), for example, showed that there were significant differences among inbred strains in spontaneous activity in an activity wheel, though it is not known whether this is related to degree of obesity.

Storer (1967) showed that there are highly significant differences between inbred strains in metabolic rate per unit of body weight, and that this variation is inversely correlated with body weight. This is shown in figure 2.3. Strain differences in liking for various flavours have also been demonstrated. Figure 2.4 shows strain differences in the preference ratio for 10 per cent alcohol in a two-bottle choice situation. In this case the strains fall into three groups; those that avoid alcohol, those that show no preference for either solution and those that show a preference for the alcohol solution. Whether such strain differences in taste preference could result in some strains consuming substantially more dietary calories, and therefore putting on more fat, is not known.

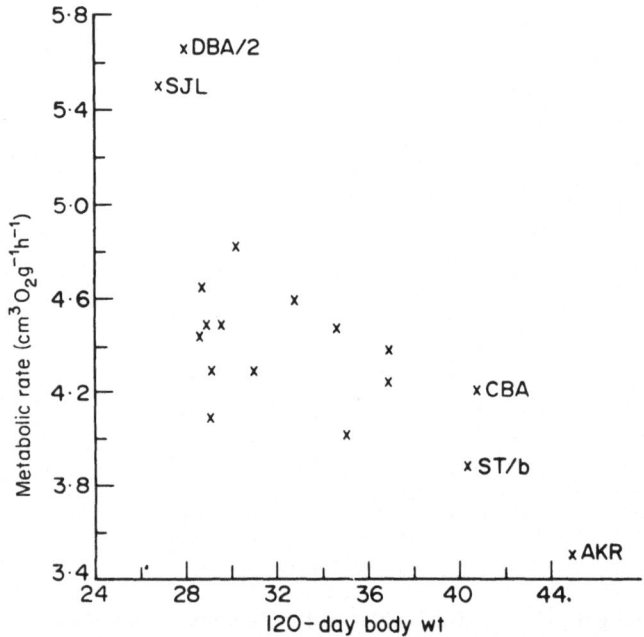

Figure 2.3   Relationship between body weight and metabolic rate per unit of body weight in 18 non-obese inbred strains of mice. Only the five extreme strains have been named on the graph. Data from Storer, 1967.

Strain differences in food consumption and food conversion, and the effect of number of mice per cage on food conversion, were described by Les (1968). Some of his data are shown in figure 2.5. In this case it is clear that AKR eats relatively little diet per gram of weight gain, and is relatively uninfluenced by the number of mice per cage, whereas the converse is true with C57BL/6. Again, the exact relationship between food consumption, weight gain, social reactivity and fatness

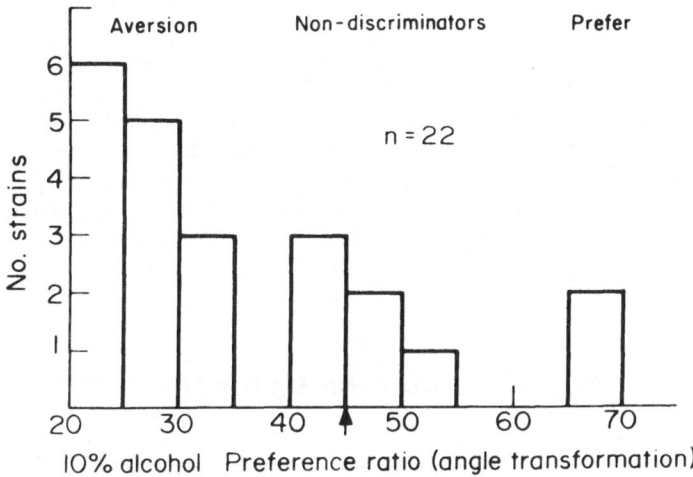

Figure 2.4   Alcohol preference ratio in 22 inbred strains of mice. Data are the percentage (subjected to an arcsine transformation) of fluid consumed as 10 per cent alcohol in a two-bottle (water or 10 per cent alcohol) choice test. The arrow indicates equal preference for water and alcohol. Unpublished data of E. Fawdington and M. F. W. Festing.

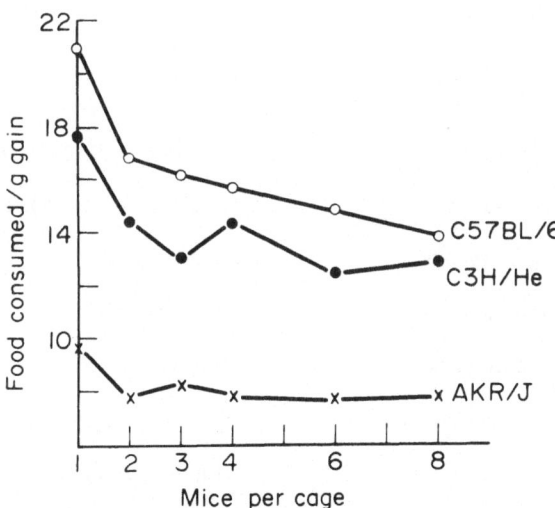

Figure 2.5   Relationship between the number of mice per cage and the food consumption per gram of weight gain in three non-obese inbred strains of mice. Note strain differences both in food conversion and in reaction to number of animals per cage. Data abstracted from Les, 1968.

is not clear, though there is little doubt that some of these variables are related to one another. From these examples, it is clear that body fat (within the normal range) and many of the underlying characteristics, such as metabolic rate, which are known to influence body fat are genetically determined. Those strains which happen through chance, or as a result of selection during inbreeding, to have many 'fatness' genes will become obese. Those strains which have slightly fewer fatness genes will presumably be near the threshold, and may become obese after the provision of a high-calorie or highly palatable diet, such as the cafeteria diet described by Stock and Rothwell (1978), or as a result of castration. Conversely, obese strains near the threshold may be cured by diet restriction.

SPECIFIC MODELS OF POLYGENICALLY DETERMINED OBESITY

Inbred strains and F_1 hybrids of the mouse in which obesity has been reported are listed in table 2.2. The obesity is classified as being either spontaneous or induced, either by diet or by some other means.

Table 2.2   Polygenic models of obesity in the mouse

Strain	Type	Reference(s)
NZO	spontaneous	Herberg and Coleman (1977) Bielschowsky and Bielschowsky (1953)
KK-A^y	spontaneous	Iwatsuka et al. (1970) Herberg and Coleman (1977)
PBB	spontaneous	Hunt et al. (1972, 1976)
BRSUNT/N	spontaneous	Hansen et al. (1973)
FL/IRe	spontaneous	Staats (1976)
C3Hfx IF$_1$	spontaneous	Jones (1964)
A/St	induced (diet)	Fenton and Dowling (1953)
C3H	induced (diet)	Fenton and Dowling (1953)
KK	induced (diet)	Kondo et al. (1957) Butler and Gerritsen (1970)
NH	induced (castration)	
LAF$_1$	induced (ACTH-secreting transplantable tumours)	Furth et al. (1953)

The NZO strain

Obesity in this strain was reported by Bielschowsky and Bielschowsky (1953), who developed the stock from the same base population as the other 'NZ' strains (Bielschowsky and Goodall, 1970). By the twelfth generation of brother × sister mating it was noted that the animals were obese, and subsequently selection was used to develop an obese strain, which was named NZO. Herberg and Coleman (1977) warn that this strain is maintained in a number of laboratories throughout the world, and some subline differentiation may have occurred. Breeding performance is poor, presumably owing at least in part to the obesity. Although the degree of obesity is comparable with that of yellow obese (A^y/-) , obese (*ob/ob*) and diabetes (*db/db*), it can be modified (though not eliminated) by dietary manipulation (Herberg and Coleman, 1977). The fat mainly accumulates in the body cavity, in contrast with other models of obesity (Bray and York, 1971).

The KK, KK-A^y and related strains

The KK inbred strain was developed by Kondo in 1944 from Japanese dealer stock. It was bred for large body size, and develops moderate obesity, hyperphagia and polyuria. The degree of obesity can be increased by dietary manipulation, by gold thioglucose administration or by the presence of the A^y gene, which has been backcrossed on to the strain (Iwatsuka *et al.*, 1970). Additional obese strains have been developed more recently from a cross involving KK and C57BL/6. One of these is known as the Toronto-KK strain (in violation of the Rules for Nomenclature of Inbred Strains of Mice: Staats, 1976) (Herberg and Coleman, 1977). The mode of inheritance of obesity is clearly polygenic, though there may be a relatively small number of genes segregating (Butler and Gerritsen, 1970).

The PBB strain

One of the newest models of obesity is the PBB mouse strain, inbred from multi-coloured pet shop mice, and first described by Hunt *et al.* (1972), with a more detailed description by Hunt *et al.* (1976). These mice become obese on a standard diet, reaching 65-72 g by 12 months of age. Blood glucose is normal, but glucose tolerance is abnormal, and serum insulin is elevated. No histological abnormalities of the pancreas have been detected so far. The condition has been compared with mature onset obesity in man (Hunt *et al.*, 1976). The mode of inheritance has not been studied, but the condition is presumed to have a polygenic mode of inheritance.

BRSUNT/N

This inbred strain was developed by L. C. Strong as a branch of BRS (now presumed to be extinct), which in turn was derived from strain NH. It is reported to become obese (Staats, 1976), but this does not appear to have been studied.

FL/1Re

This strain has a mixed ancestry involving largely strains WB/Re and C3H, and it carries the flexed-tail gene *f*. It is reported as having '...good breeding when young, later impaired by obesity...' (Staats, 1976), but the obesity does not appear to have been investigated in any detail.

C3H× IF$_1$

Obesity in the F$_1$ hybrid between strains C3H and I was first reported by Jones (1964) as a result of a systematic study of tumour development in inbred strains and hybrids. It has been named the 'Wellesley' mouse. There is a moderate degree of obesity at 3–4 months of age, with glycosuria in about 50 per cent of the males, but only 5 per cent of females (Bray and York, 1971), and there are histological abnormalities of the pancreas.

The mode of inheritance of this model of obesity does not appear to have been investigated. Although one parental strain, C3H, has a tendency to develop obesity, particularly on a high-fat diet (see table 2.1), strain I is both non-obese and resistant to the dietary induction of obesity (Fenton and Dowling, 1953). On the other hand, strain I clearly has some metabolic abnormalities. It is acutely sensitive to vitamin B6 deficiency (Bell and Haskell, 1971), and it also carries *Phk*, a sex-linked gene leading to a deficiency of phosphorylase kinase. This gives rise to a 3–4-fold elevation of skeletal muscle glycogen content (Gross *et al.*, 1975). Presumably, such a gene could account for the observed sex differences in glycosuria in the F$_1$ hybrid. It would be of great interest to know whether F$_1$ hybrids involving strain I and other genetic models of obesity such as NZO or PBB would result in obese progeny.

DIET-INDUCED MODELS OF OBESITY: C3H AND A

Early studies by Fenton and Carr (1951) showed that there were strain differences in response to high-fat (up to 47.5 per cent) diets. Later, Fenton and Dowling (1953) showed that on such high-fat diets obesity was easily induced in strains C3H and A, but not in C57BL and I. Unfortunately, studies seem to have been limited to these four strains. A survey of, say, 15–20 inbred strains would almost certainly show that obesity can be induced in many more of the common inbred strains. It would be interesting to know whether this would correlate with percentage body fat on a normal diet, as might be expected.

OTHER MODELS OF OBESITY IN MICE

Obesity can be induced in some strains of mice by other manipulations, such as castration (Hausberger and Hausberger, 1966), or some other treatment which upsets the hormone balance, such as transplantation of ACTH-secreting tumours in LAF$_1$ mice (Furth *et al.*, 1953). Castration has been recorded as inducing obe-

Table 2.3   Effects of diet, castration and genotype on the weight of epididymal adipose tissue in the mouse (Yen *et al.*, 1970)

Genotype (YS)	Treatment	Diet fat (g) 7.5%	11%
A^y/a	intact	940	1500
	castrated	2500	2400
a/a	intact	270	500
	castrated	1290	1200

sity in NH mice (which are in any case reported to develop obesity to some extent spontaneously, according to Staats, 1976 and Liebelt and Lyle, 1978). Castration will also increase obesity in yellow obese mice, though its effects are not additive with the effects of diet (Yen *et al.*, 1970; table 2.3), but there does not seem to have been any systematic study which records the degree of castration-induced obesity in a range of more common inbred strains of mice, even though it is highly probable that many strains would respond to such treatment (Chai and Dickie, 1966).

POLYGENIC MODELS IN SPECIES OTHER THAN THE MOUSE

Polygenically determined models of obesity in species other than the mouse are listed in table 2.4.

Table 2.4   Polygenic models of obesity in species other than the mouse

Species	Strain	Type	Reference(s)
Rat	BHE	spontaneous/diet	Marshall and Hildebrand (1963) Marshall *et al.* (1971)
	OM	induced (diet)	Schemmel *et al.* (1970)
Hamster	4.24	spontaneous	Homburger (1972)
Chicken	OS	spontaneous (auto-immune thyroiditis)	Wick *et al.* (1974)
Sand rat (*Psammomys obesus*)		spontaneous (environment)	Haines *et al.* (1965)
Spiny mouse (*Acomys cahirinus*)		spontaneous (environment)	Gonet *et al.* (1965)
Tuco-tuco (*Ctenomys talarum*)		spontaneous (environment)	Wise *et al.* (1968)
Djungarian hamster (*Phodopus sungorus*)		spontaneous (environment)	Herberg and Coleman (1977)

The BHE rat

At present there do not appear to be any inbred strains of rats which may be regarded as models of spontaneous obesity. However, Marshall and Hildebrand (1963) and Marshall *et al.* (1971) described the BHE stock of rats, which has been described as 'carbohydrate sensitive'. According to Berdanier (1974), the stock is the result of a cross between the Osborne-Mendel (also known as the Yale) strain and the Pennsylvania State College strain. After many years of breeding, it became apparent that BHE rats gained more body weight and had more carcass and liver lipids than similarly fed Wistar rats. These differences could not be accounted for by differences in food intake, and were particularly apparent when the diet contained large amounts of purified carbohydrates (Berdanier, 1974). The stock also has a high incidence (30 per cent) of hydronephrosis as well as a high incidence of nephrosis. The relationship between the nephrosis and the metabolic abnormalities is not clear. The stock also exhibits hyperinsulinaemia and insulin insensitivity, but a normal glucose tolerance (Berdanier, 1974). Two inbred strains are being developed from the BHE stock, which may help to identify in more detail the genetic abnormalities in the stock (Marshall *et al.*, 1971).

The OM rat

There are a number of inbred and outbred colonies of OM (Osborne-Mendel) rats, some of which are reputed to become obese on high-fat diets. Schemmel *et al.* (1970) maintained seven strains or stocks of rats on a control and on a 60 per cent fat diet and found that on the high-fat diet at 20 weeks of age the carcass fat accounted for 39 per cent of total body weight in males and 40 per cent in females in the OM strain. In contrast, in the most resistant strain, S5B, the carcass fat only accounted for 14 per cent of the total body weight in both sexes. The 'Wistar-Lewis' (more correctly known as LEW: Festing, 1978) also had a tendency to accumulate body fat under these conditions.

The BIO 4.24 strain of Syrian hamster

Homburger (1972) noted that females of the inbred B10 4.24 strain of Syrian hamsters are larger than the males, and tend to become obese at maturity. Few studies of this model appear to have been conducted so far, though it was used by Ashwell and Mead (1978) in studying the effects of transplantation on adipose tissue. The Syrian hamster is, of course, unusual in that allografts are frequently not rejected, making it a useful model for studies involving transplantation of tissues (Palm *et al.*, 1967).

The obese OS chicken

The OS chickens were first reported by van Tienhoven and Cole (1962) as a strain with hereditary hypothyroidism. The characteristic clinical symptoms are apparent by 3-4 weeks of age, and consist of a small body size, excessive amounts of subcutaneous and abdominal fat, lipaemic serum, soft pliable skin, long silky feathers, small combs, delayed sexual maturity, low reproductive performance and sensitivity to temperatures below $20°C$ (Wick *et al.*, 1974). It is now known that the condition is due to a hereditary autoimmune thyroiditis, which closely resembles Hashimoto's disease in humans. The genetics of the condition is not clear at this time, and it is debatable whether the OS chicken should be listed as a model of obesity, when the obesity is almost certainly secondary to the autoimmune disease.

Obesity in other small rodent species

Obesity and/or diabetes-like syndromes have been reported in laboratory colonies of the sand rat (*Psammomys obesus*) (Haines *et al.*, 1965), the spiny mouse (*Acomys cahirinus*) (Gonet *et al.*, 1965), the tuco-tuco (*Ctenomys talarum*) (Wise *et al.*, 1968) and the Djungarian hamster (*Phodpus sungorus*) (Herberg and Coleman, 1977). Although under laboratory conditions these syndromes occur spontaneously, they are almost certainly a result of the drastic change in the environment from that found in the wild. In the tuco-tuco there are high blood glucose, high plasma insulin levels and marked obesity at least for one period of the developing syndrome. In contrast, in the sand rat, the spiny mouse and the Djungarian hamster a condition resembling human insulin-deficient juvenile-onset diabetes is often observed. Only in some cases does obesity develop, in which case the syndrome resembles human late-onset diabetes mellitus (Herberg and Coleman, 1977). This individual variation suggests that within each species there may be genetic variation controlling the obesity, but very little work has been reported on the genetics of obesity in these species.

EFFECTS OF THE GENETIC BACKGROUND ON EXPRESSION OF THE OBESE MUTANTS

The expression of many mutant genes can be strongly influenced by genes at other loci, or the 'genetic background'. Thus, according to Herberg and Coleman (1977): 'If one keeps in mind that many, if not all, of the metabolic abnormalities of obese mice result not from the presence of the mutant gene itself but from its interaction with the background genome, it is understandable that the severity and the developmental pattern of the metabolic disturbances vary in mice in which the mutation is maintained on different inbred (or outbred) backgrounds. This variable expression has caused considerable confusion in the literature on the obese mouse.'

The effects of the genetic background on expression of mutants at a number of loci controlling obesity have been studied. Table 2.5, for example, shows that the body weight of A^{vy}/a animals on an inbred VY genetic background was 45 per cent greater than normal (*a/a*) sibs, whereas it was only 24 per cent greater on the YS × VYF$_1$ hybrid genetic background. A range of metabolic abnormalities associated with the 'yellow mouse syndrome' also varied according to the genetic background (Wolff and Pitot, 1973). Moreover, on some genetic backgrounds A^y/a mice may not become obese (Fenton and Chase, 1951).

Table 2.5   Effect of genetic background on obesity in A^y/a and A^{vy}/a mice
(Wolff and Reichard, 1970)

Genetic background	Genotype	% Excess wt. over a/a sibs
YS	A^y/a	+26
YS × VY	A^{vy}/a	+24
VY	A^{vy}/a	+45
VY × YS	A^y/a	+35

Variation in the expression of the obese (ob/ob) mutation has also been extensively studied. Thus, according to Herberg and Coleman (1977), there are some important differences among ob/ob mice maintained in various laboratories throughout the world. A Swedish colony originating from The Jackson Laboratory V stock (in which the mutation originally occurred) has characteristics similar to that of the C57BL/6-ob/ob mouse, except that there are massive islets of Langerhans. A colony maintained at the CNRS, Orleans, again derived from the V stock, has a more severe form of diabetes-like syndrome than C57BL/6-ob/ob, which may be regarded as the 'standard' ob/ob mouse. The obese stock of the Birmingham (Aston) colony has been mixed with local undefined mice, and also has a more severe diabetes type of syndrome than the C57BL/6-ob/ob mouse.

The most comprehensive studies of the effect of the genetic background on expression of mutations at both the obese and diabetes locus are those of Hummel *et al.* (1972) and Coleman and Hummel (1973, 1975). These authors have back crossed mutations at the diabetes locus (db and db^{2J}) and the obese locus (ob) to the C57BL/6J and the C57BL/Ks inbred strains. Their findings may be briefly summarised as follows.

(1) On the C57BL/Ks genetic background db/db animals are characterised by obesity, hyperphagia and a severe diabetes with marked hyperglycaemia, temporarily elevated plasma insulin levels, degenerative changes in the islets of Langerhans and a shortened lifespan.

(2) On the C57BL/6J genetic background db/db animals are characterised by marked obesity, hyperphagia, transient hyperglycaemia and markedly elevated plasma insulin concentrations associated with marked hypertrophy of the islets and increased proliferative capacity of the beta-cells.

(3) The various mutants at the db locus (table 2.1) are indistinguishable from each other on the same genetic background.

(4) On the same genetic background ob/ob mice are phenotypically indistinguishable from db/db mice.

(5) The genetic factors that modify the expression of mutants at these two loci are inherited as polygenes.

These findings emphasise the importance of specifying the genetic background in studies involving genetic models of obesity if results are to be repeatable in the future. It is worth re-emphasising that one of the basic assumptions of science is that results should be repeatable. Experiments may not be repeatable if the experimental methods, including the animals used, are not adequately specified.

FUTURE ROLE AND DEVELOPMENT OF GENETIC MODELS OF OBESITY

Scientific advance is characterised both by more detailed investigations of the
nature of observed phenomena and by studies of the generality of such phenomena.
In the case of biomedical investigations of obesity, detailed biochemical and physiological investigations will eventually reveal the extent of the biochemical abnormalities in various obese models. Assuming that a primary biochemical lesion
is eventually found for a model such as the *ob/ob* mouse, it might be possible
through some sort of genetic engineering to develop a method of 'curing' such animals. However, whether such studies have any relevance to obesity in humans is
questionable. It seem highly probable that in humans, as in the mouse, there is no
single cause, and therefore (presumably) no single cure for obesity. However, in
attempting to develop an understanding of obesity, what is needed is not only
in-depth studies, but also in-breadth studies. The comparative approach has already
been successful in showing that phenomena such as abnormalities in temperature
regulation, absence of hyperphagia, hyperinsulinaemia and insulin resistance are
general abnormalities which are common to many different models of obesity.
However, such comparisons should cover as wide a range of genetic models as possible. The identification of models which are exceptions to the general rule is also
useful. Although there are already a reasonable number of obese models, genetic
tools are already available to develop many new ones.

New polygenic models of obesity could easily be developed by a few generations
of deliberate selection for carcass fat, holding fat-free mass constant. Not only
would such an experiment develop new models, but also the correlated responses
in characters such as thermoregulation, energy consumption, fat distribution, hormone balance, activity and structure of the endocrine glands would undoubtedly
reveal some of the fundamental relationships between these factors and obesity,
particularly if replicated selection lines were used. The success of artificial selection
in altering carcass composition and the value of studying correlated responses is
clearly shown by McCarthy (1978).

Another way of developing new models of obesity would be to make use of
genetic modifiers already present in inbred strains by crossing some of the obese
mutants to several different inbred backgrounds, as has already been done with *ob*
and *db* by Coleman, Hummel and co-workers. This would be easy in the case of
the dominant mutants, but is admittedly somewhat laborious in the case of recessive genes. However, the variation in the characteristics of *ob/ob* animals with
different genetic backgrounds suggests that such a procedure would result in new
models with a wide range of different characteristics. Hybrid models of obesity in
which one parent carries the A^y or A^{vy} gene would be easy to construct. It would
be interesting to study the characteristics of, for example, the 'Wellesley' mouse
in which one parent was C3H-A^{vy} rather than just C3H.

Other interesting models could be constructed by combining obese models
with other genetic abnormalities. Thus, Wolff (1965) developed the dwarf yellow
obese mouse (dw/dw, $A^y/$-), and showed that the particular pituitary defect that
caused dwarfism did not interfere with the development of obesity. However, the
mutants Ames dwarf, pigmy, miniature, diminutive and little also cause dwarfing
of various sorts, and their effects in combination with obese genes are unknown
(Bulfield, 1977). Ashwell and Mead (1978) report on the nude obese (*nu/nu,*

ob/ob) mouse, which may have some value in studying various aspects of the immunology of obesity, and the transplantation of tissues into such animals. No doubt there are many other genetic combinations that would be of some value in studying obesity.

Finally, two other types of genetic manipulations that can be used in studying obesity should be mentioned. Coleman and Hummel (1969) used parabiosis to study the effects of humoral factors on normal and diabetic (*db/db*) mice. They found that union with a normal (+/+) mouse did not cure diabetes in the *db/db* animals, but did lead to weight loss, hypoglycaemia and death of the normal partner. Parabiosis is only feasible if the mutation is maintained on an inbred genetic background, as a congenic or segregating inbred strain, and it is obviously useful in deciding whether humoral factors are relevant in genetically determined diseases. Embryo fusion is another technique that has recently become available (McLaren, 1976). If embryos of various models of obesity (either single-gene or polygenic) could be fused with embryos of lean mice, it should be possible to determine whether the obesity is cured, and what factors determine whether a cure is achieved. This technique does not appear to have been used yet in the study of genetically determined obesity.

In conclusion, genetic techniques such as selection, the synthesis of new genetic types, and techniques such as parabiosis, tissue transplantation and embryo fusion seem to offer possibilities for the study of obesity in the future.

REFERENCES

Ashwell, M. and Meade, C. J. (1978). Adipose tissue in genetically obese rodents. This volume.

Bell, R. R. and Haskell, B. E. (1971). Metabolism of vitamin B6 in the I-strain mouse. I. Absorption, excretion and conversion of vitamin to enzyme co-factor. *Arch. Biochem. Biophys.*, 147, 588-601

Beloff-Chain, A., Hawthorn, J. and Green, D. (1975). Influence of the pituitary gland from homozygote (+/+) and heterozygote (*ob*/+) lean mouse on insulin secretion *in vitro*. *FEBS Letters*, 55, 72-74

Berdanier, C. D. (1974). Metabolic abnormalities in BHE rats. *Diabetologia*, 10, 691-695

Bielschowsky, M. and Bielschowsky, F. (1953). The New Zealand strain of obese mice. Their response to stilboestrol and to insulin. *Aust. J. Exp. Biol. Med. Sci.*, 31, 181-198

Bielschowsky, M. and Goodall, C. M. (1970). Origin of inbred NZ mouse strains. *Cancer Res.*, 30, 834-836

Bray, G. A. (1977). The Zucker-fatty rat: a review. *Fed. Proc.*, 36, 148-153

Bray, G. A. and York, D. A. (1971). Genetically transmitted obesity in rodents. *Physiol. Rev.*, 51, 598-646

Bulfield, G. (1977). Nutrition and animal models of inherited metabolic disease. *Proc. Nutr. Soc.*, 36, 61-67

Butler, L. and Gerritsen, E. C. (1970). A comparison of the modes of inheritance of diabetes in the Chinese hamster and the KK mouse. *Diabetologia*, 6, 163-167

Chai, C. K. and Dickie, M. M. (1966). Endocrine variations. *Biology of the Laboratory Mouse* (ed. E. L. Green), McGraw-Hill, New York, pp. 387-403

Coleman, D. L. and Hummel, K. P. (1967). Studies with the mutation, diabetes in the mouse. *Diabetologia*, 3, 238-248

Coleman, D. L. and Hummel, K. P. (1969). Effects of parabiosis of normal with genetically diabetic mice. *Am. J. Physiol.*, 217, 1298-1304

Coleman, D. L. and Hummel, K. P. (1973). The influence of genetic background on the expression of the obese (*ob*) gene in the mouse. *Diabetologia*, 9, 287-293

Coleman, D. L. and Hummel, K. P. (1975). Influence of genetic background on the expression of mutations at the diabetes locus in the mouse. II. Studies on background modifiers. *Israel. J. Med. Sci.*, 11, 708-713

Dickie, M. M. (1969). Mutations at the agouti locus in the mouse. *J. Hered.*, 60, 20-25

Falconer, D. S. (1960). *Introduction to Quantitative Genetics*, Oliver and Boyd, Edinburgh

Falconer, D. S. and Isaacson, J. H. (1959). Adipose, a new inherited obesity of the mouse. *J. Hered.*, 50, 290-292

Fenton, P. F. and Carr, C. J. (1951). The nutrition of the mouse. XI. Response of four strains to diets differing in fat content. *J. Nutr.*, 45, 225-233

Fenton, P. F. and Chase, H. B. (1951). Effect of diet on obesity of yellow mice in inbred lines. *Proc. Soc. Exp. Biol. Med.*, 77, 420-422

Fenton, P. F. and Dowling, M. T. (1953). Studies on obesity. I. Nutritional obesity in mice. *J. Nutr.*, 49, 319-331

Festing, M. F. W. (1977). Wheel activity in 26 strains of mouse. *Lab. Animals*, 11, 257-258

Festing, M. F. W. (1978). Inbred strains of rats. *Rat News Letter*, 3, 13-56

Furth, J., Gadsden, E. L. and Upton, A. C. (1953). ACTH secreting transplantable pituitary tumours. *Proc. Soc. Exp. Biol. Med.*, 84, 253-254

Garrow, J. S. (1974). *Energy Balance and Obesity in Man*, North-Holland, Amsterdam

Gonet, A. E., Stauffacher, W., Pictet, R. and Renold, A. E. (1965). Obesity and diabetes mellitus with striking congenital hyperplasia of the islets of Langerhans in spiny mice (*Acomys cahirinus*). I. Histological findings and preliminary metabolic observations. *Diabetologia*, 1, 162-171

Grewal, T., Schemmel, R., Cress, C. E. and Mickelsen, O. (1973). Prediction of total body fat in rats from individual fat depot weights. *Growth*, 37, 111-126

Gross, S. R., Longshore, M. A. and Pangburn, S. (1975). The phosphorylasekinase deficiency (Phk) locus in the mouse: evidence that the mutant allele codes for an enzyme with abnormal structure. *Biochem. Genet.*, 13, 567-584

Haines, E. S., Hackel, D. B. and Schmidt-Nielsen, K. (1965). Experimental diabetes mellitus induced by diet in the sand rat (*Psammomys obesus*). *Am. J. Physiol.*, 208, 297-300

Hansen, C. T., Judge, F. J. and Whitney, R. A. (1973). *Catalogue of NIH Rodents*. DHEW Publications No. (NIH) 74-606, National Institutes of Health, Bethesda, Maryland

Hausberger, F. X. and Hausberger, B. C. (1966). Castration-induced obesity in mice. Body composition, histology of adrenal cortex and islets of Langerhans in castrated mice. *Acta Endocrinol.*, 53, 571

Herberg, L. and Coleman, D. L. (1977). Laboratory animals exhibiting obesity and diabetes syndromes. *Metab. Clin. Exp.*, 26, 59-99

Homburger, F. (1972). Disease models in Syrian hamsters. *Prog. Exp. Tumor Res.*, 16, 69-86

Hummel, K. P. and Coleman, D. L. (1974). 'New mutants' *fat*. *Mouse News Letter*, 50, 43

Hummel, K. P., Coleman, D. L. and Lane, P. W. P. (1972). Influence of the genetic background on expression of mutations at the diabetes locus in the mouse. I. C57BL/KsJ and C57BL/6J strains. *Biochem. Genet.*, 7, 1-13

Hummel, K. P., Dickie, M. M. and Coleman, D. L. (1966). Diabetes, a new mutation in the mouse. *Science*, 153, 1127-1128

Hunt, C. E., Lindsey, J. R., Maxfield, L. M. and Fox, O. J. (1972). Obesity in a new strain of mouse. *Fed. Proc.*, 31, 244 (abstract)

Hunt, C. E., Lindsey, J. R. and Walkley, S. U. (1976). Animal models of diabetes and obesity, including the PBB/Ld mouse. *Fed. Proc.*, 35, 1206-1217

Ingalls, A. M., Dickie, M. M. and Snell, G. D. (1950). Obese, a new mutation in the house mouse. *J. Hered.*, 41, 317-318

Iwatsuka, H., Shino, A. and Suzuoki, Z. (1970). General survey of diabetic features of yellow KK mice. *Endocrinol. Japon.*, 17, 25-35

Jones, E. (1964). Spontaneous hyperplasia of the pancreatic islets associated with glucosuria in hybrid mice. In *The Structure and Metabolism of the Pancreatic Islets* (ed. S. E. Brolin, B. Hellman and H. Knutson), Pergamon Press, Oxford, pp. 189-191

Koletsky, S. (1973). Obese spontaneously hypertensive rats – a model for study of atherosclerosis. *Exp. Mol. Pathol.*, 19, 53-60

Kondo, K., Nozawa, K., Tomita, T. and Esaki, K. (1957). Inbred strains resulting from Japanese mice. *Bull. Exp. Animals*, 6, 107-112

Lane, P. W. (1973). Contribution. *Mouse News Letter*, 48, 34

Les, E. P. (1968). Cage population density and efficiency of feed utilization in inbred mice. *Lab. Animal Care*, 18, 305-313

Liebelt, A. D. and Lyle, P. D. (1978). Animal models of obesity. *I.L.A.R. News*, 21, 19-20

McCarthy, J. C. (1978). Normal variation in body fat and its inheritance. This volume

McLaren, A. (1976). *Mammalian Chimaeras*, Developmental and Cell Biology Series, No. 4, Cambridge University Press

Marshall, M. W., Durand, A. M. A. and Adams, M. (1971). Different characteristics of rat strains. Lipid metabolism and response to diet. In Proc. IV I.C.L.A. Symp., *Defining the Laboratory Animal* (ed. H. A. Schneider), National Academy of Sciences, Washington, DC, pp. 381-413

Marshall, M. W. and Hildebrand, H. E. (1963). Differences in rat strain response to three diets of different composition. *J. Nutr.*, 79, 227-238

Miller, D. S. and Mumford, P. (1966). Obesity: physical activity and nutrition. *Proc. Nutr. Soc.*, 25, 100-186

Palm, J., Silvers, W. K. and Billingham, R. E. (1967). The problem of histocompatibility in wild hamsters. *J. Hered.*, 58, 40-44

Schemmel, R., Mickelsen, O. and Gill, J. L. (1970). Dietary obesity in rats: body weight and body fat accretion in seven strains of rats. *J. Nutr.*, 100, 1041-1048

Staats, J. (1975). Diabetes in the mouse due to two mutant genes – a bibliography. *Diabetologia*, 11, 325-327

Staats, J. (1976). Standardized nomenclature for inbred strains of mice: sixth listing. *Cancer Res.*, 36, 4333-4377

Stauffacher, W., Orci, L., Cameron, D. P., Burr, I. M. and Rendd, A. E. (1971). Spontaneous hyperglycemia and/or obesity in laboratory rodents: an example of the possible usefulness of animal disease models with both genetic and environmental components. *Rec. Prog. Hormone Res.*, 27, 41-91

Stock, M. J. and Rothwell, N. J. (1978). Energy balance in reversible obesity. This volume

Storer, J. B. (1967). Relation of lifespan to brain weight, body weight, and metabolic rate among inbred mouse strains. *Exp. Gerontol.*, 2, 173-182

van Tienhoven, A. and Cole, R. K. (1962). Endocrine disturbances in obese chickens. *Anat. Rec.*, 142, 111-122

Wallace, M. E. (1975). Contribution. *Mouse News Letter*, 53, 20

Wick, G., Sundick, S. and Albini, B. (1974). The obese strain (*os*) of chickens: an animal model with spontaneous autoimmune thyroiditis. *Clin. Immunol. Immunopathol.*, 3, 272-300

Wise, P. H., Weir, B. J., Hime, J. M. and Forrest, E. (1968). Implications of hyperglycaemia and cataract in a colony of tuco-tucos (*Ctenomys talarum*). *Nature*, 219, 1374-1376

Wolff, G. L. (1965). Hereditary obesity and hormone deficiencies in yellow dwarf mice. *Am. J. Physiol.*, 209, 632-636

Wolff, G. L. (1971). Metabolic regulation by genes in the laboratory mouse. In *Defining the Laboratory Animal* (ed. H. A. Schneider), Proc. IV Symposium, International Committee on Laboratory Animals, National Academy of Sciences, Washington DC, pp. 205-217

Wolff, G. L. and Pitot, H. C. (1973). Influence of background genome on enzymatic characteristics of yellow ($A^y/$-, A^{vy}) mice. *Genetics*, 73, 109-123

Wolff, G. L. and Reichard, G. A. Jr. (1970). Response of serum insulin concentration to tumor growth in different genetic systems. *Horm. Metab. Res.*, **2**, 68–71

Wright, S. (1934). The results of crosses between inbred strains of guinea-pigs differing in number of digits. *Genetics*, **19**, 537–551

Yen, T. T., Shaw, W. N. and Pao-Lo, Yu (1977). Genetics of obesity in Zucker rats and Koletsky rats. *Heredity*, **38**, 373–378

Yen, T. T. T., Steinmetz, J. and Wolff, G. L. (1970). Lipolysis in genetically obese and diabetes-prone mice. *Horm. Metab. Res.*, **2**, 200–203

Zucker, L. M. and Zucker, T. F. (1961). Fatty, a new mutation in the rat. *J. Hered.*, **52**, 275–278

3

The characteristics of
genetically obese mutants

David A. York (Department of Nutrition, School of Biochemical and
Physiological Sciences, University of Southampton, Southampton, UK)

SUMMARY

A number of characteristics are common to the differing forms of rodent obesity.
An increased efficiency of energy utilisation is accompanied by, but not dependent upon, hyperphagia and hypoactivity. An impairment in thermogenic processes may be responsible for the increase in efficiency in a number of the obese
models. The developmental sequence of the obesity varies between the differing
models, recessively inherited obesities being associated with early pre-weaning
excess fat deposition, whereas obesity in other forms of rodent obesity develops
at a later age and is of a more moderate type. The rodent models also differ in
the type of adipose tissue expansion between hypertrophic and hypertrophic-
hyperplastic growth. The hyperinsulinaemia characteristic of all obesities is together with the hyperphagia the major stimulus for the enhanced hepatic and
adipose tissue lipogenesis. The obese-hyperinsulinaemic state produces secondary
effects which include changes in pancreatic morphology, insulin resistance, enhanced gluconeogenesis, increased basal lipolysis and loss of responsiveness to
lipolytic hormones. Hyperadrenoçortism and sterility observed in the recessively
inherited obesities may result from hypothalamic dysfunction. The possible metabolic basis of the obesities is discussed.

INTRODUCTION

The animal models of obesity provide a spectrum of diseases which may be helpful to our understanding of the multifactorial nature of obesity/diabetes syndromes in man. The varying animal models allow us to study the interrelationships
between and the roles played by genetic background, diet, activity and environment. It is not surprising therefore that the metabolic abnormalities which have
been reported in the various animal models of obesity are extremely variable.
Notwithstanding this, however, a number of metabolic disturbances have been

shown consistently in all models of obesity, although the severity of these changes and the age of their onset vary considerably. It is the aim of this review, within the limitations of the current literature, to discuss the metabolic changes which are characteristic of the animal obesities, with a view not only to describing those changes which are a consequence of the obese state, but also to review some of the possible defects which might underlie the obesities. Since the obese (C57

Table 3.1 Some genetic models of obesity in rodents*

Dominant		
(1)	yellow mouse	A^y/a
Recessive		
(1)	obese–hyperglycaemic mouse	*ob/ob*
(2)	diabetes mouse	*db/db*
(3)	fatty rat	*fa/fa*
Polygenic		
(1)	New Zealand mouse	NZO
(2)	Japanese/Toronto mouse	KK
(3)	Paul Bailey black mouse	PBB/Ld
(4)	*Psammomys obesus*	(sand rat)
(5)	*Acomys cahirinus*	(spiny mouse)
Hybrid		
(1)	Wellesley mouse	C3$_f$I F$_1$

* A more extensive list is given by Festing in this volume.

BL/6J-*ob/ob*)* and diabetic (C57 BL/Ks-*db/db*)* mice and the fatty (*fa/fa*) rat (Bray, 1977) have been used most extensively in this research, the review will concentrate primarily on these species, but similarities to and differences from other obese mutants will be discussed where information is available. A classification of the genetic models and their gene symbols are given in table 3.1. Several other more extensive reviews of this area have been published (Bray and York, 1971; Assimacopoulos-Jeannet and Jeanrenaud, 1976; Herberg and Coleman, 1977).

*The C57BL/6J-*ob/ob* and C57BL/Ks-*db/db* mice will be referred to throughout this review as *ob/ob* and *db/db* mice, respectively, except in those studies in which the *ob* and *db* genes were transferred to alternate background genomes.

ENERGY BALANCE

The regulation of energy balance is central to any discussion of obesity. Simply, energy balance may be regarded as an interplay between food intake, energy expenditure (metabolic and mechanical), energy storage and heat loss, as outlined in the formula:

energy intake = energy expenditure + energy storage + heat loss

Obesity results from an imbalance in this equation which could reflect either an increase in food intake, an increase in energy storage or a fall in either energy expenditure or heat loss. We would therefore expect that the gene defects in the various forms of inherited obesities would be expressed in a protein or proteins which is or are fundamental to one or more of these parameters.

Numerous studies on food intake of obese rodents have been reported, all of which have shown that the obese animal consumes more food than its lean sibling during the early stages of development. However, food intake often returns towards normal or even to normal during the static phase of obesity (Dickie and Wooley, 1946; Dickerson and Gowan, 1947; Coleman and Hummel, 1969; Dulin and Wyse, 1970; Bray and York, 1972; Wyse and Dulin, 1974; Hunt *et al.*, 1976). Typical values are shown in table 3.2, from which it can be seen that there is considerable variation in the degree of hyperphagia, the greatest being in the diabetes *db/db* mouse, which may eat over twice as much as its lean controls.

The precise age at which hyperphagia first appears is not clear for any of the mutants. Thus, although hyperphagia is observed shortly after weaning in *ob/ob*, *db/db*, *fa/fa* and KK obese animals, the possibility that milk intake during the suckling period is also increased has received little attention, mainly because of the difficulty of assessing this function. However, from recent studies with ^{3}H$_{2}$O-labelled maternal milk, it has been suggested that both the suckling *ob/ob* mouse (Jeanrenaud, 1978) and the *fa/fa* rat (Boulange *et al.*, 1977) have normal energy intake despite their storage of excess fat during this pre-weaning period.

Table 3.2 Average food intake of obese rodent models (g/d)

Model	Lean	Obese	Reference
A^{y}/a	4.4	5.3	Dickerson and Gowan (1947)
ob/ob	4.1	5.6	Fuller and Jacoby (1955)
db/db	4.8	11	Coleman and Hummell (1969)
KK*	2.4	3.4	Dulin and Wyse (1970)
PBB/Ld*	2.3	3.5	Hunt *et al.* (1976)
fa/fa	15	24	Bray and York (1972)

*C57BL/6 mice were used as controls.

The mechanisms underlying the hyperphagia of obese mutants are not understood. A number of reviews (see Powley, 1977) have clearly described the complex
interplay between neural, endocrine and metabolic signals which regulate satiety
and feeding in rodents. The hyperphagia of *ob/ob* mice and *fa/fa* rats is associated
with a loss of the normal diurnal variations in meal pattern (Bailey *et al.*, 1975;
Becker and Grinker, 1977). The essentially normal response of *fa/fa* rats to food
dilution, monoamine oxidase (MAO) inhibitors, high fat diets, amphetamine (Bray
and York, 1972), quinine (Cruce *et al.*, 1974) and cholecystokinin (McCloughlin
et al., 1977) suggests that the hyperphagia of *fa/fa* rats is not the result of a congenital lesion similar to that which may be imposed electrically or with gold thioglucose (Powley, 1977). In contrast, the impaired response of *ob/ob* mice to many
of these treatments (Parson *et al.*, 1954; Fuller and Jacoby, 1955; Sprott, 1972)
and their preference for sweetened diets (Sprott, 1972) has been proposed as evidence for a hypothalamic lesion in these mice. However, since there are also
several reports of conflicting experimental findings (see table 3.3) — e.g. the res-

Table 3.3 Some differences in the regulation of food intake in *ob/ob*
mice and *fa/fa* rats

	ob/ob	*fa/fa*
Normal response to		
1. Dietary dilution	x	x✓
2. Dietary fat increase	x	✓
3. Quinine or sucrose octoacetate	x✓	x
4. Amphetamine		✓
5. HFR* reinforcement	x✓	✓
6. Dietary sucrose	✓	✓
7. Day : night cycle	x	x

x   = abnormal response.
✓   = normal response.
x✓   = conflicting results in the literature.
*HFR = high fixed ratio.

For further details of these experiments, the reader is referred to Bray and York
(1971), Bray and York (1972), Fuller and Jacoby (1955), Parson *et al.* (1954),
Cruce *et al.* (1974), Sclafani (1976), Singh *et al.* (1974), Greenwood *et al.* (1974),
Fuller (1972), Becker and Grinker (1977) and Bailey *et al.* (1975).

ponse to high fixed ratio reinforcement schedules (Greenwood *et al.*, 1974;
Singh *et al.*, 1974; Sclafani, 1976) — a clear understanding of the hypothalamic
function in obese rodents is not yet possible.

Clear evidence does exist, however, to show that the hyperphagia is not a
necessary prerequisite for the deposition of excess fat in *ob/ob*, *db/db*, A^y/a mice
and *fa/fa* rats (Fenton and Chase, 1951; Alonso and Maren, 1955; Chlouverakis,

1970; Bray *et al.*, 1973; Zucker, 1975; Cox and Powley, 1977). For when these obese animals are either pair-fed or pair-gained to their respective lean litter mates on similar meal feeding regimes, they still deposit excessive quantities of fat, indicating an increased efficiency of energy utilisation. Such studies may, however, be criticised, as an obese animal might adapt more readily to the change in feeding pattern. The recent studies of Cox and Powley (1977) overcame this objection, as the *db/db* mice were pair-fed on a yoked feeding procedure by bar-pressing. In this situation the food intake of the diabetic *db/db* mouse was matched to lean controls in both quantity and frequency of feeding. Despite this, over a 6 week period from weaning, *db/db* mice gained 42 per cent more body weight and accumulated fivefold more lipid than the lean controls.

Parabiosis experiments offer further support for a defect in the hypothalamic regulation of food intake in both the *ob/ob* and the *db/db* mice. In these studies (Coleman and Hummel, 1969; Coleman, 1973) *ob/ob*, *db/db* and lean mice were parabiosed in differing combinations. When the *ob/ob* mouse was parabiosed to a lean mouse, the rapid accumulation of body fat by the *ob/ob* mouse stopped and its food intake was normalised. In contrast, when the *db/db* mouse was parabiosed to either lean or *ob/ob* mice, the *db/db* mouse survived the union, while both lean and *ob/ob* partners died, apparently from aphagia. These experiments have been interpreted as indicating that the *db/db* mouse produces excessive quantities of a satiety factor to which it cannot respond but which, after transfer through the parabiotic union, inhibits food intake in both the lean and the *ob/ob* mouse. The *ob/ob* mouse, on the other hand, while able to respond to this 'satiety factor', apparently is unable to secrete sufficient quantities to suppress its own food intake.

Although many investigations on hypothalamic regulation have been reported in the obese mutants, little research has been directed towards understanding the mechanisms underlying these regulatory defects. Thus the initial reports that noradrenaline but not dopamine concentrations are increased in the hypothalamus, telencephalon and brain stem of both *ad libitum*-fed and food-restricted *ob/ob* mice (Oltmans *et al.*, 1976; Lorden *et al.*, 1975), in the hypothalamus of *db/db* mice (Lorden *et al.*, 1975) and in the median eminence of *fa/fa* rats (Cruce *et al.*, 1976) and decreased in the paraventricular nucleus of *fa/fa* rats (Cruce *et al.*, 1976) are first indications of the possible biochemical changes underlying the defects in hypothalamic regulation.

Energy expenditure by obese rodents has been assessed either by indirect methods (oxygen consumption) or by direct calorimetry. Both methods have shown that the total energy expenditure is as high and often greater in the obese animal than in the lean control (table 3.4). However, when the data are expressed on a surface area basis (body weight$^{0.75}$), the energy expenditure in all models except NZO mice is less than in lean controls. The usefulness of expressing data by this method in obese animals must be queried, since the majority of the increased body weight is accounted for by the triacylglycerol store within adipocytes. These data suggest therefore that the metabolic rate may not be decreased in the obese animals. The demonstration of similar maintenance energy requirements from dietary studies in *fa/fa* rats (Deb and Martin, 1975) and *ob/ob* mice (Miller *et al.*, 1979) and their respective lean controls is further evidence that the total energy expenditure of these obese rodents may not be impaired. Conflicting data suggesting a

Table 3.4 Metabolic rate of obese rodent models

	per unit wt.$^{0.75}$/h*	per animal/h	Units	Reference
ob/ob	1.63 ± 0.08	15.0 ± 1.8	ml O_2	Otto *et al.* (1976)
+/?	1.98 ± 0.02	10.5 ± 1.2		
A^y/a	1.56 ± 0.7	14.6 ± 12	ml O_2	Bartke and Gorecki (1968)
a/a	1.74 ± 0.7	14.5 ± 7		
NZO	0.52 ± 0.01	8.1 ± 0.1	mmol O_2	Subrahmanyan (1960)
NZC†	0.42 ± 0.01	4.6 ± 0.1		
fa/fa	6.2 ± 0.4	348 ± 8	ml O_2	Bray (1969)
Fa/Fa	7.4 ± 0.5	298 ± 6		

* Significantly less in all obese models than in the relevant controls, with the exception of NZO mice.
† Lean control for NZO mice.

decrease in maintenance energy requirement in obese animals have been reported for *fa/fa* rats (Pullar and Webster, 1974) from direct calorimetric studies and for *ob/ob* mice (Woodward, Trayhurn and James, 1978) from dietary studies. However, the ability to maintain an increased body weight by the expenditure of normal amounts of energy implies an increase in efficiency of energy utilisation.

The inability of *ob/ob*, *db/db*, A^y/a and *fa/fa* mutants to maintain a normal body temperature at laboratory temperature is also indicative of an impairment in thermogenic processes (van der Kroon, 1966; Davis and Mayer, 1954; York *et al.*, 1972; Yen *et al.*, 1974; Trayhurn *et al.*, 1977). This thermogenic deficiency is strikingly apparent in the inability of *ob/ob* mice to withstand cold exposure. Death results not from an absence of the normal physical responses to cold but from the failure to increase thermogenesis. Both A^y/a mice and *fa/fa* rats also show an impaired response to cold exposure, but in these animals the effects are less severe. The impairment in thermogenesis in *ob/ob* mice is neither a secondary result of hypothyroidism, since they are euthyroid (Joosten and van der Kroon, 1974a; York *et al.*, 1978), nor secondary to the obese state, since hypothalamic obesity is not associated with impaired thermogenesis. Indeed the ability to demonstrate a low rectal temperature in *ob/ob* mice (Joosten and van der Kroon, 1974a; Trayhurn *et al.*, 1977) before any other known metabolic changes suggests that defective thermogenesis might be a primary reflection of the gene mutation. A similar conclusion might be appropriate for the fatty (*fa/fa*) rat, since their low rectal temperature is demonstrable before weaning and before any increase in hepatic lipogenesis or serum insulin (Godbole *et al.*, 1978).

The possibility that hypoactivity might contribute to the increased efficiency of energy utilisation in obese animals has been investigated by a number of groups. It is of no surprise that adult obese rodents faced with the prospect of carrying excess weight are less active than their lean siblings (Stern and Johnson, 1977; Yen and Acton, 1972; Joosten and van der Kroon, 1974b). However, there are very few studies in which activity in the pre-weaning/weaning period has been cri-

tically evaluated. Thus, Stern and Johnson (1977) failed to demonstrate any reduction in normal activity in pre-obese *fa/fa* rats before the onset of hyperphagia
and increased adiposity. In contrast, although one study reported normal activity
for suckling pre-obese *ob/ob* mice (Yen and Acton, 1972), Joosten and van der
Kroon (1974b), using a more sensitive technique, clearly showed that hypoactivity was first detectable by 5–10 days of age and was genetically determined,
and suggested that it made a significant contribution to pre-weaning storage of
excess energy.

ADIPOSITY

The developmental progression of obesity is very varied among the differing obese
species, as illustrated in figure 3.1. The homozygous recessively inherited obesities of *ob/ob* and *db/db* mice and *fa/fa* rats are characterised by a very early pre-
weaning deposition of excess fat, with obesity being visually obvious by 4 weeks
of age. In contrast, in the inbred NZO and, in particular, in the KK mice, the
dominant A^y/a mouse and the hybrid Wellesley C3$_f$I F$_1$ the obesity develops
somewhat later, with their growth curves often not being distinguishable from
those of lean controls until 2–4 months of age. Other species – e.g. the sand rat
and spiny mice – only develop obesity when their normal desert food is exchanged
for a laboratory diet (Bray and York, 1971; Herberg and Coleman, 1977). The
severity of the obesity also varies between the differing models, the obesity characteristic of KK, Wellesley and spiny mice and of desert rats being moderate in comparison with other species. However, the excess fat is distributed generally in all
the obese species except NZO mice, in which the fat is deposited mainly intra-
abdominally (Bray and York, 1971; Herberg and Coleman, 1977).

The obese species differ, however, in their mode of fat storage, whether it is
through enlargement of existing adipocytes (hypertrophic obesity) or through increasing both cell size and number (hypertrophic–hyperplastic obesity) (table 3.5).
Fat cell proliferation normally stops by 7–14 days in mice and 12–14 weeks in
rats. Furthermore, obesity resulting from hypothalamic damage results mainly
from hypertrophic growth of the adipose tissue (Johnson and Hirsch, 1972). Despite these observations, it is not possible to separate the obese mutants into hypertrophic and hypertrophic–hyperplastic obesities according to the age of onset of
the obesity. Thus, whereas the *ob/ob* mouse is an example of hypertrophic–
hyperplastic obesity, the *db/db* mice, which also develop obesity prior to weaning
and initially attain a severe form of obesity, accommodate their fat mainly through
hypertrophy of adipocytes, with only very moderate hyperplasia (Johnson and
Hirsch, 1972; Taljedal and Hellman, 1963). The KK mice, which only develop a
moderate form of obesity after 4–5 months of age, are, as expected, an example of
hypertrophic obesity. It is unlikely that we shall understand these differences in
adipose tissue growth until our knowledge of the mechanisms controlling adipocyte proliferation and pre-adipocyte maturation is advanced. It is known, however, that fat cell number may be increased in a number of obese mutant strains
after the normal age of proliferative activity by feeding high-fat diets (Lemonnier
and Alexiu, 1974; Herberg *et al.*, 1974). Indeed, the adult NZO mouse may be

David A. York

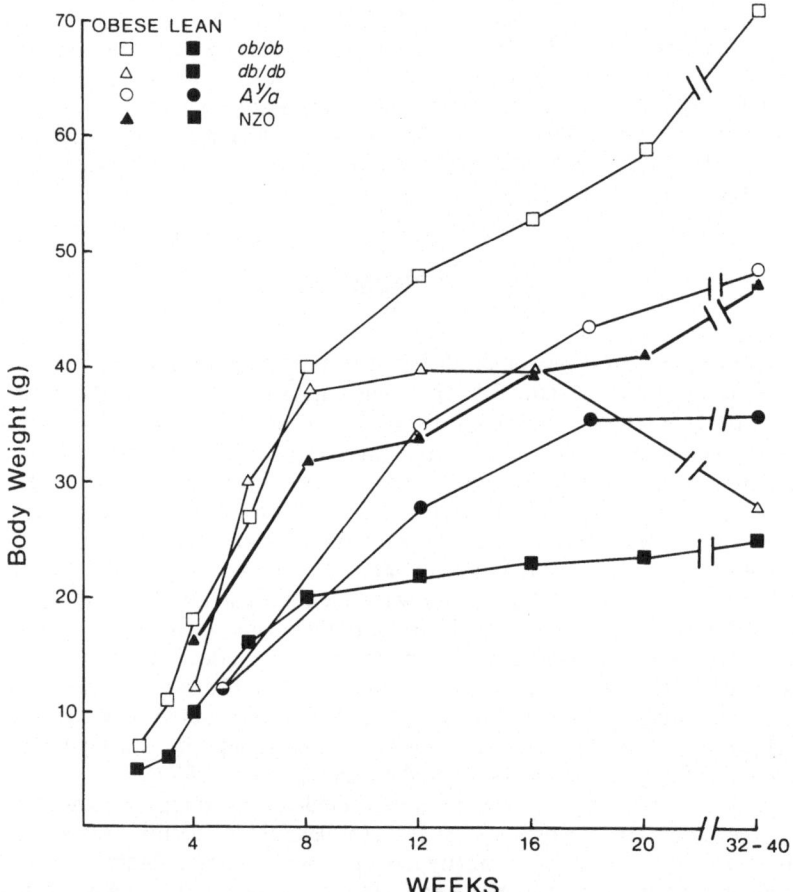

Figure 3.1   Body weight of obese mice from 2 weeks of age. As the body weights of the lean control mice for *ob/ob, db/db* and NZO mice were very similar, an average weight curve (Alonso and Maren, 1955) has been included. The figure was prepared from data in Bartke and Gorecki (1968), Wyse and Dulin (1970) and Dubuc (1977).

converted from hypertrophic to a hypertrophic–hyperplastic obesity by such a diet (Herberg *et al.*, 1974), although another report suggests that NZO mice normally exhibit hyperplastic growth of all adipose deposits (Johnson and Hirsch, 1972).

Table 3.5 Hypertrophy and hyperplasia of fat
cells in obese rodent models

Species	Fat cell no.	Fat cell size
A^y/a	o	++
ob/ob	++	++
db/db	+	++
KK	o	++
NZO	+	++
PBB/Ld	+	+
fa/fa	++	++

o = no change.
+ = increase.

For further information, the reader is referred to Johnson and Hirsch (1972),
Taljedahl and Hellman (1963), Nakamura (1962), Herberg *et al* (1970a),
Hunt *et al.* (1976) and Johnson *et al.* (1973).

PANCREATIC FUNCTION

A consistent observation in all obese mutants is that of hyperinsulinaemia
(Bray and York, 1971; Herberg and Coleman, 1977). The degree of hyperinsulinaemia not only is variable from species to species, but also varies considerably
with age in each species.

A small rise in serum insulin is one of the earliest changes in the *ob/ob* mouse,
being observed around 17 days of age 5–10 days after the onset of low body
temperature (Dubuc, 1977). After weaning from the high-fat milk diet to the
relatively high-carbohydrate mouse diet, serum insulin of *ob/ob* mice rapidly increases to reach values in excess of 500 µU/ml by 3–4 months of age (Dubuc,
1977; Herberg *et al.*, 1970b) before falling towards lean levels in older animals.
The initial hyperinsulinaemia is followed by insulin resistance which clearly expresses itself in the developing hyperglycaemia, which, again, may recede in
older animals. This picture of excessive pancreatic secretion and resultant insulin
resistance is typical of all forms of genetic obesity (Assimacopoulos-Jeannet and
Jeanrenaud, 1976); only the precise time of onset, the severity and duration of
the hyperinsulinaemia and insulin resistance vary between the differing species.
This is most clearly illustrated by comparison of C57BL/6J-*ob/ob* mice with
C57BL/Ks-*db/db* mice. The C57BL/Ks-*db/db* mouse has a small hyperinsulinaemia,
a more extreme hyperglycaemia and ketoacidosis which lasts for 3–4 months only
before the β cells fail and death ensues. However, when the background genomes
of *ob/ob* and *db/db* mice are exchanged, the C57BL/6J-*db/db* mouse develops an
hyperinsulinaemic syndrome similar to that of the C57BL/6J-*ob/ob* mouse, while
the C57BL/Ks-*ob/ob* mouse develops a syndrome similar to that of the C57BL/Ks-
db/db mouse (Coleman and Hummel, 1969; Coleman, 1973). These studies indicate the major role that background genetic material may play in the expression
of the metabolic defects which arise not only primarily from the gene defect, but
also as a consequence of obesity. Such differences in background genome probably

Table 3.6   Diabetic status in genetically inherited obesities

Type A:	Severe diabetes, insulin resistance and (ketosis)
	(1) C57BL/Ks-*db/db*
	(2) spiny mouse
	(3) C57BL/6J-*ob/ob* (no ketosis)
Type B:	Moderate hyperglycaemia, hyperinsulinaemia
	(1) A^y/a
	(2) NZO
	(3) KK
	(4) C3$_f$I F$_1$ hybrid
Type C:	Normoglycaemia, moderate hyperinsulinaemia
	(1) *fa/fa*

account for many of the differing reports in apparently similar animals that have been prevalent in the literature (Herberg and Coleman, 1977).

In general, the obese mutants may be divided into three groups (table 3.6) according to the severity of the diabetes and insulin resistance. Thus, the spiny mouse develops a severe diabetes with ketosis similar to that of CB57BL/Ks-*db/db* mice when its diet is changed from desert plants to laboratory food of high caloric density. The *ob/ob* mouse develops a severe hyperglycaemia but does not present a ketotic syndrome. The second group of obese mutants, which include A^y/a, NZO, KK and hybrid C3$_f$I F$_1$ Wellesley mice, all develop a moderate hyperglycaemia and hyperinsulinaemia which reverts towards or to normal with age. In contrast, the third group, exemplified by the fatty (*fa/fa*) rat, develops a moderate hyperinsulinaemia but has a normal or only marginally increased blood glucose (York *et al.*, 1972; Stauffacher *et al.*, 1967; Bray and York, 1971; Herberg and Coleman, 1977).

The changes in islet morphology in all models are consistent with insulin hypersecretion. Differing degrees of islet hypertrophy and hyperplasia have been reported together with increased tissue vascularisation. The number and appearance of granules within the β cells are indicative of their hyperactivity – e.g. the cells may be degranulated, packed full of dense granules or contain a mixture of dense and less densely filled granules (Shino *et al.*, 1973; Findlay *et al.*, 1973).

The morphological changes in the pancreas associated with obesity suggest that a similar sequence of events occurs in all the obese models. Thus, after an initial hypersecretion of insulin, the pancreas undergoes structural changes in an effort to synthesise more insulin. The ability to respond to the necessity for increased insulin supply varies between the obese models, resulting either in early death, stabilisation of blood glucose at a higher level or complete remission of the diabetes. Diabetic *db/db* mice and sand rats are unable to increase insulin synthesis sufficiently and severe ketosis and death result, while insulin secretion is sufficiently increased in A^y/a, NZO, Wellesley and spiny mice (Bray and York, 1971; Herberg and Coleman, 1977) to allow total or partial amelioration of the diabetes. The diabetes of KK and *ob/ob* mice, on the other hand, stabilises at a higher blood glucose level.

The hyperinsulinaemia in these obese syndromes is not associated with a defective clearance of the hormone. Studies with $[^{131}I]$-insulin suggest that the turnover of insulin is enhanced in *ob/ob* mice (Coore and Westman, 1970), while the degrading activity of adipose tissue (Westman, 1968) and of the hepatic enzyme glutathione-insulin transhydrogenase (Varandani and Nafz, 1976) are both increased. Further, in the species in which it has been studied both the biological activity and the immunological characteristics of the insulin secreted by the obese animals are normal (Laburthe *et al.*, 1975; Genuth, 1969; Stauffacher *et al.*, 1971). It is therefore unlikely that the diabetes and hyperinsulinaemia result from the secretion of a defective insulin molecule.

No consistent abnormality has been reported in α cell function of the various genetic obesities. Whereas serum glucagon is normal in *db/db* mice (Herberg *et al.*, 1976), it may be normal or slightly increased in *ob/ob* mice (Laube *et al.*, 1974; Lavine *et al.*, 1975; Mahler *et al.*, 1976) but is slightly decreased in *fa/fa* rats (Laburthe *et al.*, 1975). However, in all three animals the glucagon-to-insulin ratio is greatly reduced. Further, whereas the response of plasma glucagon to arginine stimulation is enhanced in *db/db* mice (Herberg *et al.*, 1976), it is normal in *ob/ob* mice (Beloff-Chain *et al.*, 1977) and impaired in *fa/fa* rats (Laburthe *et al.*, 1975).

Despite innumerable investigations of islet cell function, particularly in *in vitro* studies, the relationship between the primary gene defects and the altered islet function is still not clear for any of the obese mutants. Thus, the hypersecretion of insulin in response to glucose stimulation has been shown consistently in several obese types (Schade and Eaton, 1975; Herberg and Coleman, 1977; Beloff-Chain *et al.*, 1973). The hypersecretion is not rectified by reduction of body weight. Further evidence that a pancreatic defect may be central to obesity results from studies in two forms of obesity with transplanted islet tissue. In the initial report Strautz (1968, 1970) transplanted islets from lean mice into *ob/ob* mice and showed that the body weight gain ceased and the hyperinsulinaemia and hyperglycaemia normalised. However, these observations have never been repeated. In contrast, even streptozotocin-treated islets from lean mice transplanted intraperitoneally into NZO mice restored blood glucose and serum insulin (Gates *et al.*, 1972, 1974) to normal and ameliorated body weight gain. These results suggest that an additional factor which is freely diffusible and normally present in islet tissue may be important in coordinating the secretory responses of the β cells. This factor may be pancreatic polypeptide. It is absent in pancreas of NZO mice, and after injection weight gain, blood glucose and insulin normalise (Gates and Lazarus, 1977).

INSULIN RESISTANCE

The simultaneous occurrence of hyperinsulinaemia and normal or elevated blood glucose is suggestive of the insulin-resistant state which develops progressively in all of the obese models. In young animals before the development of gross obesity and severe hyperinsulinaemia, tissues show normal or even enhanced responses to insulin stimulation (Assimacopoulos-Jeannet and Jeanrenaud, 1976). An example of the changes in tissue sensitivity to insulin is shown by the response of adipose tissue from *fa/fa* rats to insulin stimulation (figure 3.2). Insulin resistance develops as a consequence of the hyperinsulinaemia and possibly acts initially as a protec-

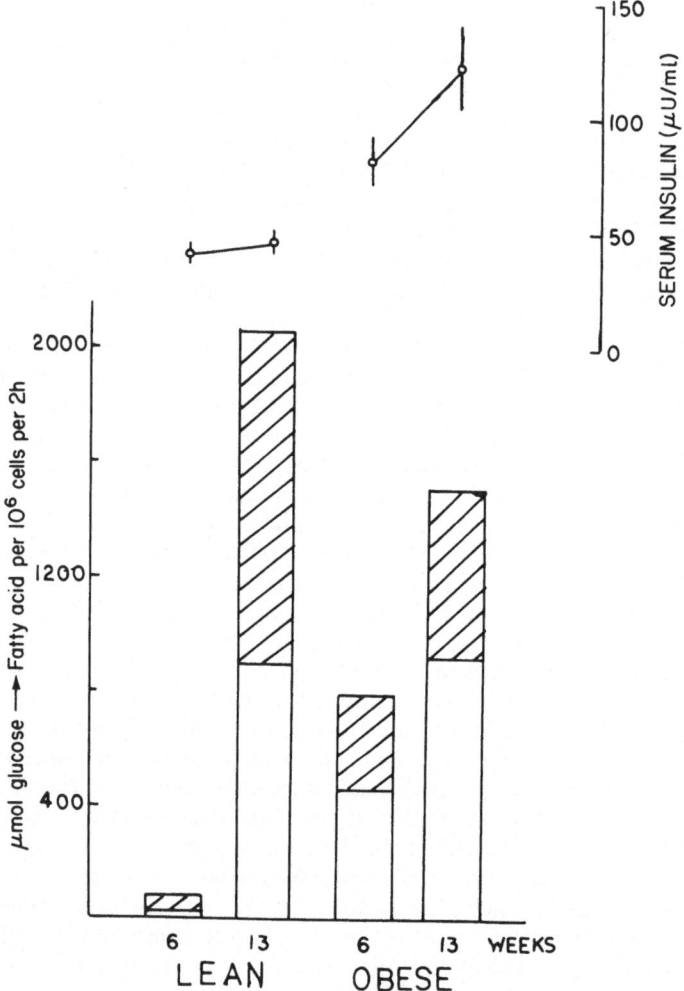

Figure 3.2 Fatty acid synthesis in adipose tissue of *ad libitum*-fed lean and obese *fa/fa* rats at 6 and 13 weeks of age. At 6 weeks of age basal fatty acid synthesis (open box) was greater in *fa/fa* rats and showed a good insulin (1 mU/ml) stimulation (cross-hatched box). At 13 weeks basal fatty acid synthesis was identical in lean and obese rats but insulin stimulation was diminished in obese rats. Serum insulin was higher at both ages in *fa/fa* rats. Data from York and Bray (1973a,b).

tive mechanism against the hypoglycaemic effects of insulin. Reduction of the hyperinsulinaemia through food restriction, adrenalectomy or drugs always improves and often restores insulin responses to normal (Chlouverakis and White, 1969; Cuendet *et al.*, 1976).

The mechanisms underlying the development of the insulin-resistant state are not fully understood (Olefsky, 1976; Czech *et al.*, 1977). The number of insulin receptors on plasma membranes of liver, adipose tissue and muscle of obese animals is decreased (Czech *et al.*, 1977; Olefsky, 1976; Soll *et al.*, 1975a, b; Chang *et al.*, 1975; Freychet *et al.*, 1972; Forque and Freychet, 1975), possibly owing to the suppressive action of high insulin concentrations on receptor number (Freychet, 1976). Although receptor number is increased in obese mutants when serum insulin and body weight are reduced, it is not possible to explain the resistant state solely by the attenuation of receptor number. Two lines of evidence argue against such a hypothesis: (1) Only a small percentage of receptors must be occupied for maximal biological response (Olefsky, 1976; Freychet, 1976). (2) The responses of different metabolic pathways within a single tissue can vary considerably. For example, in *ob/ob* mice hepatic lipogenesis is maximally stimulated in the hyper-insulinaemic state, whereas the gluconeogenic pathway is totally insensitive to insulin suppression (Assimacopoulos-Jeannet and Jeanrenaud, 1976). Thus, it is possible that a defect in the transport of glucose or its intracellular utilisation is an important factor contributing to the insulin-resistant state of these mice (Cuendet *et al.*, 1976).

LIPOGENESIS

The consequences of the hyperinsulinaemic state in these genetic mutants are profound and are reflected in (1) the overproduction of lipid by the liver and adipose tissue, (2) increased serum lipid levels and (3) changes in the control of lipolysis. The major changes in the fatty (*fa/fa*) rat which are outlined below are typical of all obese types in which these areas have been investigated. Immediately after weaning, fatty acid synthesis in both the liver and adipose tissue is greatly enhanced in the presence of the hyperinsulinaemia and hyperphagia. With increasing age the rate of hepatic fatty acid synthesis increases further, whereas the rate of fatty acid synthesis in adipose tissue falls, probably as a result of the increasing insulin insensitivity of the tissue. Despite the fall in adipose tissue fatty acid synthesis, the total amount of fatty acid synthesis in adipose tissue is increased in comparison with lean rats because of the increased mass of adipose tissue in the obese animal (Godbole and York, 1978). However, in *in situ* experiments, when a quantitative assessment of the transport of newly synthesised fatty acids from the liver to adipose tissue is made, it is clear that the liver is the major site of the enhanced lipogenesis in these animals. This is consistent with the demonstration of increased hepatic triacylglycerol synthesis (Bloxham *et al.*, 1977) and increased hepatic secretion of VLDL (Schonfeld and Pfleger, 1971; Schonfeld *et al.*, 1974) and serum hypertriglyceridaemia (Barry and Bray, 1969). This increase in serum triglyceride represents an imbalance between secretion into and removal from the blood which is found despite an increase in adipose tissue lipoprotein lipase activity (de Gasquet *et al.*, 1973). The *ob/ob* and *db/db* mice differ from this general picture in two ways. Firstly, although hepatic lipogenesis is greatly increased, it would appear that the major site of the enhanced lipogenesis is adipose tissue rather than liver (Hems *et al.*, 1975; Chan and Exton, 1977), despite the relative fall in adipose tissue fatty acid synthesis with increasing age.

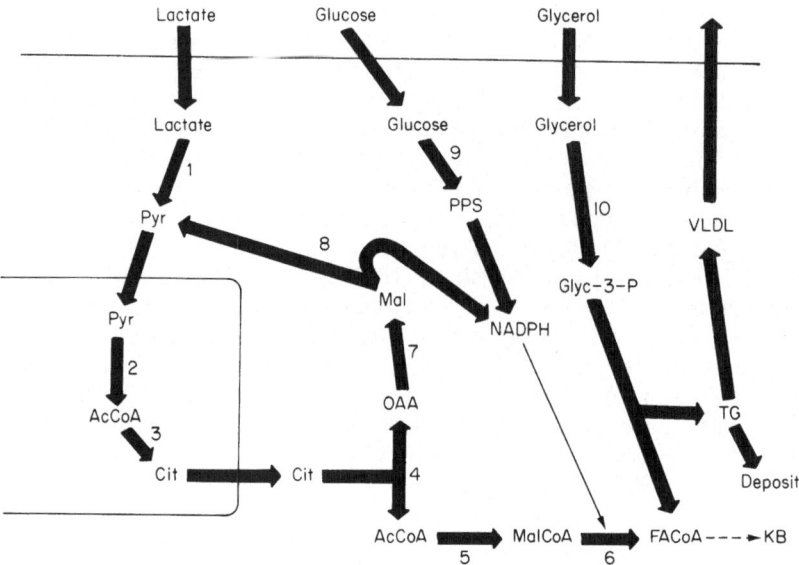

Figure 3.3   Increased lipogenesis in liver of obese mutants. Thick arrows indicate
pathways which are increased, dotted arrows pathways which are decreased compared with lean controls. Pyr = pyruvate; AcCoA = acetyl CoA; Cit = citrate;
MalCoA = malonyl CoA; FACoA = fatty acyl CoA; TG = triacylglycerol; OAA =
oxaloacetic acid; Mal = malate; PPS = pentose phosphate shunt; glyc-3-P =
glycerol-3-phosphate; KB = ketone bodies; VLDL = very low density lipoprotein.
The majority of enzymes in these pathways have been shown to have increased
activity. 1, Lactate dehydrogenase; 2, pyruvate dehydrogenase; 3, citrate synthetase; 4, ATP-citrate lyase; 5, acetyl CoA carboxylase; 6, fatty acid synthetase; 7,
NAD-dependent malic dehydrogenase; 8, malic enzyme; 9, hexokinase and
glucose-6-phosphate dehydrogenase; 10, glycerol kinase. The data are based upon
many studies, but the reader is referred to the following reviews for references:
Bray and York (1971), Herberg and Coleman (1977), Assimacopoulos-Jeannet
and Jeanrenaud (1977) and Bray (1977).

Secondly, the increase in adipose tissue and muscle lipoprotein lipase activities
is sufficient to compensate for the increased hepatic and intestinal production
of triglyceride, so that plasma triglyceride levels remain unaltered in *ob/ob* mice
(Salmon and Hems, 1973; Elliott *et al.*, 1974). Despite these small differences,
the turnover of serum triglyceride is greatly increased in all three forms of obesity.
These changes in lipogenesis are associated with enhancement in the activities of
lipogenic enzymes and in the enzymes of supporting pathways (see figure 3.3)
in young obese animals, although enzyme activity may be normal in older obese
animals in the static phase of the obesity (see Bray and York, 1971; Assimacopoulos-Jeannet and Jeanrenaud, 1976; Chan and Exton, 1977).

Several differing lines of evidence suggest that the enhancement in fatty acid
synthesis results from the hyperinsulinaemic state and the hyperphagia of these
animals. Thus, treatment of these animals with either anti-insulin serum or
streptozotocin reduces both hepatic and adipose tissue lipogenesis close to normal

values (Assimacopoulos-Jeannet *et al.*, 1977; Godbole and York, 1978). Reduction of food intake also results in lower rates of lipogenesis, although total lipogenesis remains increased because of the increased liver size and fat cell number. Before weaning it has been shown that both hepatic and adipose tissue lipogenesis in *ob/ob* mice and hepatic lipogenesis in *fa/fa* rats are normal (Jeanrenaud, 1978; Godbole *et al.*, 1978) but rise rapidly after weaning from the high-fat milk diet to the high-carbohydrate laboratory chow. The importance of weaning as the signal for hypersecretion of insulin and hyperlipogenesis was clearly illustrated in one study in which weaning was delayed (figure 3.4). In this situation the normal increase in serum insulin and hepatic lipogenesis that occurs in pre-obese *fa/fa* rats at weaning on day 21 was not observed. Although lipogenesis is normal before weaning, it can be demonstrated that both *ob/ob* mice and *fa/fa* rats

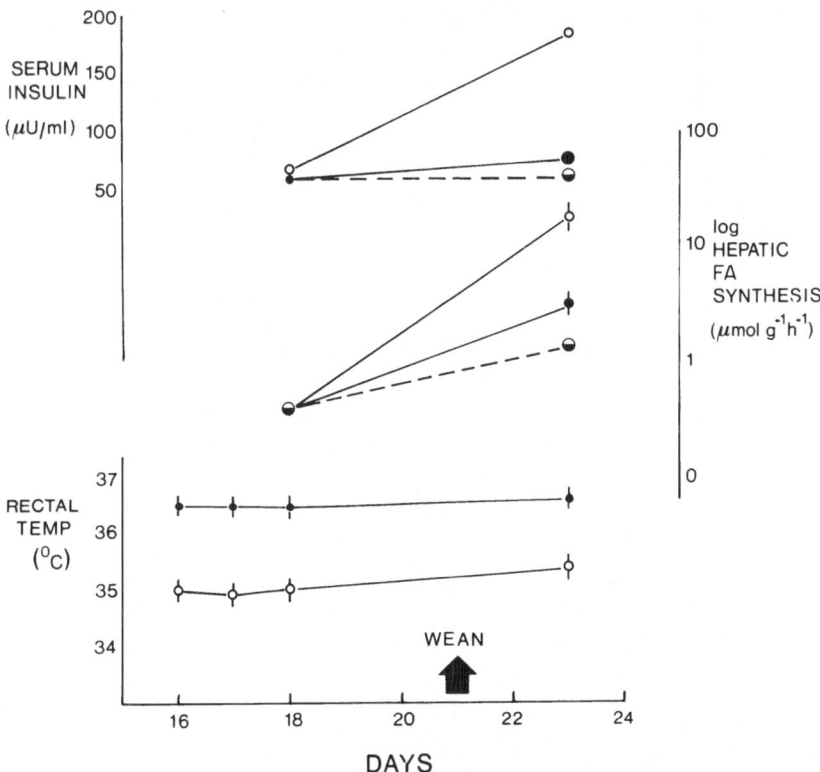

Figure 3.4 The effect of weaning on rectal temperature, serum insulin and hepatic fatty acid synthesis in lean (+/?) (●) and obese (*fa/fa*) fatty (o) rats. Weaning did not alter rectal temperatures in either group. Serum insulin and hepatic fatty acid synthesis were normal in *fa/fa* rats prior to weaning but rose to increased levels in *fa/fa* rats after weaning on day 21 (solid lines). When weaning was delayed to day 23 (dashed lines), serum insulin and hepatic lipogenesis in lean and *fa/fa* rats remained identical. Data from Godbole *et al.* (1978).

already have increased adipose tissue deposits at this age (Zucker and Antoniades, 1972; Bell and Stern, 1977).

Figure 3.5 summarises the major changes which account for the increase in both hepatic and adipose tissue lipogenesis. Although the precise relationship between liver and adipose tissue varies between the differing mutants, it is likely that the hyperinsulinaemia and hyperphagia are the major factors underlying this hyperlipogenesis.

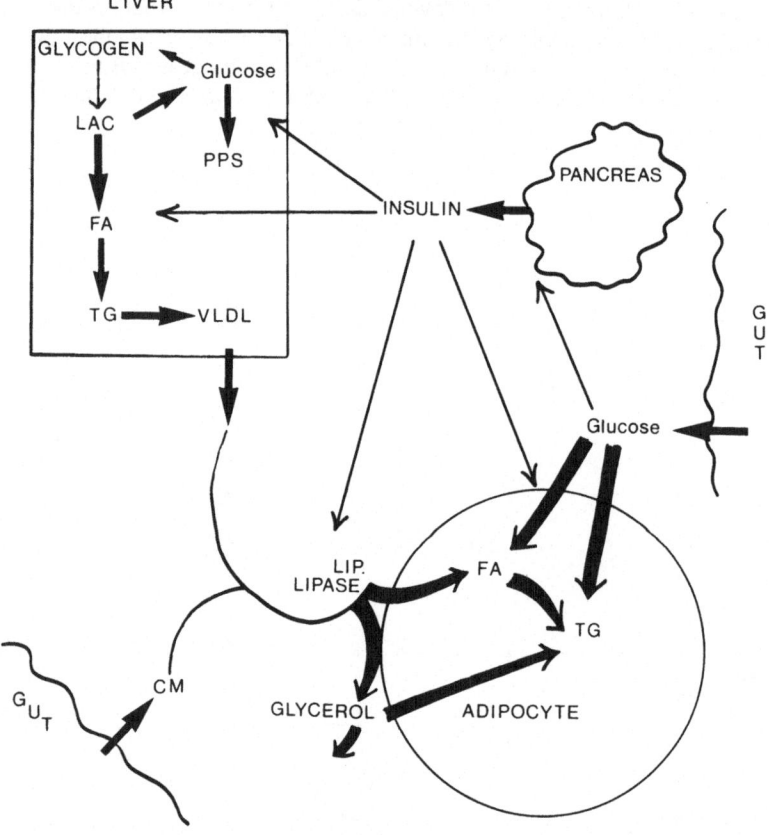

Figure 3.5 Control of lipogenesis in obese mutants. Thick arrows denote in-creased rate, whereas thin arrows denote stimulatory influences. In the presence of enhanced serum insulin both adipose tissue and hepatic fatty acid (FA) synthesis are increased. This situation is further exacerbated by an increased food intake providing increased substrate (glucose). This also further potentiates the hypersecretion of insulin. The increased very-low-density lipoprotein (VLDL) secretion from the liver and the increased fat (chylomicron, CM) absorption are counteracted by an increase in lipoprotein lipase activity under the influence of insulin. TG = triacylglycerol; Lac. = lactate; PPS = pentose phosphate shunt.

LIPOLYSIS

The response of the adipocyte lipolytic system is modified in the obese state as a result of the increased adipocyte size and the hyperinsulinaemia. The basal rate of lipolysis and glycerol release is related to the size of the fat cells (figure 3.6) despite the insulin stimulated increase in glycerokinase (E.C.2.7.1.30) activity which may account for the re-esterification of up to 20 per cent of the glycerol (Koschinsky *et al.*, 1971). Thus, with increased size and number of fat cells the obese animal releases more free fatty acid (FFA) and glycerol into the blood.

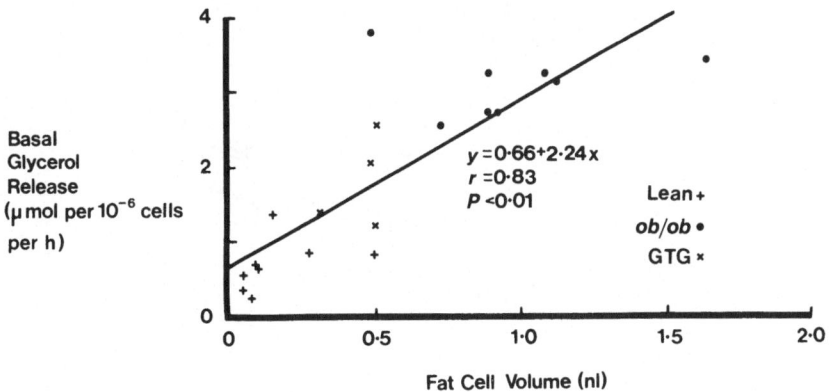

Figure 3.6   Relationship of fat cell size to basal glycerol release in lean (+/?), *ob/ob* and gold thioglucose (GTG) mice. Data reproduced from Otto *et al.* (1976), with permission of *Journal of Endocrinology.*

Much of these FFA may be reutilised in the liver for resynthesis and secretion of VLDL, a cycle (figure 3.7) which leads to increased turnover of serum FFA (Hems *et al.*, 1975). The development of enlarged fat cells also appears to be associated with some loss of sensitivity to lipolytic hormones *in vitro* (York, 1975; Otto *et al.*, 1976; York *et al.*, 1978; York and Bray, 1973a; Yen and Steinmetz, 1972; Enser, 1970). In *ob/ob* mice this loss has been attributed to either insulinisation of the tissue (Kissebah *et al.*, 1975) or a loss of thyroid hormone stimulation (Otto *et al.*, 1976; York *et al.*, 1978). However, since *in situ* studies have shown a normal rise in serum FFA in response to adrenaline (Abraham *et al.*, 1971; Zucker, 1972; Cuendet *et al.*, 1975) and since all obese mutants are able to withstand prolonged starvation, it is unlikely that these defects in the lipolytic system are of major importance to the aetiology of the obese state.

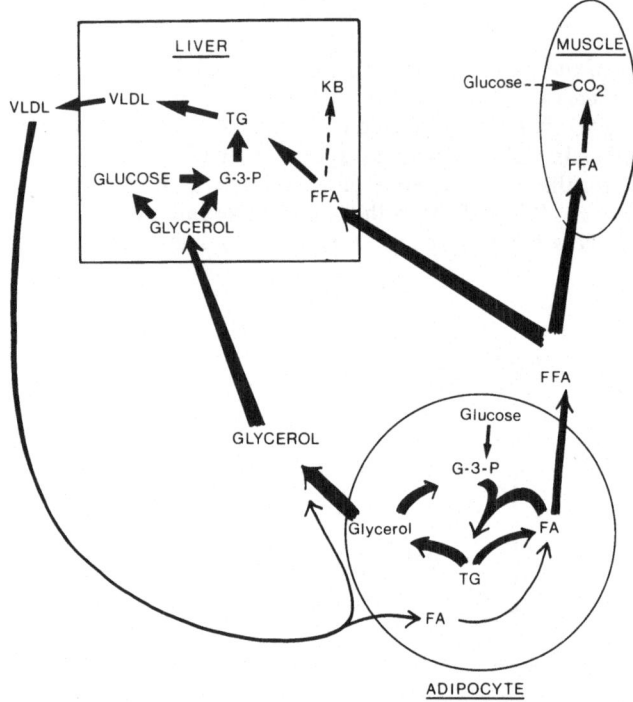

Figure 3.7   The effects of increased lipolysis in adipose tissue of obese models. The increased fat cell size results in an increase in glycerol and free fatty acid (FFA) release despite the increase in reutilisation of glycerol and re-esterification. The increased flux of free fatty acids (FFA) to the liver results in more triacylglycerol (TG) production and resecretion of very-low-density lipoprotein (VLDL), while ketone body (KB) formation is impaired. These changes lead to an increase in FFA turnover. The increased FFA may also result in a reduction of glucose utilisation by muscle, thus enhancing the insulin resistance and hyperglycaemia. Thick arrows denote increased rates; dashed lines denote reduced rates in comparison with lean controls. G-3-P = glycerol-3-phosphate.

ADRENAL FUNCTION

Some degree of adrenal hyperfunction appears to be a common characteristic of many obese mutants. Adrenal cortical hypertrophy has been described in NZO, A^y/a and fa/fa animals, whereas in KK mice the adrenals are smaller and in *ob/ob* mice adrenals are of normal weight (Bray and York, 1971). Adrenal function has been most investigated in the *ob/ob* mouse. Although serum corticosterone is first increased at day 17, it rises progressively with age and its concentration is closely related to the severity of diabetes, hyperinsulinaemia and insulin resistance (Naeser, 1974; Dubuc, 1976). Adrenalectomy results in a reduction or cessation of weight gain, a decrease in serum insulin and a fall in blood glucose when hyper-glycaemia is present in *ob/ob* (Solomon *et al.*, 1977; Solomon and Mayer, 1973; Naeser, 1973) and *A^y/a* (Jackson *et al.*, 1976) mice and *fa/fa* rats (Bray, personal communication; Hausberger and Hausberger, 1960; Godbole and York, unpublished observations). It is likely that these changes in adrenal corticosteroid secretion are (1) secondary to the obesity, since weight reduction ameliorates the adrenal hyperplasia, and are (2) responsible for the increased gluconeogenesis and its insensitivity to insulin suppression in the majority of obese mutants (see Bray and York, 1971; Herberg and Coleman, 1977; Assimacopoulos-Jeannet and Jeanrenaud, 1976). It is probable that an increased pituitary stimulation is responsible for the adrenal corticosteroid hypersecretion in *ob/ob* mice, since pituitary ACTH content is increased (Edwardson and Hough, 1975), while hypophysectomy prevents weight gain in *fa/fa* rats (Powley and Morton, 1976). It is likely that the hypersecretion of adrenal steroids plays an important role in both the hypersecretion of insulin and the insulin resistance of obese mutants.

GONADAL FUNCTION

Infertility is a common characteristic of the recessively inherited obesities in *ob/ob*, *db/db* and *fa/fa* mutants (Bray and York, 1971). Of the other inherited obesities, however, most show normal fertility, with the exception of *A^y/a*, which produce fewer litters and stop reproducing at an early age, and NZO, which have prolonged oestrus cycles. The infertility of *ob/ob* mice and *fa/fa* rats appears to involve an impaired regulation of gonadotrophin secretion (Saiduddin *et al.*, 1973; Swerdloff *et al.*, 1976). The changes are possibly secondary to or amplified by the obesity, since weight-reduced male *ob/ob* mice will reproduce with careful husbandry, while pregnant female *ob/ob* mice may be taken to term with progesterone treatment (Smithberg and Runner, 1956, 1957).

CONCLUSIONS

Although the obese models which have been described present quite variable syndromes, the major metabolic defects appear to occur in three areas: (1) the hypersecretion of insulin, (2) hyperphagia and (3) decreased thermogenesis. In the *ob/ob* mouse and *fa/fa* rat the defective thermogenesis precedes the hyper-insulinaemia and hyperphagia by several days. However, this does not mean that the latter two are a result of the impaired thermogenesis. An alternative possibility is that the primary enzyme or protein defect may be expressed in several areas, as outlined in figure 3.8. The interactions are such that a massive enhancement of lipogenesis and a massive obesity would be produced if the protein defect affected all pathways, whereas a more moderate obesity would result from impairment of fewer pathways. Such a defect in the enzyme $(Na^+ + K^+)$ ATPase has been suggested as the basis of the obesity in *ob/ob* mice (York *et al.*, 1978; Bray *et al.*, 1978).

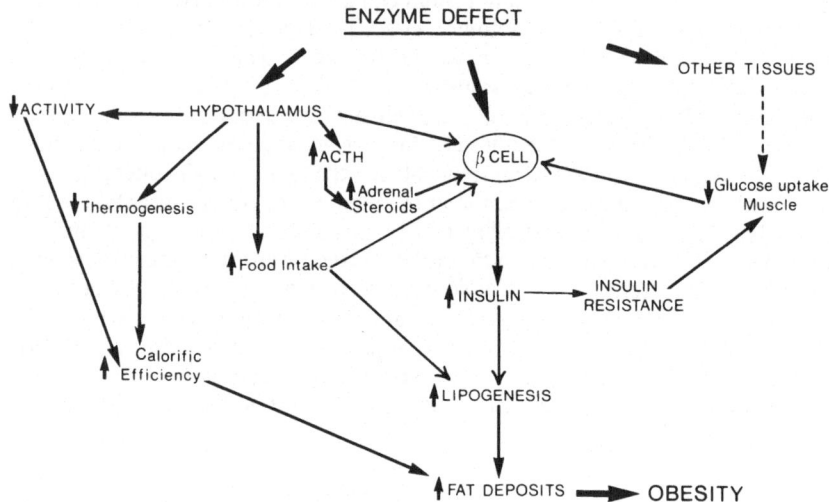

Figure 3.8   Possible locations of enzyme defect in obese mutants. The location of the enzyme defect in the hypothalamus would result in increased insulin secretion as shown. This situation would be further exaggerated if the same enzyme defect occurred in the pancreas, resulting in a hypersecretion of insulin to normal stimuli. A defective uptake of glucose by other tissues either as a result of the gene defect or as a result of insulin resistance would further increase the blood glucose level, leading to further insulin secretion.

REFERENCES

Abraham, R. R., Dade, E. Elliot, J. and Hems, D. A. (1971). Hormonal control of intermediary metabolism in obese hyperglycaemic mice. *Diabetes*, **20**, 535–541

Alonso, L. G. and Maren, T. H. (1955). Effect of food restriction on body composition of hereditary obese mice. *Am. J. Physiol.*, **183**, 284–290

Assimacopoulos-Jeannet, F. and Jeanrenaud, B. (1976). The hormonal and metabolic basis of experimental obesity. *Clin. Endocr. Metab.*, **5**, 337–365

Assimacopoulos-Jeannet, F., Karakash, C., Le Marchand, Y. and Jeanrenaud, B. (1977). Insulin effect upon lipogenesis of perfused mouse liver. *Biochim. Biophys. Acta*, **498**, 91–101

Bailey, C., Atkins, T., Conner, H., Hanley, C. and Matty, A. (1975). Diurnal variations of food consumption, plasma glucose and plasma insulin in lean and obese hyperglycemic mice. *Hormones*, **6**, 380–386

Barry, W. and Bray, G. A. (1969). Plasma triglycerides of genetically obese rats. *Metabolism*, **18**, 833–839

Bartke, A. and Goreki, A. (1968). Oxygen consumption by obese yellow mice and their normal littermates. *Am. J. Physiol.*, **214**, 1250–1252

Becker, E. E. and Grinker, J. A. (1977). Meal patterns in the genetically obese Zucker rat. *Physiol. Behav.*, **18**, 685–692

Bell, G. E. and Stern, J. S. (1977). Evaluation of body composition of young obese and lean Zucker rats. *Growth*, **41**, 63–80

Beloff-Chain, A., Newman, H. and Mansford, K. (1973. *In vitro* studies on insulin secretion in genetically obese mouse. *Diabetologia*, 9, 447–457

Beloff-Chain, A., Newman, M. and Mansford, K. (1977). Factors influencing insulin and glucagon secretion in lean and genetically obese mice. *Horm. Metab. Res.*, 9, 33–37

Bloxham, D., Fitzsimmons, S. and York, D. A. (1977). Lipogenesis in hepatocytes of genetically obese rats. *Horm. Metab. Res.*, 9, 304–309

Boulangé, A., Planche, E. and de Gasquet, P. (1977). Excess fat storage and normal energy intake in the new born Zucker rat (*fa/fa*). In *Proc. 2nd Int. Congr. Obesity, Washington, DC* (ed. G. A. Bray), Newman, London, 1978

Bray, G. A. (1969). Oxygen consumption of genetically obese rats. *Experientia*, 25, 1100–1101

Bray, G. A. (1977). The Zucker fatty rat: a review. *Fed. Proc.*, 36, 148–153

Bray, G. A. and York, D. A. (1971). Genetically transmitted obesity in rodents. *Physiol. Rev.*, 51, 598–646

Bray, G. A. and York, D. A. (1972). Studies on food intake of genetically obese rats. *Am. J. Physiol.*, 223, 176–179

Bray, G. A., York, D. A. and Swerdloff, R. (1973). Genetic obesity in rats 1. The effects of food restriction on body composition and hypothalamic function. *Metabolism*, 22, 435–442

Bray, G. A., York, D. A. and Yukimura, Y. (1978). Activity of $(Na^+ + K^+)$ ATP-ase in liver of animals with experimental obesity. *Life Sciences*, 22, 1637–1642

Chan, T. M. and Exton, J. (1977). Hepatic metabolism of genetically diabetic (*db/db*) mice. *Biochim. Biophys. Acta*, 489, 1–14

Chang, K.-J., Huang, D. and Cuatrecasas, P. (1975). Defect in insulin receptors in obese hyperglycemic mice. Probable alterations in membrane glycoproteins. *Biochem. Biophys. Res. Comm.*, 64, 566–573

Chlouverakis, C. (1970). Induction of obesity in obese-hyperglycemic (*ob/ob*) mice on normal food intake. *Experientia*, 26, 1262–1263

Chlouverakis, C. and White, P. A. (1969). Obesity and insulin resistance in the obese-hyperglycemic mouse (*ob/ob*). *Metabolism*, 18, 998–1066

Coleman, D. L. (1973). Effects of parabiosis of obese with diabetes and normal mice. *Diabetologia*, 9, 294–298

Coleman, D. L. and Hummel, K. P. (1969). The effects of parabiosis of normal with genetically diabetic mice. *Am. J. Physiol.*, 217, 1298–1304

Coleman, D. L. and Hummel, K. P. (1973). The influence of genetic background on the expression of the obese (*ob*) gene in the mouse. *Diabetologia*, 9, 287–293

Coore, H. G. and Westman, S. (1970). Disappearance of serum insulin in obese-hyperglycemic mice. *Acta Physiol. Scand.*, 78, 274–279

Cox, J. E. and Powley, T. L. (1977). Development of obesity in diabetic mice pair-fed with lean siblings. *J. Comp. Phys. Psychol.*, 91, 347–358

Cruce, J. A., Greenwood, M. R., Johnson, P. and Quatermain, D. (1974). Genetic versus hypothalamic obesity. Studies of intake and dietary manipulations in rats. *J. Comp. Physiol. Psychol.*, 87, 295–301

Cruce, J. A., Thoa, N. and Jacobwitz, D. M. (1976). Catecholamines in the brains of genetically obese rats. *Brain Res.*, 101, 165–170

Cuendet, G. S., Loten, E. G., Cameron, D., Renold, A. E. and Marliss, E. B. (1975). Hormone substrate responses to total fasting in lean and obese mice. *Am. J. Physiol.*, 228, 276–283

Cuendet, G. S., Loten, E. G., Jeanrenaud, B. and Renold, A. E. (1976). Decreased basal, non-insulin stimulated glucose uptake and metabolism by skeletal soleus muscle isolated from obese-hyperglycemic (*ob/ob*) mice. *J. Clin. Invest.*, 58, 1078–1088

Czech, M. P., Richardson, D. K. and Smith, C. J. (1977). Biochemical basis of fat cell resistance in obese rodents and man. *Metabolism*, 26, 1057–1078

Davis, T. R. and Mayer, J. (1954). Imperfect homeothermia in the hereditary obese-hyperglycemic syndrome of mice. *Am. J. Physiol.*, 177, 222–226

60   *David A. York*

Deb, S. and Martin, R. (1975). Effects of exercise and of food restriction on the development of spontaneous obesity in rats. *J. Nutr.*, **105**, 543–549

de Gasquet, P., Pequinot, R., Lemmonier, D. and Alexiu, A. (1973). Adipose tissue lipoprotein lipase activity and cellularity in genetically obese Zucker rat (*fa/fa*). *Biochem. J.*, **132**, 633–635

Dickerson, G. E. and Gowan, J. W. (1947). Hereditary obesity and efficient food utilisation in mice. *Science*, **105**, 496–498

Dickie, M. M. and Wooley, G. W. (1946). The age factor in weight of yellow mice. Weight reduction of ageing yellow and 'thin-yellows' revealed in littermate comparisons. *J. Heredity*, **37**, 365–368

Dubuc, P. U. (1976). Basal corticosterone levels in young *ob/ob* mice. *Horm. Metab. Res.*, **9**, 95–97

Dubuc, P. U. (1977). Development of obesity, hyperinsulinaemia and hyperglycemia in *ob/ob* mice. *Metabolism*, **25**, 1567–1574

Dulin, W. E. and Wyse, B. M. (1970). Diabetes in the KK mouse. *Diabetologia*, **6**, 317–323

Edwardson, J. A. and Hough, C. A. M. (1975). The pituitary adrenal system of the genetically (*ob/ob*) mouse. *J. Endocrinol.*, **65**, 99–107

Elliott, J. A., Dade, E., Salmon, D. M. W. and Hems, D. A. (1974). Hepatic metabolism in normal and genetically obese mice. *Biochim. Biophys. Acta*, **343**, 307–323

Enser, M. (1970). Fatty acid mobilisation in obese mice. *Nature*, **226**, 175–177

Fenton, P. F. and Chase, H. B. (1951). Effect of diet on obesity of yellow mice in inbred lines. *Proc. Soc. Exp. Biol. Med.*, **77**, 420–422

Findlay, J. A., Rookledge, K. A., Beloff-Chain, A. and Lever, J. D. (1973). A combined biochemical and histological study of the islets of Langerhans in the genetically obese hyperglycemic mouse and in the lean mouse, including observations on the effect of strepozotocin treatment. *J. Endocrinol*, **56**, 571–583

Forque, M. and Freychet, P. (1975). Insulin receptors in heart-muscle — demonstration of specific binding sites and impairment of insulin binding in plasma membrane of obese hyperglycemic mouse. *Diabetes*, **24**, 715–713

Freychet, P. (1976). Interaction of polypeptide hormones with cell membrane specific receptors. Studies with insulin and glucagon. *Diabetologia*, **12**, 83–100

Freychet, P., Laudat, M., Laudat, P., Rosselin, G., Kahn, C., Gordon, P. and Roth, J. (1972). Impairment of insulin binding to the fat cell plasma membrane in the obese hyperglycemic mouse. *FEBS Lett.*, **25**, 339–342

Fuller, J. L. (1972). Genetic aspects of the regulation of food intake. *Adv. Psychosom. Med.*, **1**, 2–24

Fuller, J. L. and Jacoby, G. A. (1955). Central and sensory control of food intake in genetically obese mice. *Am. J. Physiol.*, **183**, 279–283

Gates, R. J., Hunt, M. and Lazarus, N. R. (1974). Further studies on the amelioration of the characteristics of New Zealand obese (NZO) mice following implantation of islets of Langerhans. *Diabetologia*, **10**, 401–406

Gates, R. J., Hunt, M., Smith and Lazarus, N. R. (1972). Return to normal of blood glucose, plasma insulin and weight gain in New Zealand (NZO) mice after implantation of islets of Langerhans. *Lancet*, **ii**, 567–570

Gates, R. J. and Lazarus, N. R. (1977). The ability of pancreatic polypeptides (APP and BPP) to return to normal the hyperglycemia, hyperinsulinaemia and weight gain of New Zealand obese mice. *Horm. Res.*, **8**, 189–202

Genuth, S. M. (1969). Hyperinsulinism in mice with genetically determined obesity. *Endocrinology*, **84**, 386–391

Godbole, V. and York, D. A. (1978). Lipogenesis *in situ* in the genetically obese Zucker fatty rat (*fa/fa*). Role of hyperphagia and hyperinsulinaemia. *Diabetologia*, **14**, 191–198

Godbole, V., York, D. A. and Bloxham, D. P. (1978). Developmental change in the fatty (*fa/fa*) rat. Evidence for defective thermogenesis preceding the hyperlipogenesis and hyperinsulinaemia. *Diabetologia*, 15, 41–44

Greenwood, M. R. C., Quatermain, M., Johnson, P., Cruce, J. and Hirsch, J. (1974). Food motivated behaviour in genetically obese and hypothalamic-hyperphagic rats and mice. *Physiol. Behav.*, 13, 687–692

Hausberger, F. X. and Hausberger, B. G. (1960). The etiologic mechanism of some forms of hormonally induced obesity. *Am. J. Clin. Nutr.*, 8, 671–681

Hems, D. A., Rath, E. A. and Verinder, T. R. (1975). Fatty acid synthesis in liver and adipose tissue of normal and genetically obese (*ob/ob*) mice during the 24 hour cycle. *Biochem. J.*, 150, 167–173

Herberg, L., Berger, M., Buchanan, K., Gries, F. and Kern, H. (1976). Tiermodelle in der Diabetesforschung metabolische und hormonelle Besonderheiten. *Z. Versuchstierk.*, 18, 91–105

Herberg, L. and Coleman, D. (1977). Laboratory animals exhibiting obesity and diabetes syndromes. *Metabolism*, 26, 59–99

Herberg, L., Doeppen, W., Major, E. and Gries, F. (1974). Dietary induced hypertrophic-hyperplastic obesity in mice. *J. Lipid Res.*, 6, 580–585

Herberg, L., Gries, F. A. and Hesse-Wortmann, C. H. (1970a). Effect of weight and cell size on hormone-induced lipolysis in New Zealand obese mice and American obese hyperglycemic mice. *Diabetologia*, 6, 300–305

Herberg, L., Major, E., Hemmings, U., Gruneklee, D., Freytag, G. and Gries, F. A. (1970b). Differences in the development of the obese-hyperglycemic syndrome in *ob/ob* and NZO mice. *Diabetologia*, 6, 292–299

Hunt, C. E., Lindsey, J. R. and Walkley, S. U. (1976). Animal models of diabetes and obesity including PBB-Ld mouse. *Fed. Proc.*, 35, 1206–1217

Jackson, E., Stolz, D. and Martin, R. (1976). Effect of adrenalectomy on weight gain and body composition of yellow obese mice (A^y/a). *Horm. Metab. Res.*, 8, 452–454

Jeanrenaud, B. (1978). An overview of experimental models of obesity. In *Recent Advances in Obesity Research*, Vol. 2 (ed. G. A. Bray), Newman, London, p. 111–122

Johnson, P., Stern, Greenwood, H., Zucker, L., and Hirsch, J. (1973). Effect of early nutrition in adipose cellularity and pancreatic insulin release in Zucker rat. *J. Nutr.*, 103, 738–743

Johnson, P. R. and Hirsch, J. (1972). Cellularity of adipose deposits in six strains of genetically obese mice. *J. Lipid Res.*, 13, 2–11

Joosten, H. F. and van der Kroon, P. H. W. (1974a). Role of the thyroid in the development of the obese-hyperglycemic syndrome in mice (*ob/ob*). *Metabolism*, 23, 425–436

Joosten, H. F. and van der Kroon, P. H. W. (1974b). Growth pattern and behavioral traits associated with development of obese-hyperglycemic syndrome in mice (*ob/ob*). *Metabolism*, 23, 1141–1147

Kissebah, A. H., Clark, P., Vydelingum, N., Gill, F., Tulloch and Fraser, T. (1975). Mechanism of insulin resistance associated with obesity. In *Recent Advances in Obesity Research* (ed. A. Howard), Newman, London, pp. 152–154

Koschinsky, Th., Gries, F. A. and Herberg, L. (1971). Regulation of glycerol kinase by insulin in isolated fat cells and liver of Bar Harbor obese mice. *Diabetologia*, 7, 316–322

Laburthe, M., Rancon, F., Freychet, P. and Rosselin, G. (1975). Glucagon and insulin from lean rats and genetically obese rats. Studies by radio-immunoassay, radioreceptor assay and bioassay. *Diabetologia*, 11, 517–526

Laube, H., Fussgang, R. and Pfeiffer, E. (1974). Paradoxical glucagon release in obese hyperglycemic mice. *Horm. Metab. Res.*, 6, 426–429

Lavine, R. L., Voyles, N., Perrino, P. V. and Recant, L. (1975). The effect of fasting on tissue cyclic AMP and glucagon in the obese hyperglycemic mouse. *Endocrinology*, 97, 615–620

Lemonnier, D. and Alexiu, A. (1974). Nutritional, genetic and hormonal defects of adipose tissue cellularity. In *Regulation of Adipose Tissue Mass (Proc. IV Int. Meeting Endocrin.)* (ed. J. Vague and J. Boyer), Excerpta Medica, Amsterdam, pp. 158–173

Lorden, L., Oltmans, G. and Margules, D. (1975). Central catecholamine levels in genetically obese mice (*ob/ob* and *db/db*). *Brain Res.*, 96, 390–394

McCloughlin, C., Baile, C. and Chalup, N. (1977). Relative sensitivity of Zucker lean and obese rats to food intake depressing effect of cholecystokinin and D-amphetamine sulphate and stimulating effect of diazepam. *Fed. Proc.*, 36, 1150

Mahler, R., Dubuc, P., Mobley, P. and Ensink, J. (1976). Glucagon and insulin relationships in the obese hyperglycemic mouse (*ob/ob*). *Horm. Metab. Res.*, 8, 79–80

Miller, B., Otto, W., Grimble, R., York, D. A. and Taylor, T. G. (1979). The relationship between protein turnover and energy balance in lean and genetically obese (*ob/ob*) mice. *Br. J. Nutr.* (in press)

Naeser, P. (1973). Effects of adrenalectomy on the obese-hyperglycemic syndrome in mice. *Diabetologia*, 9, 376–379

Naeser, P. (1974). Function of the adrenal cortex in obese-hyperglycemic mice. *Diabetologia*, 10, 449–453

Nakamura, M. (1962). A diabetic strain of the mouse. *Proc. Japan Acad.*, 38. 348–352

Olefsky, J. (1976). Insulin receptor. Its role in insulin resistance of obesity and diabetes. *Diabetes*, 25, 1154–1162

Oltmans, G. A., Lorden, J. F. and Margules, D. L. (1976). Effects of food restriction and mutation on central catecholamine levels in genetically obese mice. *Pharm. Biochem. Behav.*, 5, 617–620

Otto, W., Taylor, T. G. and York, D. A. (1976). Glycerol release *in vitro* from adipose tissue of obese (*ob/ob*) mice treated with thyroid hormones. *J. Endocrinol.*, 71, 143–155

Parson, W., Camp, J. L. and Crispell, K. R. (1954). Dietary dilution studies in mice with gold-thioglucose-induced obesity and in mice with the hereditary obesity-diabetes syndrome of mice utilising C^{14}-acetate. *Metabolism*, 3, 351–356

Powley, T. (1977). The ventromedial hypothalamic syndrome, satiety and a cephalic phase hypothesis. *Psychol. Rev.*, 84, 89–126

Powley, T. and Morton, S. (1976). Hypophysectomy and regulation of body weight in genetically obese Zucker rat. *Am. J. Physiol.*, 230, 982–987

Pullar, J. D. and Webster, A. J. (1974). Heat loss and energy retention during growth in congenitally obese and lean rats. *Br. J. Nutr.*, 31, 377–392

Saiduddin, S., Bray, G. A., York, D. A. and Swerdloff, R. (1973). Reproductive function in genetically obese rats. *Endocrinology*, 93, 1251–1256

Salmon, M. W. and Hems, D. A. (1973). Plasma lipoproteins and the synthesis and turnover of plasma triglyceride in normal and genetically obese mice. *Biochem. J.*, 136, 551–563

Schade, D. S. and Eaton, R. P. (1975). Insulin secretion by perfused islets from Zucker obese rats. *Proc. Soc. Exp. Biol. Med.*, 149, 311–314

Schonfeld, G., Felski, C. and Howald, M. (1974). Characterisation of plasma lipoproteins of the genetically obese hyperlipoproteinemic Zucker fatty rat. *J. Lipid Res.*, 15, 457–464

Schonfeld, G. and Pfleger, B. (1971). Overproduction of very low density lipoproteins by livers of genetically obese rats. *Am. J. Physiol.*, 220, 1178–1181

Sclafani, A. (1976). Appetite and hunger in experimental obesity syndromes. In *Hunger: Basic Mechanisms and Clinical Implications* (ed. D. Novin, W. Wyrwicka and G. A. Bray), Raven Press, New York, pp. 281–295

Shino, A., Matsuo, T., Iwatsuku, H. and Suzuoki, Z. (1973). Structural changes of pancreatic islets in genetically obese rats. *Diabetologia*, 9, 413–421

Singh, D., Lakey, J. and Sanders, M. (1974). Hunger motivation in gold thioglucose-treated and genetically obese female mice. *J. Comp. Physiol. Psychol.*, 86, 890–897

Smithberg, M. and Runner, M. N. (1956). The induction and maintenance of pregnancy in prepubertal mice. *J. Exp. Zool.*, 133, 441–458

Smithberg, M. and Runner, M. N. (1957). Pregnancy induced in genetically sterile mice. *J. Heredity*, 48, 97–100

Soll, A. H., Kahn, C. R., Neville, D. and Roth, J. (1975a). Insulin binding to liver plasma membranes in obese-hyperglycemic mouse. Demonstration of a decreased number of normal receptors. *J. Biol. Chem.*, **250**, 4702–4707

Soll, A. H., Kahn, C. R., Neville, D. and Roth, J. (1975b). Insulin receptor deficiency in genetic and acquired obesity. *J. Clin. Invest.*, **56**, 769–780

Solomon, J., Bradwin, J., Cocetia, H., Coffey, D., Condor, T., Garity, W. and Grieco, W. (1977). Effects of adrenalectomy on body weight and hyperglycemia in 5 month old *ob/ob* mice. *Horm. Metab. Res.*, **9**, 152–156

Solomon, J. and Mayer, J. (1973). The effect of adrenalectomy in the development of the obese-hyperglycemic syndrome in *ob/ob* mice. *Endocrinology*, **93**, 510–513

Sprott, R. (1972). Long term studies of feeding behaviour in obese diabetic and viable yellow mutant mice under *ad lib* and operant conditions. *Psych. Rep.*, **30**, 991–1003

Stauffacher, W., Lambert, A., Vecchio, D. and Renold, A. E. (1967). Measurements of insulin activities in pancreas and serum of mice with spontaneous ('obese' and 'New Zealand obese') and induced (gold thioglucose) obesity and hyperglycemia with considerations on the pathogenesis of the spontaneous syndrome. *Diabetologia*, **3**, 230–237

Stauffacher, W., Orci, L., Cameron, D., Burr, I. and Renold, A. (1971). Hyperglycemia and/or obesity in laboratory rodents: an example of the possible usefulness of animal disease models with both genetic and environmental components. *Rec. Prog. Horm. Res.*, **27**, 41–95

Stern, J. S. and Johnson, P. (1977). Spontaneous activity and adipose cellularity in genetically obese Zucker rat (*fa/fa*). *Metabolism*, **26**, 371–380

Strautz, R. L. (1968). Islet implants: reduction of glucose levels in the hereditary obese mouse. *Endocrinology*, **83**, 975–978

Strautz, R. L. (1970). Studies of hereditary obese mice (*ob/ob*) after implantation of pancreatic islets in millipore filter capsules. *Diabetologia*, **6**, 306–312

Subrahmanyan, K. (1960). Metabolism in New Zealand strain of obese mice. *Biochem. J.*, **76**, 548–556

Swerdloff, R. S., Batt, R. A. and Bray, G. A. (1976). Reproductive hormonal function in the genetically obese mouse. *Endocrinology*, **98**, 1359–1364

Täljedahl, I.-B. and Hellman, B. (1963). Morphological characteristics of the epididymal adipose tissue in different types of hereditary obese mice. *Pathol. Microbiol.*, **26**, 149–157

Trayhurn, P., Thurlbey, P. and James, W. (1977). Thermogenic defect in preobese *ob/ob* mice. *Nature*, **266**, 60–62

van der Kroon, P. H. (1966). Hereditary obesity in mice (PhD thesis). CAtholic Univ. Nijmegen (in Dutch)

Varandani, P. and Nafz, M. A. (1976). Insulin degradation. Regulation of glutathione: insulin transhydrogenase in hyperglycemic *ob/ob* mice. *Biochim. Biophys. Acta*, **451**, 382–392

Westman, S. (1968). Degradation of insulin *in vitro* by liver and epididymal adipose tissue from obese-hyperglycemic mice. *Biochem. J.*, **106**, 543–547

Woodward, C., Trayhurn, P. and James, W. (1978). Costs of maintenance and growth in genetically obese (*ob/ob*) mice. *Proc. Nutr. Soc.*, **36**, 115A

Wyse, B. and Dulin, W. (1970). The influence of age and dietary conditions on diabetes in the *db* mouse. *Diabetologia*, **6**, 268–273

Wyse, B. and Dulin, W. (1974). Further characterisation of the diabetes-like abnormalities in the T-KK mouse. *Diabetologia*, **10**, 617–623

Yen, T. T. T. and Acton, J. (1972). Locomotor activity of various types of genetically obese mice. *Proc. Soc. Exp. Biol. Med.*, **140**, 647–650

Yen, T. T. T., Fuller, R. and Pearson, D. (1974). Response of obese (*ob/ob*) and diabetic (*db/db*) mice to treatments that influence body temperature. *Comp. Biochem. Physiol.*, **49**, 377–385

Yen, T. T. T. and Steinmetz, J. A. (1972). Lipolysis of genetically obese and/or hyperglycemic mice with reference to insulin response of adipose tissue. *Horm. Metab. Res.*, **4**, 331–337

York, D. A. (1975). Lipid metabolism in genetic models of obesity. *Proc. Nutr. Soc.*, **34**, 249–256

York, D. A. and Bray, G. A. (1973a). Genetic obesity in rats. II. The effect of food restriction on metabolism of adipose tissue. *Metabolism*, **22**, 443–454

York, D. A. and Bray, G. A. (1973b). Adipose tissue metabolism in 6 week old fatty rats. *Horm. Metab. Res.*, **5**, 355–360

York, D. A., Bray, G. A. and Yukimura, Y. (1978). An enzymatic defect in the obese (*ob/ob*) mouse. Loss of the thyroid inducible sodium-potassium dependent adenosine triphosphatase. *Proc. Nat. Acad. Sci. USA*, **75**, 477–481

York, D. A., Hershman, J. M., Utiger, R. D. and Bray, G. A. (1972). Thyrotropin secretion in genetically obese rats. *Endocrinology*, **90**, 67–72

York, D. A., Otto, W. and Taylor, T. G. (1978). Thyroid status of obese (*ob/ob*) mice and its relationship to adipose tissue metabolism. *Comp. Biochem. Physiol.*, **59B**, 59–65

York, D. A., Steinke, J. and Bray, G. A. (1972). Hyperinsulinaemia and insulin resistance in genetically obese rats. *Metabolism*, **21**, 277–284

Zucker, L. M. (1972). Fat mobilisation *in vitro* and *in vivo* in the genetically obese Zucker rat 'fatty'. *J. Lipid Res.*, **13**, 234–243

Zucker, L. M. (1975). Efficiency of energy utilisation by the Zucker hereditarily obese rat 'fatty'. *Proc. Soc. Exp. Biol. Med.*, **148**, 498–500

Zucker, L. M. and Antoniades, H. (1972). Insulin and obesity in the Zucker genetially obese rat 'fatty'. *Endocrinology*, **90**, 1320–1330

4

The husbandry
and maintenance of
genetic models of obesity

David P. Lovell (Medical Research Council, Laboratory Animals Centre,
Woodmansterne Road, Carshalton, Surrey, UK)

SUMMARY

Methods for the large-scale production of genetic models of obesity are reviewed.
Particular emphasis is placed upon a knowledge of the genetics of the model to
assist in its efficient production. The possibility of identifying heterozygotes of a
recessive mutant by non-genetic means is investigated both to assist in efficient
production and also as an aid in the search for the primary lesion in these models.

INTRODUCTION

This paper will concentrate on the maintenance and efficient production of the
various genetic models of obesity. The emphasis is on the use of genetic techniques to maximise production rather than on animal husbandry as such, which
has been adequately covered elsewhere (UFAW, 1972; Lane-Petter and Pearson,
1971), while the nutritional requirements of laboratory animals have recently been
reviewed (LAC, 1977).

Table 4.1 lists representatives of the various types of genetic models of obesity.
More complete lists are available in Herberg and Coleman (1977), Bray and York
(1971) and Festing (1978). The inheritance of obesity can be split into two main
types: polygenic and single-gene, or Mendelian. In the polygene case obesity is a
characteristic of a particular inbred strain or outbred stock and is due to the
action of a number of separate genes. In the Mendelian models obesity can have
either a dominant or recessive inheritance, though other background genes may
affect the manifestation of these single genes.

65

Table 4.1   Some genetic
                   models of obesity

Inbred strains and F_1 hybrids
      KK, NZO, LAF_1, $C3_fI$ F_1
Outbred stocks
      BHE (rat)
Dominant mutants
      A^y, A^{vy}, A^{iy}, Ad
Recessive mutants
      ob, db, db^{ad}, db^{2J}, fa (rat)

MAINTENANCE OF POLYGENIC MODELS

Most polygenic models are inbred strains or F_1 hybrids; however, the BHE rat is
an outbred stock and the background genotypes of many of the single-gene mutants are also genetically variable. Maintenance of these strains follows that for any
inbred or outbred strain. Inbred strains are maintained by continuous brother ×
sister matings (to be-defined as inbred, a strain requires 20 continuous generations of such matings), though multiplication of the colony to produce large
numbers of experimental subjects may involve other matings. One such scheme is

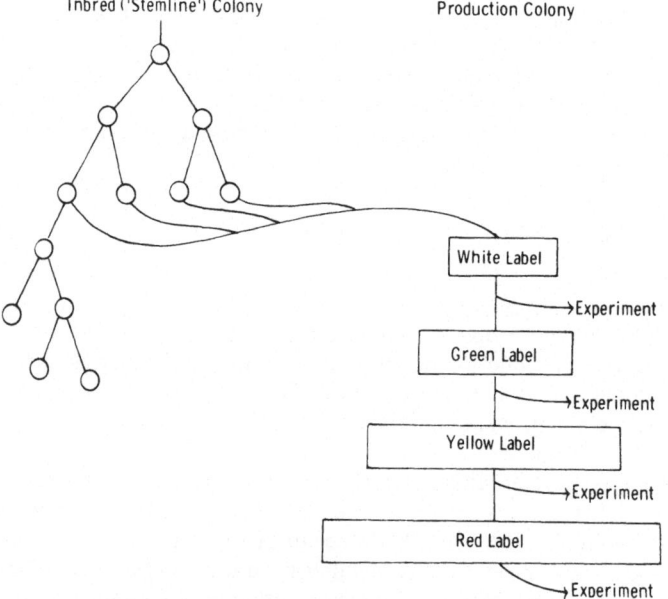

Figure 4.1   Traffic light system of breeding inbred animals. The colony on the
left is maintained by continuous brother × sister mating; the colony on the right
by random mating for three generations only.

termed the Traffic Light Scheme (Lane-Petter and Pearson, 1971; figure 4.1). In this method two colonies are maintained; an inbred colony is maintained by brother × sister matings which supplies breeding stock for a random-bred colony. However, this latter colony is bred only for up to three generations. This prevents chance mutations contributing to a serious increase in genetic variation. Different-coloured labels to designate the offspring of the various generations of random mating are used to control which animals can be used for mating. Animals with red labels are used solely for experiment and are not permitted in the breeding colony.

Inbred strains are usually less productive than outbred animals and therefore require more attention. Adequate future breeding stock must always be available and the colonies must be monitored regularly to ensure that they do not contain too many old animals. This is particularly important when one of the characteristics of the strain is increasing obesity on ageing.

In contrast, outbred stocks are maintained by a system which attempts to maintain their genetic variability. Random mating is an effective system with colony sizes of 100 pairs or more; where colony sizes are smaller, systems which minimise inbreeding are required. Falconer (1972) provides examples of some of these.

Care is necessary in maintaining outbred colonies to prevent selection of breeding stock affecting the manifestation of obesity. Such selection could result in a change in the obese syndrome. This is also the case when single-gene models are maintained on a non-inbred genetic background.

MAINTENANCE OF SINGLE-GENE MODELS

When single-gene differences control an animal's phenotype, the model is maintained by specific crosses which result in predictable phenotypic ratios in the offspring. However, the mutants are maintained on either an inbred or an outbred genetic background, so that the discussion above applies to them if their genetic background is required to be kept constant.

Table 4.2 lists all the possible crosses between animals where there is a gene with two alleles and one, A, is dominant to the other, a. Five crosses are available for the dominant gene to be present in the offspring (Types 1–5), while three are

Table 4.2   The principal types of mating available when two alleles (a, A) are present at a single locus and A is dominant over a

| | | | % offspring | | |
			AA	Aa	aa
Type 1	Incross	$AA \times AA$	100	–	–
Type 2	Backcross (test cross)	$AA \times Aa$	50	50	–
Type 3	Cross (outcross)	$AA \times aa$	–	100	–
Type 4	Intercross	$Aa \times Aa$	25	50	25
Type 5	Backcross (outcross)	$Aa \times aa$	–	50	50
Type 6	Incross	$aa \times aa$	–	–	100
$A > a$					

available for production of the recessive heterozygote (Types 4–6). However, if recessive homozygotes are sterile only one mating (Type 4) produces homo- . zygotes. The most suitable cross will depend upon consideration of whether maternal genotype might affect the results, whether control slim litter mates are required, and whether knowledge of the genotype of phenotypically similar animals is required. Maternal genotype has been reported to exert a significant effect on the weight of 100 day old rats (Jinks and Broadhurst, 1963).

Dominant genes

Examples of dominant genes controlling obesity are A^{vy} and A^y. A^{vy} is partially dominant with the heterozygote in most cases distinguishable from both homozygotes. Five crosses are available (table 4.3) to maintain the gene or produce particular ratios of genotypes. A^{vy} is, however, very variable in appearance from yellow to agouti coat colour (Wolff, 1965), so there may be some danger of misclassification. A^y is lethal when homozygous, so only two crosses are available for its maintenance (table 4.4). One cross produces homozygotes which die *in utero* with a consequent reduction in litter size. The other cross produces the same number of heterozygotes but an increase in homozygous normals and no *in utero* deaths.

Table 4.3   Matings involving the mutant A^{vy} in which both homozygotes and heterozygotes are fertile

Mating	% offspring		
	$A^{vy}A^{vy}$	$A^{vy}+$	++
$A^{vy}A^{vy} \times A^{vy}A^{vy}$	100	–	–
$A^{vy}A^{vy} \times A^{vy}+$	50	50	–
$A^{vy}+ \ \ \times A^{vy}+$	25	50	25
$A^{vy}+ \ \ \times ++$	--	50	50
$A^{vy}A^{vy} \times ++$	–	100	–

Table 4.4   Matings involving the mutant A^y, a dominant allele lethal in the homozygote

	% offspring		
Mating	A^yA^y (lethal in utero)	A^y+	++
$A^y+ \times A^y+$	25	50	25
$A^y+ \times ++$	–	50	50

Recessive genes

The majority of recessive gene models of obesity in the mouse and rat are effectively sterile. This means that only one cross, that between two heterozygotes, will produce the homozygote recessive animal. Table 4.5 gives the lifetime production for various crosses involving the recessive gene. It makes certain assumptions about the breeding performance of a colony which, when compared with an actual colony (C57BL/6-*ob*/+ strain at the LACO, are faily realistic. The table also lists the total production of various genotypes from 90 pairs. Unfortunately, when the

Table 4.5   Efficiency of various mating systems in breeding recessive
mutants which are infertile in the homozygous state. Numbers in
parentheses refer to actual production obtained in the colony of
C57BL/6-*ob*/+ maintained by the MRC Laboratory Animals Centre

Assumptions:
Litter every 8 weeks
8 young/litter
Exact Mendelian ratios
Reproductive life-span 24 weeks

i.e. Y/♀/wk   = 1.00 (1.20)
L/♀/M   = 0.50 (0.71)
Av. litter size = 8.0 (6.7)

	ob/ob	*+/ob*	*++*	*% ob/ob*
ob/ob × *ob/ob*	2160	–	–	100
ob/ob × *+/ob*	1080	1080	–	50
+/ob × *+/ob*	540	1080	540	25
2/3 +/ob × *2/3 +/ob*	240	960	960	11
†2/3 +/ob × 2/3 +/ob	216	614	467	17
(first litter assumed) test cross)				

*'Lean' offspring from cross between two heterozygotes has probability
of 2/3 of being heterozygote.
† All pairs failing to produce *ob/ob* in first litter culled.

homozygotes are effectively sterile, breeding stock has to be taken from the offspring of heterozygote crosses. Homozygous normal (+/+) and heterozygous animals (+/*ob*) are phenotypically identical. There is therefore a 2/3 probability of selecting a heterozygote when each animal is selected. The probability of the first pair mated at random both being +/*ob* is therefore 4/9. The total lifetime production of these animals mated at random is given in the table. Only 11 per cent of offspring are *ob/ob*. It is possible to discard pairs which fail to produce *ob/ob* young in their first litters, as these pairs are probably not both +/*ob*. However, 10 per cent of litters of eight from +/*ob*+/*ob* matings will fail to produce any homozygotes by chance. Therefore, taking into account that 10 per cent of heterozygote matings will be discarded in error, a new measure of total lifetime production can be obtained; this gives approximately 17 per cent production of *ob/ob* (if litter sizes of five are produced approximately 25 per cent of pairs would be discarded in error). It is, however, more efficient to set up new pairs

(44 per cent will be +/ob pairs) than to remate the old pairs at random (16 per cent).

There is thus a maximum production of homozygotes of 25 per cent or a ratio of 1:3 which cannot be improved on unless homozygous animals can be bred. A ratio of one *ob/ob* to five lean can be attained with reasonable ease if a simple breeding plan is followed; further improvement depends upon the skills and re-sources available. The reduction in costs due to a reduction of total numbers of animals bred will have to be set against the increased costs of labour involved in test mating, rematings and genotype identification necessary for improving on this ratio. The relative costs of these factors will vary from place to place and over time, making any 'optimum' breeding plan unlikely.

METHODS OF OBTAINING FERTILE RECESSIVE HOMOZYGOTES

If homozygous *ob/ob* animals were fertile, there would be no need for test crosses and the proportion of homozygotes could be increased to either 50 per cent or 100 per cent, depending upon whether one or both sexes were fertile. If one sex was fertile, a continual backcross to the heterozygote would maintain the mutant without recourse to test crosses.

Table 4.6   Techniques available for increasing efficiency of breeding
             programmes of recessive models of obesity

Homozygote males
   (1)   Diet restriction
   (2)   Hormone therapy
   (3)   Artificial insemination

Homozygote females
   (1)   Ovary transplants
   (2)   *In vitro* fertilisation

Identification of heterozygotes prior to mating
   (1)   Detectable phenotypic differences between homozygote
        normals and heterozygotes
   (2)   Linked marker genes

Table 4.6 lists the possible methods of either producing fertile homozygotes or detecting heterozygotes prior to mating. Consider first slimming the animals. No breeding success has been reported from reducing food intake in female obese mice (*ob/ob*) and obtaining litters. Male (*ob/ob*) mice (Eastcott, 1972) and *fa^k/fa^k* rats (Yen *et al.*, 1977) have sired litters after their food intake has been restricted. Eastcott succeeded in breeding 90 per cent of a group of 20 male obese mice of the genetically heterogeneous Aston strain. In all, 84 litters were produced with a pro-duction of 0.44 litters per female per month. It was possible to maintain a colony of 50 pairs of known heterozygous offspring to produce experimental animals from these matings.

Results from the LAC's colony of *ob/ob* mice on a C57BL/6 inbred back-ground were not as good: six slimmed males out of twelve sired litters. A total of

38 animals from seven litters were weaned and a production of 0.22 young per female per week was obtained from the females when the males were mated three or four times a week to females. A larger scale trial involving 24 males, slimmed at either 3 or 6 weeks of age, is in progress at present and is yielding similar results. There may be husbandry differences between the two studies but more probably the results can be explained by the genetic differences between the two colonies. Miller (personal communication) feeds his random-bred colony carrying the obese (*ob*) gene a diet which includes the drug ephedrine. Males are fertile and the colony is maintained by backcrossing to heterozygotes.

Slimming by the use of drugs or diet reduction can thus produce fertile males. While increasing the yield of homozygotes, it is more labour-intensive and production levels may be below corresponding heterozygous crosses. There is little evidence at present that these techniques can be applied efficiently to mutants maintained on inbred backgrounds.

Both sexes of the *ob/ob* mice appear to have hormonal imbalances. Smithberg and Runner (1957) have reported that female homozygous obese (*ob/ob*) mice were capable of being fertilised, maintaining pregnancies and rearing young if a suitable drug regime was used. Animals were injected with gonadotrophin to obtain pregnancies which were maintained with progesterone. Unfortunately, no reference was given to the background genotype and also the technique appears very labour-intensive. It might be feasible to apply hormonal therapy to *ob/ob* males to see whether this will improve their breeding performances.

Artificial insemination may be a feasible procedure, as males have fertile sperm in the vas deferens even if this is in reduced quantities (Lane and Dickie, 1954), but owing to technical difficulties AI could only be used for special studies requiring, for example, known *ob/ob* embryos.

Ovary transplants are used by The Jackson Laboratory to maintain their stock of obese mice (Hummel, 1957). Ovaries from *ob/ob* females are removed from the animal and transplanted to a female mouse of the C57BL/6 inbred strain. This female is then mated to a male from the C57BL/6 strain to obtain heterozygotes. This technique is only available if both the donor and the recipient animals share the same histocompatibility loci. It does, however, have the advantage of maintaining the gene by repeated backcrossing into an inbred background. The Jackson Laboratory regularly breed obese mice by this method, producing homozygous animals from crosses between the heterozygotes produced by animals which have undergone an ovary transplant. The Jackson Laboratory produce about 5000–6000 obese animals a year and have now reached the thirty-sixth backcross (N36) (Jackson Laboratory, 1977).

Other techniques, such as *in vitro* fertilisation, ovum and embryo culture, and *in vitro* fertilisation and embryo transfer, are now well worked out (Daniels, 1971) and would allow production of litters of *ob/ob* animals in foster females. Alternatively, combination of either slimmed males or artificial insemination of females which have received ovary transplants could yield 100 per cent *ob/ob* litters. However, all these techniques require skilled technical work and some practice before they work efficiently, and these extra costs must be balanced against any savings in increased proportion of obese homozygotes produced. They do, however, have possibilities for the production of certain specific animals such as obese animals reared by obese mothers.

IDENTIFICATION OF HETEROZYGOTE ANIMALS BEFORE BREEDING AGE

Returning to the problem of identification of heterozygotes: If a method existed for identifying heterozygous animals from the homozygous normals (+/+) before or at breeding age, all crosses could be between known heterozygotes and the ratio of homozygote obese would immediately rise to 25 per cent without any decrease in general productivity. Two methods exist in practice: direct measurement of some differences between the two genotypes and an indirect method relying on linked marker genes on the same chromosome as the mutant of interest.

Direct measurement of differences between heterozygotes and homozygote normals

Direct measurement of differences between the two genotypes has to be almost 100 per cent successful in discriminating between the two genotypes to be worthwhile. It must be quick, cheap and, of course, leave the animal unharmed. Trayhurn *et al.* (1977) used the response of obese and lean mice to a temperature of 4°C to distinguish pre-obese *ob/ob* mice from their lean (+/*ob* or +/+) litter mates.

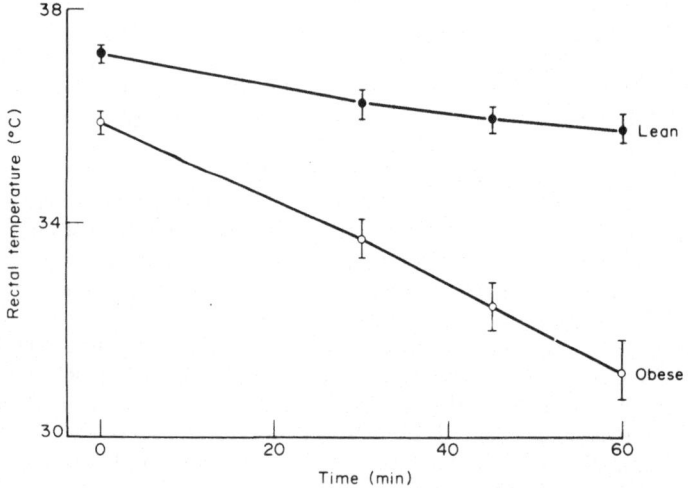

Figure 4.2   Identification of pre-obese *ob/ob* mice from 'lean' litter mates by measuring rectal temperatures at 17 days of age in an environment maintained at 4°C. From Trayhurn *et al.* (1977).

The body temperature of the *ob/ob* mouse fell much more rapidly than that of the lean animals (figure 4.2). However, it appears unlikely that in this case there were differences between homozygous normals (+/+) and heterozygote animals (+/*ob*), as the standard errors of the presumably genetically variable animals were less than those of the homozygote *ob/ob*.

The length of anaesthesia after injection with pentobarbitone sodium is a test of liver enzyme function and is affected by genotype, temperature and body composition of the animal (Lovell, in preparation). It was therefore decided to see whether this test could be used to discriminate between +/ob and +/+. Table 4.7 shows the results. Obese (ob/ob) mice slept shorter than either homozygous C57BL/6-+/+ or animals which had a 2/3 probability of being heterozygous (+/ob).

Table 4.7   Sleeping time under pentobarbitone anaesthesia (log min)

Experiment	Genotype	Number	Age (weeks)	Mean sleeping time ($\pm$ S.E.)
1	+/+	5	18–31	2.04 $\pm$ 0.038
	+/ob	5	18–31	2.01 $\pm$ 0.057
2	+/+	10	7–11	2.12 $\pm$ 0.021
	+/?	30	7–11	2.14 $\pm$ 0.021
	ob/ob	10	7–11	2.04 $\pm$ 0.025

Note: In experiment 1 there was no significant difference between animals of the two gentoypes. In experiment 2 the ob/ob mice slept for a significantly shorter time than either of the other genotypes ($P < 0.05$), which did not differ significantly from each other ($P > 0.05$). However, the variance of sleeping time of the +/? was significantly greater than that of the +/+ mice ($P < 0.05$).

However, there were no significant differences between +/+ and +/ob animals that were about 6 months old. There was an indication ($P < 0.05$) that the genetically variable mixture of +/+ and +/ob were slightly more variable in response than the +/+ normal inbred C57BL/6J animals. The test, however, does not look particularly useful for detecting +/ob

Yen *et al.* (1968) reported that +/ob mice differ significantly from both ob/ob and +/+ in the oxidation of glucose by adipose tissue as measured by radioactive-labelled CO_2 release (figure 4.3). The test was performed *in vitro* and required 100 mg of adipose tissue. Here is perhaps a test which could be used, as figure 4.3 seems to indicate a clear distinction between the genotypes and a biopsy could be carried out to test potential breeders prior to breeding age. No other measurements showed significant differences between +/ob and +/+ animals (such as body weight, incorporation of glucose into lipids or blood glucose levels).

Joosten and van der Kroon (1974) claim that the three genotypes ob/ob, +/ob and +/+ can be distinguished at 13 days of age by their fat cell diameters. Only a small number of animals were examined and it is not clear whether environmental factors between litters would decrease the discrimination.

Chick *et al.* (1970) report that +/db male animals have a reduced glucose tolerance compared with +/+ animals. The finding of a gene dosage effect with the heterozygote intermediate compared with the two homozygotes is of interest in itself; it implies an action of the mutant gene which can be distinguished from the normal effect and yet does not produce the characteristic pattern seen in the double mutant homozygote. It indicates either an insignificant effect of the mutant gene on the overall phenotype or some compensatory effect preventing the

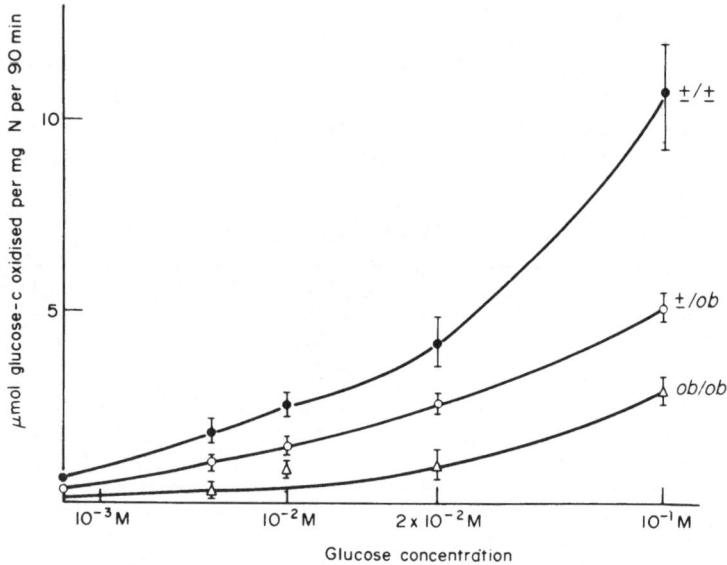

Figure 4.3 A possible technique for distinguishing between *ob*/+ and +/+ mice. The oxidation of glucose by adipose tissue as measured by the release of labelled CO_2 in animals of the three genotypes. From Yen *et al.* (1968).

full effects of gene showing. It would be of interest to find more differences between homozygous normals and the heterozygotes both for easy distinction of the genotypes and as a means of investigating the primary lesion in these mutants without secondary effects complicating the results.

Use of marker genes

Marker genes have been used for distinguishing heterozygotes for animals which are difficult to breed in the homozygote state. The criterion for a good marker gene is that it is closely linked to the gene of interest and is easily identifiable. The gene misty (*m*) has been used in the maintenance of the diabetic model *db*. Figure 4.4 lists two possible crosses. The first, where both recessive genes are carried on the same chromosome (termed coupling), is used to identify *db*/*db* animals by their coat colour before other symptoms become apparent. All animals that are misty (have a grey coat) will be *db*/*db* as well. This particular system does not improve selection of heterozygotes for future breeding, as the remaining 75 per cent of animals consist of a ratio of 2 +/*m*, +/*db* to 1 +/+, +/+, all of which appear normal. The second system, where the mutant genes are maintained on different chromosomes (called 'repulsion'), provides three types of offspring: one misty but lean (*m*/*m* +/+), two black lean (+/*m* +/*db*) and one black diabetic (+/+ *db*/*db*). The +/*m*, +/*db* (black–lean) animals provide future breeding stock, as they are heterozygous for both genes. The misty animals are discarded.

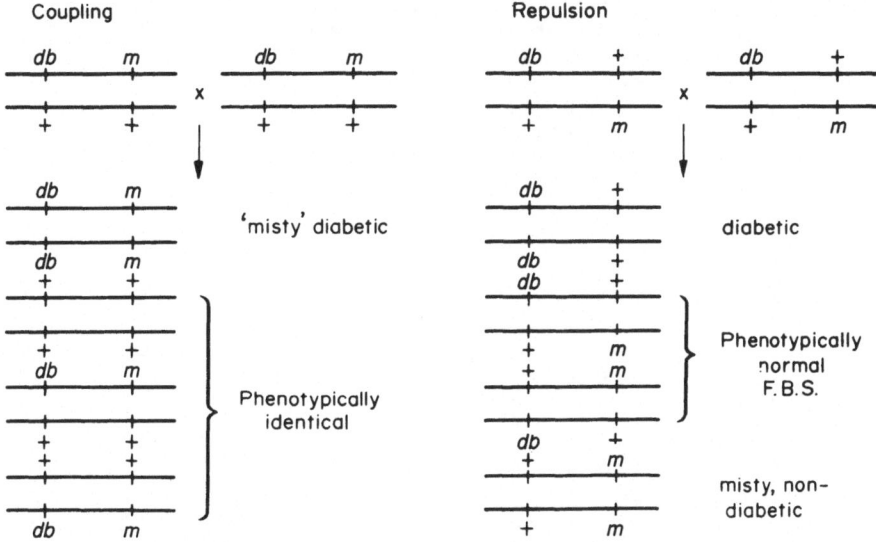

Figure 4.4   Breeding of *db/db* mice using the linked marker gene misty (*m*). Matings of coupling double heterozygotes (on the left) lead to the production of mice homozygous for both *db* and *m*, so may be used for the early identification of homozygotes. Matings of repulsion double heterozygotes (on the right) in which the mutants *db* and *m* are on different homologous chromosomes lead to the production of non-misty diabetics, phenotypically normal mice used as breeding stock and homozygous misty mice which are discarded. Note that in this case the linked recessive marker could be infertile or even lethal in the homozygous state without losing its utility. Crossover types have not been included in this diagram.

The previous discussions assume that no crossovers occur between the two genes during meiosis. In fact, *m* and *db* are one crossover unit apart on chromosome 4. In one in a hundred gametes the genes for *m* and *db* are recombined and the breeding system breaks down, as it is, in effect, treating the two genes as if they were one. Table 4.8 shows the effect of 20 per cent recombination between two genes. In this situation approximately one-third of the supposedly repulsion heterozygotes have in fact lost or had their mutant genes rearranged. This means that the mutant used as a marker is giving only limited information about the mutant under investigation, and in extreme cases will result in poorer prediction of the heterozygote than if animals were selected at random.

In practice, linkage of less than 5 per cent should give reasonable prediction. Looser linkage than this results in an increasing amount of supervisory time required to ensure that the system is performing satisfactorily.

Another problem with linked marker genes is that the development of these special stocks can cause complications in the interpretation of the results. Reports on diabetic misty animals at first were interpreted as meaning that misty altered the manifestation of the diabetic gene. Later it was clarified that the background

David P. Lovell

Table 4.8 The use of linked markers when crossovers occur, and the mating is between repulsion double heterozygote animals, a +‖+ b

| Offspring | Expected frequency when crossover % is: | | Use* |
	0	20%	
Non-crossover			
$\dfrac{a\ \ +}{a\ \ +}$	0.25	0.16	E
$\dfrac{a\ \ +}{+\ \ b}$	0.50	0.32	B
$\dfrac{+\ \ b}{+\ \ b}$	0.25	0.16	D
Crossover			
$\dfrac{a\ \ b}{a\ \ +}$	0	0.08	E
$\dfrac{a\ \ b}{+\ \ b}$	0	0.08	D
$\dfrac{+\ \ +}{a\ \ +}$	0	0.08	B†
$\dfrac{+\ \ +}{+\ \ b}$	0	0.08	B†
$\dfrac{a\ \ b}{a\ \ b}$	0	0.01	D
$\dfrac{a\ \ b}{+\ \ +}$	0	0.02	B†
$\dfrac{+\ \ +}{+\ \ +}$	0	0.01	B†

*E = experimental animals homozygous a/a.
 B = phenotypically normal, used for breeding.
 D = homozygotes at b locus discarded.
 † This breeding stock will give a reduced output of homozygous a/a offspring if used for breeding.

on which the misty gene had been maintained was in fact interacting with the diabetic gene (Coleman and Hummel, 1975).

This illustrates the fact that background genetic variation may interfere quite considerably with the manifestation of a mutant gene. This must be borne in mind at all times when new genetic material is introduced into a colony for whatever reason.

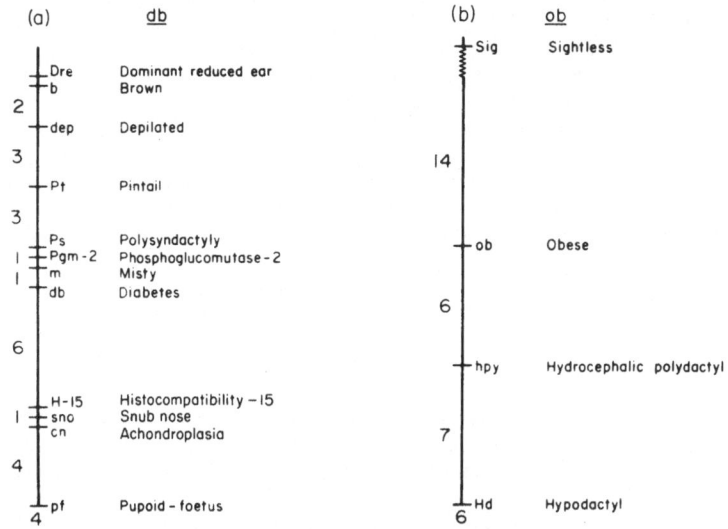

Figure 4.5   Known linkage relationships of genes causing obesity in the mouse: (a) chromosome 4 carrying *db* and (b) chromosome 6 carrying *ob*.

Figure 4.5 gives the known linkages for two recessive genes in the mouse, *db* and *ob*. The gene *fa* in the rat has not yet been mapped. As *db* is linked by one map unit to *m*, which is easily recognised, it is unlikely that other linked markers would be an improvement.

Obese (*ob*) is, however, isolated to some extent with only a small number of known linkages. The mutant *hyp*, which causes hydrocephalic polydactyl, is only six crossover units away and is on the edge of being a possible marker gene. The possibility of using this particular gene in a repulsion cross should perhaps be considered; the fact that it is deleterious is not a disadvantage for use as a marker in a repulsion cross.

REFERENCES

Bray, G. A. and York, D. A. (1971). Genetically transmitted obesity in rodents. *Physiol. Rev.*, 51, 598–646

Chick, W. L., Lavine, R. L. and Like, A. A. (1970). Studies in the diabetic mutant mouse. V. Glucose tolerance in mice homozygous and heterozygous for the diabetes (*db*) gene. *Diabetologia*, 6, 257–262

Coleman, D. L. and Hummel, K. P. (1975). Influence of genetic background on the expression of mutations at the diabetes locus in the mouse. II. Studies on background modifiers. *Israel J. Med. Sci.*, **11**, 708-713

Daniels, J. C. (Ed.) (1971). *Methods in Mammalian Embryology*, Freeman, San Francisco

Eastcott, A. (1972). Influence of restricting the diet on breeding in genetically obese mice. *Lab. Anim.*, **6**, 9-18

Falconer, D. S. (1972). Genetic aspects of breeding methods. In *The UFAW Handbook on the Care and Management of Laboratory Animals*, 4th edn, Churchill Livingstone, Edinburgh and London, pp. 5-25

Festing, M. F. W. (1978). This symposium

Herberg, L. and Coleman, D. L. (1977). Laboratory animals exhibiting obesity and diabetes syndromes. *Metab. Clin. Exp.*, **26**, 59-100

Hummel, K. P. (1957). Transplantation of ovaries of the obese mouse (abstr). *Anat. Rec.*, **128**, 569

Jackson Laboratory (1977). 48th Annual Report. The Jackson Laboratory, Bar Harbor, Maine, USA

Jinks, J. L. and Broadhurst, P. L. (1963). Diallel analysis of litter size and body weight in rats. *Heredity*, **18**, 319-336

Joosten, H. F. P. and van der Kroon, P. H. W. (1974). Enlargement of epididymal adipocytes in relation to hyperinsulinaemia in obese hyperglycemic mice (*ob/ob*). *Metabolism*, **23**, 59-66

LAC (1977). Dietary Standards for Laboratory Animals. Report of the Laboratory Animals Centre Diets Advisory Committee. *Lab. Animals*, **11**, 1-28

Lane, P. W. and Dickie, M. M. (1954). Fertile obese male mice. Relative sterility in obese males corrected by dietary reductions. *J. Hered.*, 56-58

Lane-Petter, W. and Pearson, A. E. G. (1971). *The Laboratory Animal – Principles and Practice*, Academic Press, London

Smithberg, M. and Runner, M. N. (1957). Pregnancy induced in genetically sterile mice. *J. Hered.*, **48**, 97-100

Trayhurn, P., Thurlby, P. L. and James, W. P. T. (1977). Thermogenic defect in pre-obese *ob/ob* mice. *Nature*, **266**, 60-61

UFAW (1972). *The UFAW Handbook on the Care and Management of Laboratory Animals* (ed. UFAW), Churchill Livingstone, Edinburgh and London

Wolff, G. L. (1965). Body composition and coat colour correlation in different phenotypes of 'Viable Yellow' mice. *Science*, **147**, 1145-1147

Yen, T. T., Lowry, L. and Steinmetz, J. (1968). Obese locus in *Mus musculus*. A gene dosage effect. *Biochem. Biophys. Res. Comm.*, **33**, 883-887

Yen, T. T., Shaw, W. N. and Pao-Lo, Yu (1977). Genetics of obesity in Zucker rats and Koletsky rats. *Heredity*, **38**, 373-378

5

The use of animal models in the detection and evaluation of compounds for the treatment of obesity

M. A. Cawthorne (Beecham Pharmaceuticals, Research Division, Walton Oaks, Tadworth, Surrey, UK)

SUMMARY

Obesity is one of the most important medical and public health problems of our time. In order to develop new treatments for obesity, animal models are required for detecting and evaluating drugs that affect food intake and energy expenditure. Further, models are needed for studying overall energy balance.

The models for detecting anorexic agents have been largely designed around the ability to detect activity in known anorexic drugs. Thus, the main advances in therapy have been an improvement in side-effects rather than the development of different types of anorexigenics.

The idea of searching for thermogenic drugs has developed over the last decade and screening procedures are being developed for such drugs.

It is suggested that the genetically obese (*ob/ob*) mouse is a suitable model for studying overall energy balance. These mice have a very small variability in their body composition and this aids the evaluation of anti-obesity effects. Further, reduction in food intake of young *ob/ob* mice reduces the growth rate, but the mice still continue to lay down fat, albeit at a reduced rate. Thus, the *ob/ob* mouse can be used to separate substances acting solely on food intake from those having a thermogenic or mixed action.

INTRODUCTION

A recent publication on research on obesity by a joint Department of Health and Social Security and Medical Research Council group stated: 'We are unanimous in our belief that obesity is a hazard to health and a detriment to well-being. It is common enough to constitute one of the most important medical and public health problems of our time, whether we judge importance by a shorter expectation of life, increased morbidity or cost to the community in terms of both money and anxiety.' However, it is difficult to measure the statistical importance

79

of obesity. It very rarely appears on a death certificate and only merits a one-line entry in the International Classification of Diseases. Nevertheless, actuarial evidence from such bodies as life insurance companies shows that with increasing obesity there is an increasing risk of an early death. It is now generally accepted that, if a man is 27 per cent, or greater, overweight before he is 40, he has an extremely good chance of an earlier than average death. An insight of the size of the disease problem is that 20 per cent of the relevant male population of the USA come into this category (Bray, 1978).

Obesity is associated with an increase in the normal mortality caused by diabetes, nephritis, digestive system diseases and circulatory diseases, including coronary and cerebral thrombosis. In the UK the great increase in the incidence of deaths from ischaemic heart disease has been described as an epidemic. While it is difficult to assess the contribution of obesity to the increase in deaths from ischaemic heart disease because of the many interrelated risk factors, many official bodies in different countries have concluded that obesity does predispose to ischaemic heart disease. Perhaps the strongest statement comes from the Framingham Study, which concluded that '. . . we might expect a 20% reduction in weight in the obese to result in a 40% reduction in the chances of coronary heart disease.' (Kannel and Gordon, 1974).

There are two extreme schools of thought on the origins of obesity. First, there is the view that '. . . since the immediate causes of obesity are overeating and underexercising, the remedies are available to all, but many patients require much help in using them.' (Davidson and Passmore, 1969). The other view, which was suggested by Astwood (1962), is as follows: 'I wish to propose that obesity is an inherited disorder and due to a genetically determined defect in an enzyme, in other words that people who are fat are born fat and nothing much can be done about it.' Both of these views probably have some truth in them. Although many people would not subscribe to the fatalistic view of Astwood, there is clearly a genetic element in obesity, as Shields (1962) found. He showed that the body weights of monozygotic twins brought up apart were closer than those of dizygotic twins reared together. On the other hand, unless man is to disobey the laws of thermodynamics, then to become overweight he must have eaten more than his requirement. It seems likely that obesity is neither one condition nor one disease, and a complex multi-factorial aetiology is usually implicated. However, whatever the origin of obesity, the excessive body weight arises from a surfeit of calorie intake over calorie expenditure. The major approach to the treatment of obesity has been to reduce the intake of food, either by a low-calorie diet or by the use of anorexic agents to help the patient to reduce his food intake. However, there is an equally logical treatment of obesity, and that is to increase energy expenditure. One approach to this is to undertake an exercise programme, which can be very successful. However, as Miller (1974) has stated, there seems to be no real reason why one should not seek drugs to aid heat loss.

This paper examines the utility of animal models for detecting drugs that affect food intake and those that affect energy expenditure, together with those in which an attempt is made to explore the balance between the complicated energy input side of the equation with the incompletely understood energy output side of the equation. The latter system is important, since it seems to be almost impossible to affect one of these parameters without affecting the other to some extent.

ANIMAL MODELS FOR DETECTING ANOREXIC AGENTS

Nathanson (1937) first reported that amphetamine produced anorexia in man, and Lesses and Myerson (1938) showed that it was effective in the management of obesity. Unfortunately, amphetamine has potent central stimulant activity; thus, over the intervening years much chemical and pharmacological research has been orientated towards the finding of phenylethylamine derivatives which, while retaining the anorexic effect of amphetamine, have a reduced central stimulant effect.

In setting up an animal model, one wishes to simulate as closely as possible the human situation. However, in spite of much research, our ignorance of the critical mechanism for the control of food intake is almost total. We have little knowledge of the mechanisms that initiate a meal, sustain a meal, terminate a meal or maintain a normal inter-meal interval. As a result of this paucity of knowledge, the test systems used have been designed around the ability to detect activity in known anorexic agents. This has restricted research and resulted in the discovery of novel chemicals that have a similar action to those already available.

The primary test must allow a chemist to develop a structure–activity relationship within the series of chemicals that he is synthesising. Thus, the test should be reproducible from week to week and even from year to year. The most common species used for the primary screening test are rats and mice, but since these animals eat mainly at night, they must be either fasted overnight or meal-trained in order to obtain a consistent and measurable food consumption during the test period. In our own studies we have found that rats give more reproducible results than mice, although the latter have the advantage that less of the chemical is required for the test. Rats are housed in pairs and trained over a period of 10 days to eat their daily food during a 4 h period. On the eleventh day the compound under study is given 30 min before the meal and the food eaten after 2 h and 4 h intervals measured. By making measurements at the two separate times, one can get some measure of the duration of action of the compound. It is important in this sort of test to measure the food wastage, which can be significant, and also to observe the animals during the test. The more obvious side-effects affecting the central nervous system and neuromuscular coordination can usually be easily detected.

Many of the marketed anorexic agents are derivatives of phenylethylamine. The most potent of these is amphetamine, but this is now rarely used in the UK for the treatment of obesity, because it is a controlled drug as a result of its dependence and stimulant properties. Some of the other agents produce some stimulation in animal tests but not to the same degree as amphetamine. With this background, the effect of any potential anorexic agent on locomotor activity must be measured. However, it is not only strongly stimulant activity that is undesirable, for any depressant effect on mood is also unsuitable. Several systems of measuring locomotor activity have been developed over the last 15 years. The early systems were based on light beam detectors (actophotometers), but these were highly inaccurate, since it is not possible to cover the whole area of a cage with such a system. More recent systems have been based on ultrasonic probes (C. F. Palmer, High Wycombe, UK) or disturbances in an electromagnetic field (Animex, LKB, Croydon, UK). The latter system has the advantage that it is possible to tune the

instrument to obtain selected measurements of large (e.g. walking) and small
(e.g. grooming, washing) movements simultaneously.

At some stage during the early development of a new anorexic agent it becomes necessary to determine its anorexic activity in a non-rodent species, since
toxicity testing in such a species will be required. The usual choices for these
tests are dogs or monkeys. However, both present problems in terms of anorexic
testing. The laboratory dog (and even the domestic pet) is possibly the most extreme form of a meal-trained animal that we know — the daily meal often being
consumed in as little as 30 s. It is a matter of speculation as to whether the normal pharmacological control systems apply to this situation. The technique that
has usually been adopted to measure the anorexic activity of compounds in the
dog is to present small amounts of food at regular intervals. Abdallah and White
(1970) presented the compounds under study in a capsule embedded in a meat
ball. Each dog was then offered a tablespoonful of canned dog food every 30 min.
Any food not eaten at the end of each 30 min period was removed and replaced
with a fresh offering. Thus, the onset and duration of anorexia can be calculated.
Macko et al. (1972) preselected dogs for their ability to consume 50 g of food
every 30 min for 6 h. Their criterion for anorexia was a refusal to take food for
two consecutive 30 min periods. Krapcho and High (1967) adopted a different
technique and starved dogs for 18 h before offering food ad libitum for 1 h,
15 min after an oral dose of the compound under study.

The problems of anorexic testing in monkeys seem at first sight to be even
greater than those in dogs. Most workers have circumvented the problem by
studying the effect of anorexic agents on the intake of orange juice in monkeys
trained under operant conditions to lever press. However, in a small trial (unpublished) we have been able to measure voluntary food intake in squirrel monkeys and have shown that it can be inhibited by anorexic agents.

The earlier anorexic agents were marketed with virtually no knowledge of
their mechanism of action. However, as a result of recent experiments, most
anorexic agents are now known to act on brain monoamines. Nevertheless, with
the possible exception of amphetamine, there is little information available on
the effect of anorexic agents on the various aspects of monoamine uptake, turnover, release and activity. Further, while most anorectics show a preferential
action for a particular monoaminergic system, they can also affect other systems
in the brain. Some insight can be gained as to which amine(s) is(are) involved by
measuring the anorexic effect in animals that have been treated so as to affect
the action of particular amines. Using this type of procedure, it has been shown
that the anorexic activity of amphetamine is antagonised by α-methyl-p-tyrosine,
which is an inhibitor of catecholamine synthesis at the level of tyrosine hydroxylase (Abdallah, 1971; Weissman et al., 1966). A similar antagonism of the anorexic
action of amphetamine occurs in rats that have received the neurotoxin 6-hydroxy-
dopamine intraventricularly (Garattini et al., 1975). However, while these results
suggest an involvement of catecholamines, they do not clarify the separate role of
noradrenaline and dopamine. The uses of the dopamine receptor blockers pimo-
zide (Kruk, 1973) and haloperidol (Frey and Schulz, 1973) show that the anorexia
induced by amphetamine has a dopaminergic component. The effects of blockers
of α- and β-adrenergic receptors such as phentolamine and propanolol on the
anorexic effects of amphetamine are less clear-cut. The discrepancies found in

the literature may arise from the differences in drug effects depending on species investigated, dosage and time of drug pretreatment or effects on the metabolism and distribution of amphetamine (Garattini and Samanin, 1976). In contrast to amphetamine, the anorexia induced by fenfluramine is largely unaffected by procedures affecting the catecholaminergic system and leads to the conclusion that amphetamine and fenfluramine induce anorexia by quite separate mechanisms (table 5.1).

Table 5.1   Effect of various procedures affecting brain catecholamines on the activity of amphetamine and fenfluramine in the rat

Experimental procedure	Effect on catecholamines	Anorexia by	
		amphetamine	fenfluramine
α-Methyl p-tyrosine	inhibition of dopamine and noradrenaline synthesis	reduced	unchanged
6-Hydroxydopamine	destruction of catecholamine-containing neurons	reduced	unchanged or enhanced
Haloperidol	blockade of dopamine receptors	reduced	unchanged
Pimozide	blockade of dopamine receptors	reduced	unchanged
1-Propanolol	blockade of β-noradrenergic receptors	reduced, unchanged or enhanced	unchanged
Phentolamine	blockade of α-noradrenergic receptors	unchanged or reduced	unchanged

Similar experiments to those described for the catecholaminergic system can also be carried out on the serotoninergic system (table 5.2). There is now considerable evidence that fenfluramine-induced anorexia may be due to an interaction with brain 5-HT. Of particular interest is the finding that anorexia induced by fenfluramine, but not by amphetamine, can be inhibited by the 5-HT-receptor-blocking drugs methergoline, cyproheptidene (Garattini *et al.*, 1975) and pizotifen (M. A. Cawthorne, unpublished).

It is now apparent that the mechanism of action of phentermine, diethylpropion and mazindol have similarities to that of amphetamine. It is suggested that it is likely that the mechanism of any novel compound can be approximately typed between catecholaminergic and serotoninergic using the procedures indicated in tables 5.1 and 5.2, but one should be aware that neither the pharmacological tools nor the anorexic drugs are likely to exert a pure action, and a complex interaction between the catecholamine and serotonin neurons can be expected. Such an interaction has been long predicted as a result of lesion experiments and has evoked the concept of a ventromedial 'satiety' region and a lateral hypothalamic 'feeding' region. The elegant studies of Ungerstedt (1971a) have demonstrated that bilateral lesions in the substantia nigra, which contains the cell bodies for all dopaminergic neurons, produces aphagia and appears to be equivalent to a lesion of the lateral hypothalamus. Lesioning of the substantia nigra can be used to sub-

Table 5.2   Effect of various procedures affecting brain serotonin on the anorexic
activity of amphetamine and fenfluramine in the rat

Experimental procedure	Effect on serotonin	Anorexia by	
		amphetamine	fenfluramine
Lesion of the mid-brain Raphé nucleus	destruction of 5-HT neurons	unchanged	reduced or unchanged
5,6-Dihydroxytryptamine	destruction of 5-HT terminals	unchanged	reduced or unchanged
p-Chlorophenylalanine	inhibition of 5-HT synthesis	reduced	unchanged
Methergoline	blockade of 5-HT receptors	unchanged	reduced
Cyproheptadine	blockade of 5-HT receptors	unchanged	reduced
Pizotifen	blockade of 5-HT receptors	unchanged	reduced
Chlorimipramine	inhibition of 5-HT uptake	unchanged or enhanced	reduced

stantiate a dopaminergic action in a novel agent, for if a lesion is made on only
one side of the brain, there will be an imbalance in the dopaminergic output. Thus,
drugs such as amphetamine that cause the release of dopamine will only produce
an effect on the unlesioned side of the brain and the rat will turn in circles, towards the side of the lesion (Ungerstedt, 1971b).

At the beginning of this section it was noted that, in general, the anorexic
drugs that are available currently stem from amphetamine. While the dependence
faults of amphetamine have been overcome, few would regard the weight-reducing
performance of currently available anorexic agents as adequate. Thus, the challenge
to the pharmaceutical industry is to find better anorexic drugs for the future. To
achieve this end, it may be necessary to develop new animal models and test
systems. One new type of test system is the pellet-detecting eatometer (Kisseleff,
1970). It delivers a new 45 mg food pellet each time one is eaten and provides a
means of recording every single pellet consumed during the experimental period.
Blundell and Lesham (1975) have shown that rats eat in meals with discrete inter-
meal intervals. It is therefore possible to assess the effect of drugs on meal frequency, meal size, inter-meal intervals, latency of first meal and velocity of eating.
Blundell et al. (1976) showed that amphetamine acts mainly by delaying the
taking of the first meal. However, its duration of action is short and later the
inter-meal interval is decreased and the period of anorexia is followed by a period
of hyperphagia. Fenfluramine acts differently: it reduces the meal size and increases the inter-meal interval. This experimental feeding system may well be a
much more satisfactory system for detecting future drugs useful in controlling
obesity. In particular, it may lead to new classes of drugs — for example, those
having a specific satiety effect or those that reduce the velocity of eating. Unfortunately, as yet, the rat experiments that have been carried out by Blundell and
his colleagues have not been completely replicated in man, although Silverstone
and Fincham (1977) have described an apparatus that may be useful for the human
studies. Until these human studies are complete, the discovery of new anorexic
agents will probably remain empirical.

ANIMAL MODELS FOR DETECTING THERMOGENIC AGENTS

There is only one group of agents that are currently used clinically to raise metabolic rate in obese man. These are the thyroid hormones thyroxine and tri-iodothyronine. However, these agents have a very limited use. Certainly the administration of moderate doses of the drugs to patients with a normal metabolic rate merely leads to suppression of endogenous thyroxine production (Garrow, 1974) and there is no net effect on metabolic rate or on energy expenditure. However, thyroid preparations have a place in treating those refractory obese patients who have been described by Garrow (1974), because of their extremely low metabolic rate, as hibernators. The number of people with this type of obesity are relatively few and they can be treated in a metabolic ward.

Many authorities would argue that one should not seek agents to increase metabolic rate, and by implication suggest that obesity is always achieved by gluttony. However, the work of Payne and Griffiths (1976) has shown that in a group of children of identical weight and height some of the children were only maintaining normal body weight by maintaining a below-average food intake. It seems very likely that there is a general variation in the basal energy expenditure in man. In order to lose weight, subjects require a treatment that raises the energy expenditure by about 10 per cent. Such an increase would still result in an energy expenditure within the normal range.

Miller and his colleagues have led the search for thermogenic agents. They note that 250 mg caffeine, which is approximately equivalent to two cups of coffee, will raise the metabolic rate of man by about 10 per cent (Miller *et al.*, 1974). Miller (1974) concludes that '. . . the possibility of manipulating heat production in the treatment of obesity is clearly viable. Certainly the obese would prefer to eat as normal people rather than suffer the rigours of dieting with its poor prognosis. Thermogenic drugs would seem to provide a more positive solution to their faulty homoeostatic regulation, and at the same time, allow them the hedonic pleasures of gastronomy.'

The search for thermogenic agents has been hindered by the lack of a suitable primary screen. Miller and Stock (1969) suggested that one could detect thermogenic agents by doing energy balance experiments in rats. Thus, rats were fed on a diet containing a drug for 7 days and the total amount of food eaten was measured. During the experiment, faeces and urine were collected and at the end of the experiment the animals were killed. The calorific values of the ingested food, the faeces, the urine and the carcass were all measured by bomb calorimetry. In addition, the calorific values of control animals, which had been killed at the beginning of the experiment, were also measured. From these data Miller was able to calculate the amount of missing energy, and clearly any drug that decreases the efficiency of food utilisation is potentially an anti-obesity agent. However, the method is rather too time-consuming and labour-intensive for a primary screen, and a more rapid screening procedure is required.

One can get a measure of total energy output by measuring the amount of oxygen consumed by animals. The simplest system is that designed by Haldane (1892), where the oxygen consumed by animals is estimated by weight. The method is rather insensitive and it is only possible to obtain a single measure over a period of several hours. Two different types of continuous recording systems have been developed. The first is a closed-circuit system in which animals are placed in an airtight box equipped with an absorbent for carbon dioxide and a pressure detector. As the carbon dioxide is produced by the animal and absorbed, the gas pressure in the chamber falls. This fall in pressure is detected and via a transducer activates a gas-tight syringe to deliver pure oxygen until atmospheric pressure is restored. The whole system can be automated and the data (time and number of deliveries of oxygen) recorded on a chart recorder (Stock, 1966). In the second system the oxygen content of expired air of animals is measured by a paramagnetic oxygen analyser (Taylor Servomex, Crowborough, Sussex, UK). Thus, by a knowledge of flow rate of air through the animals' cage, one can monitor oxygen consumption constantly. Boroumand and Miller (1977) have suggested that the latter system is a suitable screen for thermogenic drugs.

ANIMAL MODELS FOR MEASURING ENERGY BALANCE

Neither the anorexic nor the thermogenic screens described above actually measure the essential thing for obese people — has any weight (fat) been lost? Further, while it is useful to treat the two sides of the energy balance equation separately in order to discover new drugs, it is important to realise that one cannot affect one side of the energy equation without affecting the other side to some extent. In simple terms, reducing the food intake of an obese person will result in weight loss. However, the decreasing body mass will result in a reducing basal metabolic rate, so that the point will eventually be reached when energy intake and output are once again in balance.

There are many types of animal models of obesity, all of which have some merits and demerits. One of the most popular models is the genetically obese (*ob/ob*) mouse, which has a low energy expenditure with only a slight increase in food intake in the growing phase. The mature *ob/ob* mouse has a low food intake if due compensation is made for the increased body size. Therefore, it is a

Table 5.3 Body weight and lipid content of various models of obesity

	Weight	Lipid	
		(%)	(g)
C57BL/6J-*ob/ob*	39.9 ± 1.8	55.5 ± 0.6	22.2 ± 1.1
C57BL/6J-*ob/+*	18.7 ± 0.7	23.9 ± 2.9	4.7 ± 0.7
CFLP	42.1 ± 0.8	10.3 ± 0.7	4.4 ± 0.4
CFLP — gold thioglucose	51.1 ± 2.2	29.5 ± 3.0	15.8 ± 2.0
CFLP — high-sucrose diet	36.4 ± 1.0	22.8 ± 1.6	8.3 ± 0.7

Results are given as mean ±S.E. of 10 mice

model for the obese person with a low metabolic rate. A major disadvantage of the model is that the obesity is the result of a single genetic defect and it is likely that very few people, if any, will have the same primary lesion as the mouse. However, the secondary effects of the lesion may well be very similar to the changes that take place in a much larger number of obese people.

A major merit of the C57BL/6J-*ob/ob* mouse model is the extremely constant body composition. Table 5.3 compares the body composition of the various types of mice and shows that the variation of the body composition of the *ob/ob* mouse is much smaller than that of the other models. This narrow range in body composition aids the statistical evaluation of the effect of a drug. Indeed, the body composition is so constant that it has been possible to establish equations for the relationship of the various components of body composition with body weight for obese mice fed on Oxoid breeders' diet (table 5.4). It was found that the C57BL/6J-*ob/ob* mice from Bar Harbor had slightly different growth incre-

Table 5.4   Correlates for components of body composition of
The Jackson Laboratory C57BL/6J-*ob/ob* mice

Upper limit of body weight (g)   = 53.7
Lower limit of body weight (g)   = 30.3
$n = 182$

Lipid content (g) = (body wt. × 0.721) – 6.146
Correlation of lipid wt. with body wt., $r = 0.967$

$$\text{Water content (g)} = \frac{(\text{body wt.} - 0.400)}{3.061}$$

Correlation of water wt. with body wt., $r = 0.817$

$$\text{Lean body fat-free mass (g)} = \frac{(\text{body wt.} - 7.6299)}{7.6845}$$

Correlation of LBFFM with body wt., $r = 0.6562$

---

ment characteristics from C57BL/6J-*ob/ob* mice from the MRC Laboratory Animals Centre, presumably as a result of environmental differences in the breeding colonies.

The *ob/ob* mouse has a tremendous capacity for laying down fat and when presented with a high-fat diet will increase its lipid burden considerably. Further, if one reduces the food intake of the young *ob/ob* mice, they show a reduced growth rate but still continue to lay down fat, albeit at a reduced rate. In contrast, in all the non-genetically inherited models of obesity reducing the food intake will selectively affect the accumulation of lipid rather than the other components of the body mass. Thus, the *ob/ob* mouse can be used to separate substances acting solely on food intake from those having a thermogenic or mixed action. For example, it is well known that amphetamine causes the release of noradrenaline, which in turn will stimulate thermogenesis (Stirling and Stock, 1968). In the *ob/ob* mouse amphetamine at the dietary level of 0.05 per cent will only reduce food in-

Table 5.5   Effect of amphetamine and fenfluramine on the lipid content of genetically
obese mice

	Control	Fenfluramine (0.05%)	Control	Amphetamine (0.05%)
Wt. gain (g per mouse per 4 weeks)	9.5 ± 0.6	6.2 ± 0.6	6.2 ± 0.6	0.5 ± 1.2*
Food eaten (g per 5 mice per week)	129	117	127	137
Carcass lipid (%)	50.9 ± 0.6	51.1 ± 0.8	55.6 ± 0.8	47.0 ± 1.3
Lipid per mouse (g)	15.9 ± 1.0	14.4 ± 0.5	21.0 ± 1.1	15.2 ± 0.5*

Female C57BL/6J-*ob/ob* mice were fed on the Oxoid powdered breeders' diet supplemented
with either fenfluramine (0.05 per cent) or amphetamine (0.05 per cent) for 4 weeks. Results
are the mean ± S.E. of 10 mice.
* Significantly different from controls, $P < 0.001$.

take for a few days, after which food intake is normal or even slightly elevated.
However, there is a significant change in the body composition of the animals
(table 5.5). Fenfluramine, on the other hand, does not affect body composition
significantly, although it reduces food intake and weight gain.

As stated earlier, the metabolic anti-obesity agent in current use is tri-iodothyronine. However, Bray has shown that in man '. . . the effect of thyroid hormones on
weight loss reflects primarily losses of lean body tissue with a smaller component
representing the metabolism of adipose tissue per se.' In our experiments it has
been found that tri-iodothyronine produces a similar effect in the *ob/ob* mouse
(table 5.6). However, in a free-feeding situation thyroid hormones will also increase
appetite, presumably as a compensatory measure. A major effect on lipid loss only
occurs when food intake is controlled. It is not clear whether thyroid preparations
increase appetite when used clinically in man, since the subjects studied are

Table 5.6   Effect of tri-iodothyronine (T_3) on the body composition of genetically obese
mice

	Ad libitum fed		Pair – fed	
	Control	T_3	Control	T_3
Wt. gain (g/mouse/4 weeks)	9.2 ± 1.0	7.0 ± 0.5	7.5 ± 0.8	4.0 ± 0.3
Food eaten (g/5 mice/week)	135 ± 4	161 ± 5*	115 ± 5	115 ± 5
Carcass lipid (%)	54.0 ± 0.7	52.5 ± 0.5	53.4 ± 0.6	50.0 ± 0.5*
Lipid (g)	19.5 ± 0.5	18.7 ± 0.3	15.8 ± 0.6	13.0 ± 0.4*

Female C67BL/6J-*ob/ob* mice were treated with saline or T_3 (30 μg/kg i.p.) daily for 4 weeks.
The results are given as mean ± S.E. of 10 values.
* Significantly different from control mice, $P < 0.001$.

normally restricted to a low-calorie diet. However, the clear aim for the future should be anti-obesity agents that can cause a specific loss of fat without promoting appetite. It is suggested that the genetically obese *ob/ob* mouse is a suitable model for evaluating such agents.

REFERENCES

Abdallah, A. H. (1971). On the role of norepinephrine in the anorectic effect of d-amphetamine in mice. *Arch. Int. Pharmacodyn.*, **192**, 72–77

Abdallah, A. and White, H. D. (1970). Comparative study of the anorectic activity of phenindamine, d-amphetamine and fenfluramine in different species. *Arch. Int. Pharmacodyn.*, **188**, 271–283

Astwood, E. B. (1962). The heritage of corpulence. *Endocrinology*, **71**, 337–341

Blundell, J. E., Latham, C. J. and Leshem, M. B. (1976). Differences between the anorexic actions of amphetamine and fenfluramine—possible effects on hunger and satiety. *J. Pharm. Pharmacol.*, **28**, 471–477

Blundell, J. E. and Leshem, M. B. (1974). Analysis of the mode of action of anorexic drugs. In *Recent Advances in Obesity Research*, Vol. 1 (ed. A. Howard), Newman, London, pp. 368–371

Boroumand, M. and Miller, D. S. (1977). An automated apparatus for measuring daily energy expenditure in laboratory animals. *Proc. Nutr. Soc.*, **36**, 14A

Bray, G. A. (1978). *Proceedings of the Second International Conference in Obesity, Washington, D.C.*, Newman, London

Davidson, S. and Passmore, R. (1969). *Human Nutrition and Dietetics*, 4th edn, Livingstone, Edinburgh, p. 385

Frey, H. H. and Schulz, R. (1973). On the central mediation of anorexigenic drug effects. *Biochem. Pharmacol.*, **22**, 3041–3049

Garattini, S., Bizzi, A., de Gaetano, G., Jori, A. and Samanin, R. (1975). Recent advances in the pharmacology of anorectic agents. In *Recent Advances in Obesity Research*, Vol. 1 (ed. A. Howard), Newman, London, pp. 354–367

Garattini, S. and Samanin, R. (1976). Anorectic drugs and brain neurotransmitters. In *Appetite and Food Intake* (ed. T. Silverstone), Dahlen Konferenzen, pp. 83–103

Garrow, J. (1974). *Energy Balance and Obesity in Man*, North-Holland, Amsterdam and London, p. 266

Haldane, J. (1892). A new form of apparatus for measuring the respiratory exchange of animals. *J. Physiol. (London)*, **13**, 419–430

Kannel, W. B. and Gordon, T. (1974). Obesity and cardiovascular disease: the Framingham Study. In *Obesity* (ed. W. L. Burland, P. D. Samuel and J. Yudkin), Churchill Livingstone, London, pp. 24–51

Kisseleff, H. R. (1970). Free feeding in normal and 'recovered lateral' rats monitored by a pellet-detecting eatometer. *Physiol. Behav.*, **5**, 163–173

Krapcho, J. and High, J. P. (1967). 2-Acylimino-1,1-dimethylphenethylamines and related compounds. Anorectic agents. *J. Med. Chem.*, **10**, 495–497

Kruk, Z. L. (1973). Dopamine and 5-hydroxytryptamine inhibit feeding in rats. *Nature*, **246**, 52–53

Lesses, M. F. and Myerson, A. (1938). Human autonomic pharmacology: benzedrine sulphate as an aid to the treatment of obesity. *New Engl. J. Med.*, **218**, 119–124

Macko, E., Saunders, H., Heil, G., Fowler, P. and Reichard, G. (1972). Pharmacology of an halogenated aralkylamine with anorectic properties. *Arch. Int. Pharmacodyn.*, **200**, 102–117

Miller, D. S. (1974). Thermogenesis in everyday life. In *Regulation of Energy Balance in Man* (ed. E. Jequier), Éditions Médicine et Hygiène, Genève, 1975, pp. 199–208

Miller, D. S. and Stock, M. J. (1969). A rapid method for the estimation of thermic energy in rats. *Proc. Nutr. Soc.*, **28**, 70A

Miller, D. S., Stock, M. J. and Stuart, J. A. (1974). The effects of caffeine and carnitine on the oxygen consumption of fed and fasted subjects. *Proc. Nutr. Soc.*, **33**, 28A

Nathanson, M. H. (1937). The central action of beta-aminopropylbenzene (benzedrine). *J. Am. Med. Assoc.*, **108**, 528–531

Payne, P. R. and Griffiths, M. (1976). Energy expenditure in small children of obese and non-obese parents. *Nature*, **260**, 698–700

Shields, J. (1962). *Monozygotic Twins Brought up Apart and Brought up Together*, Oxford University Press, London

Silverstone, T. and Fincham, J. (1977). Techniques for evaluating anorectic drugs in man. In *Central Mechanism of Anorectic Drugs* (ed. S. Garattini and R. Samanin), Raven Press, New York

Stirling, J. L. and Stock, M. J. (1968). Metabolic origins of thermogenesis induced by diet. *Nature*, **220**, 801–802

Stock, M. J. (1966). Determination of the oxygen consumption and the carbon dioxide production of the rat. *Proc. Nutr. Soc.*, **25**, xli–xlii

Ungerstedt, U. (1971a). Adipsia and aphagia after 6-hydroxydopamine-induced degeneration of the nigro-striatal dopamine system. *Acta Physiol. Scand.*, **82**, Suppl. No. 367, 95–122

Ungerstedt, U. (1971b). Striatal dopamine release after amphetamine or nerve degeneration revealed by rotational behaviour. *Acta Physiol. Scand.*, **82**, Suppl. No. 367, 49–68

Weissman, A., Doe, B.K. and Tenen, S. A. (1966). Anti-amphetamine effects following inhibition of tyrosine hydroxylase. *J. Pharmacol. Exp. Therap.*, **151**, 339–352

6

Abnormal function of the endocrine pancreas in genetic and experimentally induced obesity in rodents

Anne Beloff-Chain (Department of Biochemistry, Imperial College of Science and Technology, London, UK)

SUMMARY

Hyperinsulinaemia is a consistent feature of obesity; studies on the hyperactivity of the β-cell function of the endocrine pancreas in experimental and genetic obesity in rodents are described. Evidence that the hypersecretion of insulin may be one of the primary defects in many forms of obesity is discussed. Adrenal cortical hyperplasia and increased circulating glucocorticosteroids, as well as high pituitary ACTH levels, are among the multiple endocrine abnormalities described in the *ob/ob* mice.

Experiments are reported which provide evidence for a physiological relationship between hyperinsulinaemia and hyperactivity of the adrenal cortex in the *ob/ob* mouse as well as in other forms of obesity.

The experimental work described has demonstrated that the neuro-intermediate lobe of the pituitary gland secretes a peptide with insulin-releasing properties and there is a hyperactivity of this insulin secretagogue in *ob/ob* mice as well as in other forms of obesity. Evidence that this factor is CLIP, the corticotrophin-like intermediate lobe peptide, consisting of the 18–39 fragment of the ACTH molecule, or a closely related peptide, is given.

The possible role of such an insulin-releasing pituitary factor in the development of hyperinsulinaemia prior to the onset of obesity, hyperphagia, hyperglycaemia and insulin resistance is discussed in relation to observations made on young 3 week old *ob/ob* mice.

These findings could represent a common hypothalamic origin in obesity of both pancreatic and adrenocortical dysfunction.

91

Hyperinsulinaemia is one of the most consistent features of obesity encountered both in the clinical condition and in genetic and experimentally induced obesity in rodents. The degree of hyperinsulinaemia reported in the literature is very varied not only in different types of obesity, but also in any given type, even of the same age group. Insulin levels will, of course, vary, depending on the time at which blood samples are taken in relation to the feeding habits of the animals. However, in our laboratory, working with genetically obese mice (*ob/ob*) under controlled conditions, the plasma insulin levels covered a wide range (Abraham *et al.*, 1971; Findlay *et al.*, 1973).

The obese hyperglycaemic mouse colony originated from The Jackson Memorial Laboratory, Bar Harbor, Maine, USA, and was introduced into local mixed colonies (not pure strains) in Edinburgh and Birmingham, from which our original stock was obtained. The animals used were from a colony bred from these stocks at Imperial College over a period of about ten years.

Although it is generally accepted that hyperinsulinaemia is a characteristic of the obese syndrome, there is a good deal of controversy as to whether this is a primary defect in the pancreatic islet cell function preceding hyperphagia and obesity, or whether it is a secondary defect appearing as a consequence of the two latter abnormalities. It is generally agreed that hyperphagia promotes insulin secretion, and obesity results in insulin resistance for which the pancreas partially compensates by hypersecretion of insulin. On the other hand, it is also true that hyperinsulinaemia itself can promote hyperphagia and obesity.

One form of experimental obesity which has been widely studied is that produced by lesioning the ventromedial nucleus of the hypothalamus. Frohman and Bernardis (1968) carried out this operation on weanling rats and demonstrated increased circulating insulin levels, although these animals were not hyperphagic or obese. Further evidence that hyperinsulinaemia may be dissociated from hyperphagia and obesity was obtained by Martin *et al.* (1974), who demonstrated decreased insulin content of the pancreas, islet β-cell degranulation, a tendency to enlargement of islets and increased responsiveness of the latter to insulin secretagogues in lesioned rats maintained on a restricted diet. Both these groups of investigators have suggested that hyperinsulinaemia develops independently and prior to hyperphagia, hyperglycaemia and obesity. Rohner *et al.* (1977) have demonstrated a marked increase in insulin secretion following glucose administration *in vivo* within 10 min after electrolytic lesioning of the ventromedial hypothalamic area in anaesthetised rats. They suggest that these experiments show acute regulation of insulin secretion by the hypothalamus.

Furthermore, York and Bray (1972) reported that in rats rendered insulin-deficient by the β-cell toxin 'streptozotocin', hypothalamic lesioning did not produce obesity or hyperphagia. When these animals were subsequently treated with insulin, hyperphagia developed and abnormal weight gain occurred. Boozer and Mayer (1976), however, who treated young obese (*ob/ob*) mice (4–5 weeks of age) with streptozotocin to maintain normal insulin levels, showed that over a 16 week period these animals still became obese and insulin-resistant. They concluded that early and persistent hyperphagia rather than hyperinsulinaemia is the cause of these abnormalities in the *ob/ob* mouse. Recently Inoue *et al.* (1977) have reported that when foetal rat pancreas was transplanted under the kidney capsule in streptozotocin-diabetic rats, hypothalamic lesioning failed to produce

obesity or hyperphagia. From these experiments they conclude that the neural
control of the pancreas plays an essential role in the development of obesity.

This feature of high circulating insulin levels associated with obesity has led
a number of investigators to study the histology of the endocrine pancreas,
particularly in rodents with genetic obesity. Hyperplasia of the islets of Langerhans
and degranulation of the β-cells was described in early work on the genetically
obese mouse *(ob/ob)* (Bleisch *et al.*, 1952; Wrenshall *et al.*, 1955). Gonet *et al.* (1965)
have carried out studies on the Spiny mouse *(Acomys cahirinus)*, a group of small
rodents indigenous to the Eastern Mediterranean and Africa, which when bred in
the laboratory frequently develop striking obesity and a diabetic syndrome. They
demonstrated a considerable degree of hyperplasia and hypertrophy of the islets of
Langerhans in these animals. They measured the percentage of pancreatic mass
occupied by the islets in these mice as well as in a number of other forms of genetic obesity (figure 6.1) and compared this with the value obtained for normal lean

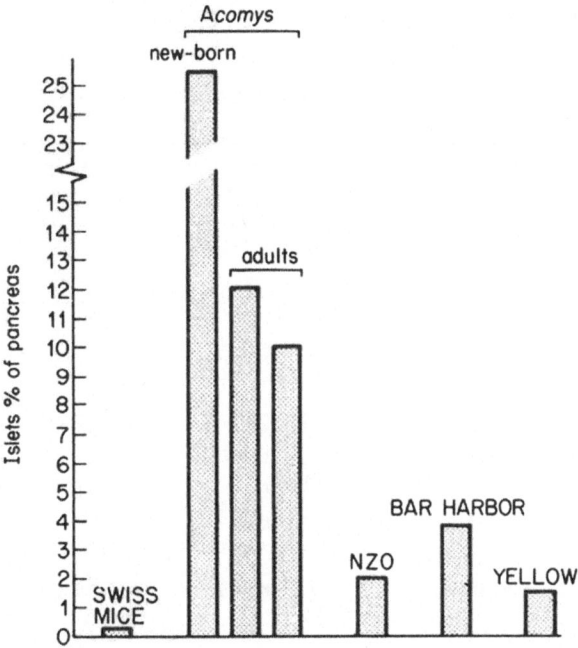

Figure 6.1   Percentage of pancreatic mass occupied by the islets of Langerhans in
Acomys cahirinus and several strains of *Mus musculus*. NZO = New Zealand
Obese; Bar Harbor = C57BL/6J-*ob/ob*; yellow = $A^y/-$. From Gonet *et al.*, 1965.

mice *(Mus musculus)*. They concluded that some degree of hyperplasia and hypertrophy of the islets is characteristic of all types of genetic obesity in rodents. Shino
et al. (1973) have examined the structural changes in genetically obese *(fa/fa)* rats from
5 to 52 weeks of age. They observed hypertrophy of islets from a very early age,
and also a highly developed endoplasmic reticulum and Golgi complex, suggesting
active insulin synthesis and cytological evidence for increased insulin secretion.

Zucker and Antoniades (1972) showed that relatively high concentrations of insulin persist in fatty rats during fasting, and suggest that this is due to excessive insulin output, which can also be demonstrated *in vitro* from isolated pancreatic islets of these animals (Stern *et al.,* 1972).

The results of studies carried out to correlate histological findings in the pancreas with certain biochemical parameters in the obese (*ob/ob*) mouse are shown in table 6.1. Findlay *et al.* (1973), in agreement with the findings of other investigators, showed that the islets of Langerhans were significantly enlarged in the *ob/ob* mice and that there was some increase in the total pancreatic insulin content. This increase was, however, much less marked than the very high concentration of circulating insulin, which was increased by a factor of almost 50. The histological studies showed pancreatic endocrine hyperfunction in the obese animals by highly vascular as well as enlarged islets. Furthermore, gland hyperplasia was suggested not only by the presence of the mega islets, but also by duct islet cell metaplasia, which was not evident in the lean mouse pancreas.

Although there is general agreement concerning insulin hypersecretion in genetic

Table 6.1 Statistical evaluation of parameters in obese (*ob/ob*) and lean (+/*ob* or +/+) control mice (Findlay *et al.,* 1973)

	Lean mice	Obese mice	*P*-value
Body weight (g)	41.7 ± 0.8 (30)	80.0 ± 2.8 (15)	< 0.001
Blood glucose (mg/100 ml)	189 ± 5 (30)	272 ± 27 (15)	< 0.001
Serum immunoreactive insulin (μu/ml)	55 ± 6 (24)	2602 ± 710 (14)	< 0.001
Pancreas weight (mg)	451 ± 15	515 ± 34	< 0.001
Mean major axis in μm of islet profiles	138 ± 9.73 (119)	330 ± 21.95 (133)	0.000 03*
Pancreatic insulin (mu/100 mg tissue)	469 ± 80 (5)	625 ± 60 (10)	< 0.2
Total pancreatic insulin (mu)	1923 ± 341 (5)	3142 ± 246 (5)	< 0.02
Pancreatic glucagon (ng/100 mg tissue)	100 ± 10 (5)	165 ± 27 (10)	< 0.2
Total pancreatic glucagon (ng)	357 ± 51 (5)	818 ± 115 (10)	< 0.01
α-cells (%)	8.98 ± 2.00 (6328 cells in 6 mice)	1.54 ± 0.71 (4118 cells in 3 mice)	< 0.05

Number of observations in parentheses. *P*-values were calculated by Student's *t*-test except for the one marked with an asterisk, where a non-parametric method was employed. Means ± S.E.M.

and experimental obesity, the significance of differences in glucagon secretion which have been reported are much less consistent. In our experiments (Findlay *et al.*, 1973; table 6.1) the total glucagon content of the pancreas was significantly higher in the obese mouse, although the percentage of α-cells of the islets is probably reduced owing to the increase in the number of β-cells, and the peripheral distribution of α-cells, characteristic of normal islets, was not found in the obese mouse islets in which these cells appeared to have a random distribution.

Dubuc *et al.* (1977) have reported high circulating glucagon levels in *ob/ob* mice of 8 weeks of age, and suggest that this is maintained in spite of hyperglycaemia and hyperinsulinaemia. Lavine *et al.* (1975) had also shown that in 8 week old *ob/ob* mice the circulating glucagon levels were high but, unlike the lean mice, these did not increase further by 48 h fasting (table 6.2). They have suggested that this could be due to the lack in response of the α-cell to suppression of glucagon secretion by carbohydrate feeding.

Table 6.2   Effect of 48 h fast on plasma glucagon in C57BL/6J-*ob/ob* and +/+ mice (Lavine *et al.*, 1975)

	Plasma glucagon (ng/ml)		Statistical significance
	ob/ob	+/+	(*P*)
Fed	0.74 ± 0.03 (6)*	0.42 ± 0.05 (6)	< 0.001
48 h fast	0.89 ± 0.07 (6)	1.06 ± 0.10 (6)	> 0.05
Statistical significance (*P*)	> 0.05	< 0.001	

*Mean ± S.E.M. and the number of animals.

Table 6.3   Factors influencing glucagon secretion from islets of lean (*ob/+* and +/+) and obese (*ob/ob*) mice (Beloff-Chain *et al.*, 1977)

Additions	Lean mice	Obese mice
None	2.23 ± 0.27 (5)	2.23 ± 0.48 (5)
Arginine (10 mM)	4.73* ± 0.94 (6)	4.14* ± 0.38 (7)
Leucine (10 mM)	4.11 ± 0.92 (6)	3.39 ± 0.65 (6)
Adrenaline (43 μM)	2.56 ± 0.75 (5)	10.86* ± 2.06 (4)
Anti-insulin serum (final concentration, 1%)	2.90 ± 0.70 (6)	3.65* ± 0.88 (5)

Results expressed as ng glucagon per 5 islets per 30 min. Mean values ± S.E.M. Number of experiments in parentheses. 100 mg% glucose present in all experiments.

*$P < 0.05$ compared with values with no addition (paired *t*-test).

We were unable to show abnormal circulating glucagon levels in 3–4 month old *ob/ob* mice or abnormal glucagon secretion from isolated islets (Beloff-Chain *et al.*, 1977). Furthermore, arginine and leucine, which are known to stimulate glucagon secretion, had the same effect on islets of lean and obese mice but adrenaline had a much larger effect on glucagon secretion in the *ob/ob* mice (table 6.3). Anti-insulin serum tended to increase glucagon secretion *in vitro* (table 6.3) and to a much greater extent *in vivo* (table 6.4). These results suggest that the *ob/ob* mouse islets are dependent on higher concentrations of insulin for the normal suppressive effect of glucose on glucagon secretion, which could indicate an insulin resistance in the α-cells.

Table 6.4   Influence of anti-insulin serum injection on blood glucose levels and plasma glucagon in lean (+/+ or +/*ob*) and *ob/ob* mice (Beloff-Chain *et al.*, 1977)

Animals	Blood glucose		Plasma glucagon	
	GPS-injected	AIS-injected	GPS-injected	AIS-injected
+/+ or +/*ob*	161 ± 12 (6)	296* ± 17 (6)	339 ± 91 (10)	355 ± 76 (11)
ob/ob	249* ± 26 (5)	330* ± 26 (6)	341 ± 44 (6)	628† ± 90 (6)

Results expressed as mg glucose per 100 ml blood and pg glucagon per ml of plasma. Mean values ± S.E.M. Number of experiments in parentheses. Animals fed *ad libitum*. Controls injected with 0.5 ml guinea-pig serum (GPS) and treated animals with 0.5 ml anti-insulin serum (AIS) by tail vein injection 1 h prior to sacrifice.
*$P < 0.02$ compared with GPS-injected lean mice.
†$P < 0.02$ compared with AIS-injected lean mice.

Eaton *et al.* (1976) have reported reduced circulating glucagon levels in the Zucker rats, rather than the increased levels reported in the *ob/ob* mice and reduced glucagon secretion *in vivo* following administration of arginine. They suggest that the suppression of glucagon is due to the high levels of glucose and insulin in the plasma, and possibly also the high plasma-free fatty acids.

However, none of the work reported would suggest that abnormal glucagon levels play an important role in genetic or experimental obesity, whereas the role of insulin would appear to be crucial.

Genuth (1969), working with the *ob/ob* mice, first pointed out that the degree of hyperinsulinaemia is disproportionate to the concurrent glucose stimulus and to the pancreatic insulin content. We concluded that excessive insulin stimulation (Beloff-Chain *et al.*, 1973) could be due to greater sensitivity of the β-cells in the *ob/ob* mouse to a normal biological stimulus or to excessive stimulation by some unknown factor or, as the evidence we have now accumulated would suggest, to a combination of the two.

Our first studies were on the effect of known insulin secretagogues on the pancreas of +/+ and *ob/ob* mice. In order to separate the effects of hyperphagia and concomitant hyperglycaemia from primary differences in the response of the pancreas, *ob/ob* mice fed *ad libitum* were compared with those maintained on

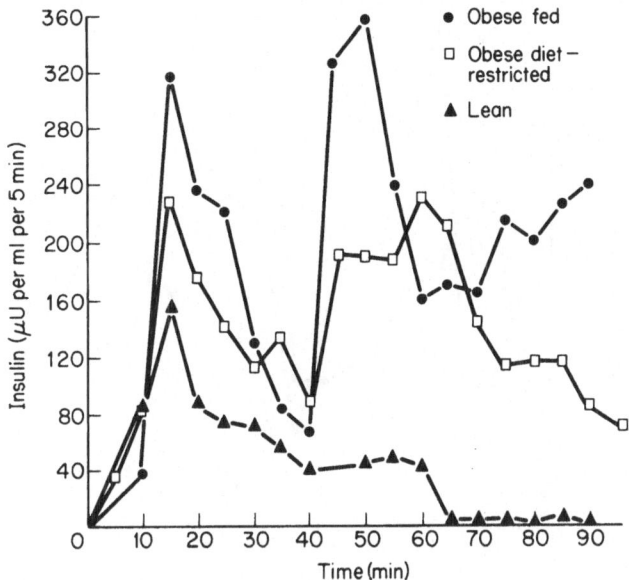

Figure 6.2 Influence of increasing glucose concentration on secretion of insulin in perfused pieces of pancreas from obese (*ob/ob*) fed, obese (*ob/ob*) diet-restricted and lean (*ob/+* or *+/+*) mice. Perfusion fluid contained 150 mg% glucose at the start of the experiment, and this was increased to 300 mg% after 40 min. From Beloff-Chain *et al.*, 1973.

a restricted diet. Figures 6.2 and 6.3 show the effect of high glucose concentration and of leucine, respectively, on insulin secretion, and demonstrate a marked increase in response from the *ob/ob* mouse islets whether fed *ad libitum* or maintained on a restricted diet. This would suggest that the increased response of the islets is independent of hyperphagia and, furthermore, is independent of the insulin content of the pancreas, which is normal in *ob/ob* mice maintained on a restricted diet (table 6.5). Newman (1977) demonstrated an elevated glycogen content in pancreatic islets of *ob/ob* mice and also increased phosphorylase and synthetase activity, indicating an increase in glycogen turnover (table 6.6). Atkins and Matty (1971) showed increased activity of phosphodiesterase and adenyl cyclase in islets of *ob/ob* mice and suggested an increased turnover of cyclic AMP in obese islets. Newman supports this conclusion by showing a greater effect on cyclic AMP levels with a phosphodiesterase inhibitor and with glucagon in islets of *ob/ob* mouse as compared with lean mice. He suggests that increased turnover of both glycogen and cyclic AMP may represent part of the altered pattern of metabolism contributing to increased insulin secretion in obesity.

There is thus considerable evidence for the hyper-responsiveness of the β-cells in the pancreatic islets of obese (*ob/ob*) mice to stimulation of insulin secretion. In the second part of this article evidence is given for the presence and nature of an insulin secretagogue in the pituitary gland, which is elevated in the *ob/ob* mouse.

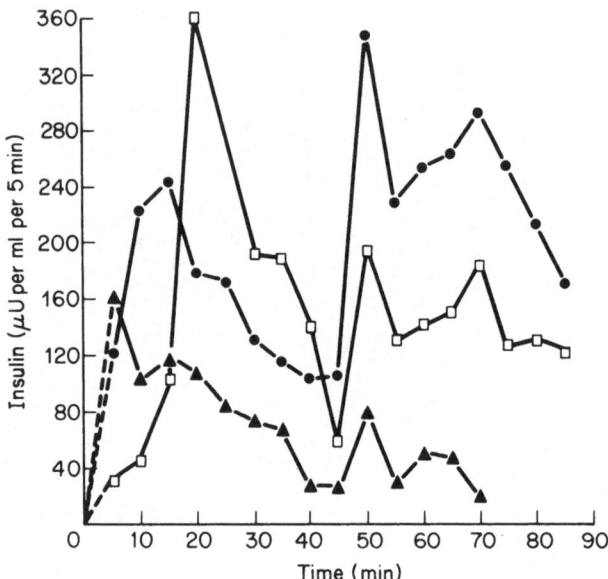

Figure 6.3 Influence of leucine on the secretion of insulin in perfused pieces of pancreas from obese fed, obese diet-restriced and lean mice. Perfusion fluid contained 150 mg% glucose at start of experiment and 10 mM leucine was added after 40 min. Key as in figure 6.2. From Beloff-Chain *et al.*, 1973.

One of the multiple endocrine abnormalities in the *ob/ob* mouse is adrenal cortical hyperplasia (Hellerstrom *et al.*, 1962) and increased circulating glucocorticosteroids reported by Naeser (1974). Simultaneous occurrence of elevated plasma insulin levels and increased adrenal cortical function has been reported in most forms of genetic and experimental obesity in rodents and is frequently encountered in clinical obesity. The physiological relationship of hyperinsulinaemia and hyperactivity of the adrenal cortex has so far not been elucidated. There is now evidence that the intermediate lobe of the pituitary gland secretes a corticotrophin-like peptide with insulin-releasing properties and that the hyperactivity of this compound

Table 6.5 Insulin content of pancreases from *ob/ob*, *ob/ob* RD (restricted diet) and lean (+/+ or +/*ob*) mice (Beloff-Chain *et al.*, 1973)

Animal	Weight of pancreas (mg)	P	mU insulin per pancreas		mU insulin per mg wet wt. pancreas	P
Lean	298 ± 18		982 ± 161		3.49 ± 0.27	
ob/ob	282 ± 23	N.S.	2249 ± 316	<0.01	8.46 ± 2.04	<0.05
ob/ob RD	272 ± 17	N.S.	1213 ± 254	N.S.	4.35 ± 0.63	N.S.

Results are the mean of six animals in each group, ± S.E.M.
N.S. = not statistically significant at *P* = 0.05.

Table 6.6 Incorporation of glucose and enzyme activities (pmol per islet per min) (Newman, 1977)

Activity	Lean (*ob*/+ or +/+)	Obese (*ob*/*ob*)	P
Incorporation of [C^{14}]-glucose	1.78 ± 0.20 (6)	3.85 ± 0.94 (7)	≤0.05
Phosphorylase	34 ± 9 (4)	73 ± 26 (4)	≤0.1
Glycogen synthetase (I + D)*	41 ± 10 (11)	38 ± 7 (14)	NS
Glycogen synthetase (I)*	17 ± 1 (3)	40 ± 9 (7)	≤0.05

Mean values ± S.E.M. Number of experiments in parentheses. *P*-Values determined by Student's *t*-test.
*Glucose-6-phosphate dependent and independent I and D form, respectively.

in the *ob*/*ob* mouse points to its possible involvement in the hyperinsulinaemia associated with obesity (Beloff-Chain *et al.*, 1976; Beevor *et al.*, 1977).

Edwardson and Hough (1975) reported elevated levels and secretion of ACTH from the pituitary gland of the *ob*/*ob* mouse *in vitro*. A direct effect of ACTH on insulin secretion has been reported by a number of investigators, although the concentrations of the hormone studied were far in excess of normal biological levels. Therefore experiments were carried out in which the pituitary gland of the *ob*/*ob* and +/? mice were perfused in series with microdissected islets (Beloff-Chain *et al.*, 1975). It was shown that the pituitary gland rapidly stimulated insulin release. Further investigations showed (figure 6.4) that when the lean mice were separated

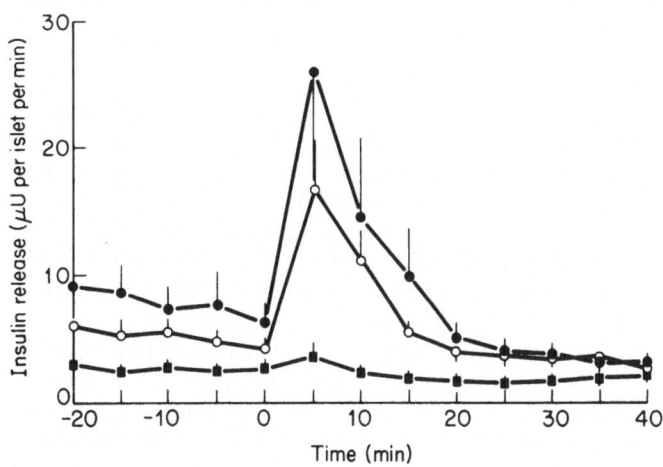

Figure 6.4 The effect of the perfusate from pituitary glands of *ob*/*ob* mice, +/+ lean mice and *ob*/+ lean mice on isolated islets from +/+ lean mice. —○—○—, *ob*/*ob* (*n* = 8); —●—●—, *ob*/+ (*n* = 5); —■—■—, +/+ (*n* = 5). Vertical bars indicate S.E.M. Pituitary glands introduced at zero time. From Beloff-Chain *et al.*, 1975.

into homozygotes (+/+) and heterozygotes (*ob*/+), the pituitary glands from the *ob*/*ob* and *ob*/+ stimulated insulin secretion, whereas the glands from +/+ mice had no effect (Beloff-Chain *et al.*, 1975).

The pattern of insulin secretion suggests that the rapid stimulation may come from a labile insulin pool, and further evidence for this was obtained by experiments in which a fresh lot of islets was introduced into the system after 25 min and a second peak was obtained (figure 6.5; Beloff-Chain and Hawthorn, 1976). When the same islet preparation was restimulated by a fresh pituitary gland, however, no further stimulation was obtained.

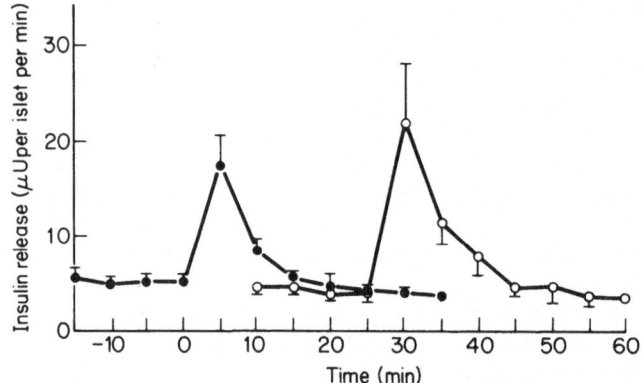

Figure 6.5 Influence of the perfusate from pituitary glands of *ob*/*ob* mice on insulin secretion from two lots of isolated lean mouse pancreatic islets. —●—●— Insulin secretion from first lot of islets. Perfused pituitaries transferred to fresh lot of islets at 25 min. —○—○— Insulin secretion from second lot of islets. Average of five experiments ± S.E.M.

In order to ascertain the nature of the pituitary factor, experiments were carried out in which pure porcine ACTH at a concentration of approximately threefold that present in the perfusate of the *ob*/*ob* glands was tested and compared with Synacthen, the synthetic 1-24 NH_2 terminal fragment of the ACTH molecule with full steroidogenic properties (figure 6.6). The pure ACTH gave some stimulation, whereas Synacthen had no effect (Beloff-Chain *et al.*, 1976). This suggested that the effect might be due to the —COOH terminal region of ACTH. Scott *et al.* (1973) described a peptide consisting of the 18–39 —COOH terminal of ACTH to be present in the neuro-intermediate lobe of rat pituitary gland. The corticotrophin-like intermediate lobe peptide 'CLIP' was tested in our system and shown to stimulate insulin secretion, although only sufficient material for one experiment was available (figure 6.6). Further experiments carried out with a synthetic peptide corresponding to the 17–39 fragment of human ACTH have shown variable results. Out of a total of 33 perfusions, 15 perfusions gave a stimulation exceeding 50 per cent, which can be considered a significant response; the lack of response of the remaining 18 perfusions must be due to a variation which we have periodically encountered in the islet preparations and for which

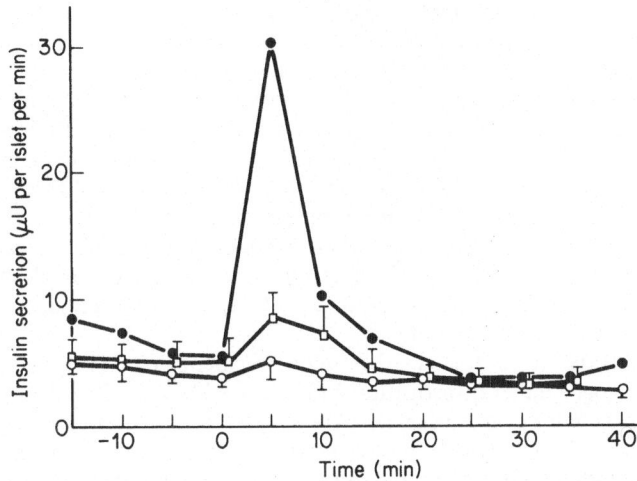

Figure 6.6   Influence of pure porcine ACTH, Synacthen and CLIP on insulin secretion from isolated islets of lean mice. —□—□—, ACTH 20 ng/ml (*n* = 5); —○—○—, Synacthen 25 ng/ml (*n* = 5); —●—●—, CLIP (corticotrophin-like intermediate lobe peptide) 20 ng/ml (*n* = 1). Peptide introduced at time zero: mean results for ACTH and Synacthen. Vertical bars indicate S.E.M. From Beloff-Chain *et al.*, 1976.

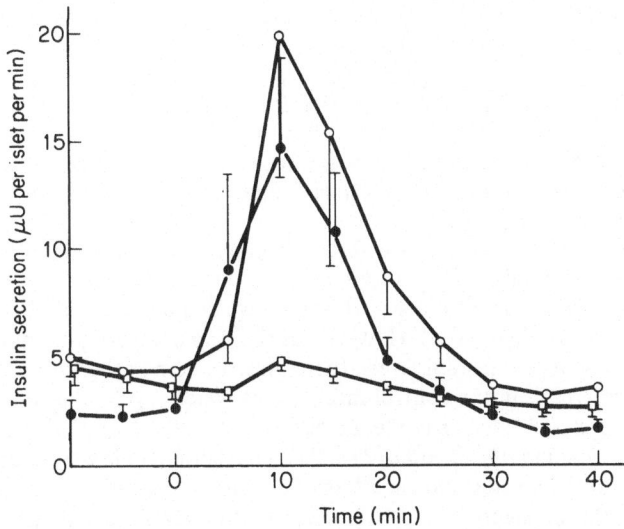

Figure 6.7   Influence of the whole pituitary gland, the neurointermediate lobe and the adenohypophysis on insulin secretion from isolated islets of lean mice. —●—●—, Whole pituitary (*n* = 8); —○—○—, neurointermediate lobe (*n* = 12); —□—□—, adenohypophysis (*n* = 14). Mean results; vertical bars indicate S.E.M. From Beloff-Chain *et al.*, 1976.

we have not yet found a satisfactory explanation. However, in a number of perfusions carried out with growth hormone, prolactin, vasopressin and α-MSH (α-melanocyte-stimulating hormone) no effect on insulin secretion was obtained. The hypothesis that 'CLIP' or a related peptide is responsible for the pituitary effect on insulin secretion is supported by experiments carried out with dissected pituitary glands (figure 6.7; Beloff-Chain *et al.*, 1976). These experiments demonstrate that the activity is in the neuro-intermediate lobe of the gland and that the adenohypophysis which contains nearly all the ACTH is inactive.

As Dr Edwardson demonstrates in his article published in this Symposium, the neuro-intermediate lobe of the pituitary gland of *ob/ob* mice has an elevated level of an α-MSH and CLIP (or a closely related peptide), as well as a high concentration of ACTH in the adenohypophysis.

All these results suggest that the C-terminal fragment of ACTH may be the insulin secretagogue under investigation and this would be the first demonstration of a biological activity of this peptide. Experiments are at present in progress to test whether the insulin-releasing properties of the pituitary gland perfusate can be removed by treating with antisera raised to the −COOH terminal moiety of the ACTH molecule.

It is of interest that Chieri *et al.* (1976) have proposed the presence of a hypophyseal factor that stimulates insulin secretion. In dogs given a continuous infusion of glucose through the carotid artery they observed a rapid increase in insulin secretion in the pancreatic duodenal vein, not due to increased circulating glucose levels. Similar experiments in hypophysectomised animals failed to stimulate insulin secretion. They also reported the presence of an insulin-stimulating factor in plasma samples from the jugular vein of dogs receiving glucose via the carotid artery and re-injected into a second dog via the pancreatic duodenal artery. Again the plasma collected from the jugular vein of hypophysectomised dogs was inactive. It is of interest in this connection that it has been shown that pituitary glands taken from lean mice, 15–20 min following a glucose load (0.5 ml of a 400 mg/ml glucose solution given intraperitoneally), showed insulin-releasing properties which were not present in glands from lean mice similarly treated with saline (Beloff-Chain *et al.*, 1978).

In the experiments already described the pituitary gland from the *ob/+* as well as the *ob/ob* stimulated insulin secretion, although in later experiments in our laboratory dilutions of the pituitary perfusates have demonstrated that there is a gene–dosage response (figure 6.8). However, as the lean heterozygote mice have normal circulating plasma insulin levels, there must be some other differences between these two groups in the control mechanisms which regulate insulin secretion *in vivo*. There could, for example, be differences in the hypothalamic control of the pituitary gland in *ob/ob* and *ob/+* mice. Another possible explanation could be a difference in the response of the islets of *ob/ob*, +/+ and *ob/+* to pituitary stimulation. In this connection, it is of interest that in *ob/ob* mice of 12 weeks of age in which obesity and hyperinsulinaemia are fully developed the islets from obese mice responded slightly but significantly less than islets from lean mice. However, in very young animals of 3 weeks of age the response of the *ob/ob* mice islets was greater than that of the +/+ islets (figure 6.9). It has also been established that pituitary glands from the young obese mice (3 weeks of age) stimulate insulin secretion. These results would suggest that the pituitary insulin secretagogue may contri-

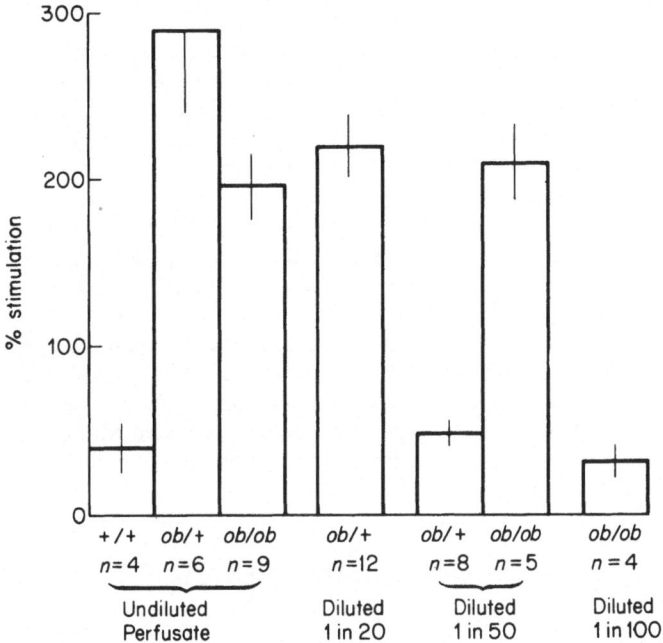

Figure 6.8 Influence of pituitary perfusate at different dilutions on stimulation of insulin secretion from lean mouse islets. From Beloff-Chain *et al.*, 1978.

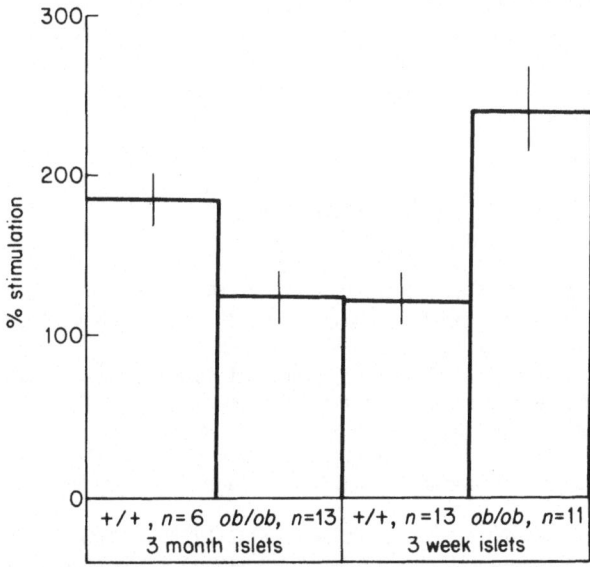

Figure 6.9 Response of pancreatic islets from young and adult lean and obese mice to stimulation by the pituitary glands of adult obese mice. From Beloff-Chain *et al.*, 1978.

bute to the development of hyperinsulinaemia in young animals prior to the development of obesity, hyperglycaemia and insulin resistance and later the islets develop some resistance to the pituitary-stimulating factor. No differences have been observed in the response of adult heterozygote and homozygote lean mice islets, but experiments on islets from young heterozygote and homozygote lean animals have not yet been carried out.

Recent experiments have shown that the pituitary glands from mice rendered obese by gold thioglucose treatment and by dietary manipulation exhibit insulin-releasing activity similar to that described above for the pituitary glands of genetically obese mice (Beloff-Chain *et al.*, 1978).

Our work would thus at present suggest that a corticotrophin-like peptide of the intermediate lobe of the pituitary with insulin-releasing properties may represent a component in the multiple control mechanism which regulates insulin secretion and may be involved in the hyperinsulinaemia associated with obesity. This could represent a common hypothalamic origin in obesity of both pancreatic and adrenocortical dysfunction.

ACKNOWLEDGEMENTS

I am grateful to the Wellcome Trust and the Herbert E. Dunhill Trust for financial support and to the Science Research Council and Beecham Pharmaceutical Research Division for CASE awards (M. Newman and S. Bogdanovic). I should also like to acknowledge the contribution of Mr D. Green and his staff for their expert handling of the obese mouse colony at Imperial College: and to thank Mr H. Dalal, Mr C. J. Hardcastle and Mr E. D'Souza for valuable technical assistance, and Miss S. Beevor for help in the preparation of the manuscript and for reading the latter at the Meeting in my absence.

REFERENCES

Abraham, R. R., Dade, E., Elliott, J. and Hems, D. A. (1971). Hormonal control of intermediary metabolism in obese hyperglycemic mice. II. Levels of plasma free fatty acids, immunoactive insulin and liver glycogen. *Diabetes,* **20**, 535–541

Atkins, T. and Matty, A. J. (1971). Adenyl cyclase and phosphodiesterase activity in the isolated islets of Langerhans of obese mice and their lean littermates; the effect of glucose, adrenaline and drugs on adenyl cyclase activity. *J. Endocrinol.,* **51**, 67–78

Beevor, S., Beloff-Chain, A., Donaldson, A. and Edwardson, J. A. (1977). Pituitary intermediate lobe function in genetically obese (*ob/ob*) and lean mice. *Proc. Physiol. Soc.,* November, 73P

Beloff-Chain, A., Bogdanovic, S. and Cawthorne, M. A. (1979). *J. Endocrinol.* (in press)

Beloff-Chain, A., Edwardson, J. A. and Hawthorn, J. (1975). The influence of the pituitary gland on insulin secretion in genetically obese (*ob/ob*) mice. *J. Endocrinol.,* **65**, 109–116

Beloff-Chain, A., Edwardson, J. A. and Hawthorn, J. (1976). Corticotrophin-like intermediate lobe peptide as an insulin secretagogue. *J. Endocrinol.,* **73**, 28P–29P

Beloff-Chain, A. and Hawthorn, J. (1976). The release of insulin from pancreatic islets of lean and obese mice stimulated in vitro by pituitary glands from obese mice and by high glucose concentrations. *FEBS Letters*, **64**, 214–217

Beloff-Chain, A., Hawthorn, J. and Green, D. (1975). Influence of the pituitary gland from the homozygote (+/+) and heterozygote (*ob*/+) lean mouse on insulin secretion *in vitro*. *FEBS Letters*, **55**, 72–74

Beloff-Chain, A., Newman, M. E. and Mansford, K. R. L. (1973). *In vitro* studies on insulin secretion in the genetically obese mouse. *Diabetologia*, **9**, 447–452·

Beloff-Chain, A., Newman, M. E. and Mansford, K. R. L. (1977). Factors influencing insulin and glucagon secretion in lean and genetically obese mice. *Horm. Metab. Res.*, **9**, 33–36

Bleisch, V. B., Mayer, J. and Dickie, M. M. (1952). Familial diabetes mellitus and insulin resistance associated with hyperplasia of the islets of Langerhans in mice. *Am. J. Pathol.*, **28**, 369–385

Boozer, C. N. and Mayer, J. (1976). Effect of long term restricted insulin production in obese hyperglycemic (genotype *ob*/*ob*) mice. *Diabetologia*, **12**, 181–187

Chieri, R. A., Basabe, J. C. and Farina, J. M. S. (1976). Evidence for a hypophyseal factor that stimulates insulin secretion by the pancreas (Insulinotrophine). *Horm. Metab. Res.*, **8**, 329–332

Dubuc, D. U., Mobley, D. W., Mahler, R. J. and Ensinek, J. M. (1977). Immunoreactive glucagon levels in obese hyperglycaemic (*ob*/*ob*) mice. *Diabetes*, **26**, 841–846

Eaton, P. R., Conway, M. and Schade, D. S. (1976). Endocrine glucagon regulation in genetically hyperlipemic obese rats. *Am. J. Physiol.*, **230**, 1336–1341

Edwardson, J. A. and Hough, C. A. M. (1975). The pituitary adrenal system of the genetically obese (*ob*/*ob*) mouse. *J. Endocrinol.*, **65**, 99–107

Findlay, J. A., Rookledge, K. A., Beloff-Chain, A. and Lever, J. D. (1973). A combined biochemical and histological study on the islets of Langerhans in the genetically obese hyperglycemic mouse and in the lean mouse, including observations on the effects of streptozotocin treatment. *J. Endocrinol.*, **56**, 571–583

Frohman, L. A. and Bernardis, L. L. (1968). Growth hormone and insulin levels in weanling rats with ventromedial hypothalamic lesions. *Endocrinology*, **82**, 1125–1132

Genuth, S. M. (1969). Hyperinsulinism in mice with genetically determined obesity. *Endocrinology*, **84**, 386–391

Gonet, A. E., Stauffacher, W., Pictet, R. and Renold, A. E. (1965). Obesity and diabetes mellitus with striking hyperplasia of the islets of Langerhans in Spiny Mice (*Acomys cahirinus*). 1. Histological findings and preliminary metabolic observations. *Diabetologia*, **1**, 162–171

Hellerstrom, C., Hellman, B. and Larsson, S. (1962). Some aspects of the structure and histochemistry of the adrenals in obese hyperglycemic mice. *Acta Pathol. Microbiol. Scand.*, **54**, 365–372

Inoue, S., Bray, G. A. and Muller, Y. S. (1977). Effect of transplantation of pancreas on development of hypothalamic obesity. *Nature*, **266**, 732–744

Lavine, R. L., Voyles, N., Perrino, P. and Recant, L. (1975). The effect of fasting on tissue cyclic c AMP and plasma glucagon in the obese hyperglycemic mouse. *Endocrinology*, **97**, 615–620

Martin, J. M., Konijnendijk, W. and Bouman, P. R. (1974). Insulin and growth hormone secretion in rats with ventromedial hypothalamic lesions maintained on restricted food intake. *Diabetes*, **23**, 203–208

Naeser, P. (1974). Function of the adrenal cortex in obese hyperglycemic mice (gene symbol *ob*). *Diabetologia*, **10**, 449–453

Newman, M. E. (1977). Glycogen metabolism and cyclic AMP levels in isolated islets of lean and genetically obese mice. *Horm. Metab. Res.*, **9**, 358–361

Rohner, R., Dufour, A., Karakash, C., Le Marchand, Y., Ruf, B. and Jeanrenaud, B. (1977). Immediate effect of lesion of the ventromedial hypothalamic area upon glucose induced insulin secretion in anaesthetized rats. *Diabetes*, **13**, 239–242

Scott, A. P., Ratcliffe, J. G., Reese, L. H., Landon, J., Bennett, H. P. J., Lowry, P. J. and McMartin, C. (1973). Pituitary peptides. *Nature New Biol.*, **244**, 65–67

Shino, A., Matsuo, T., Iwatsuka, H. and Suzuoki, Z. (1973). Structural changes of pancreatic islets in genetically obese rats. *Diabetologia*, **9**, 413–421

Stern, J., Johnson, P. R., Greenwood, M. R. C., Zucker, L. M. and Hirsch, J. (1972). Insulin resistance and pancreatic insulin release in the genetically obese Zucker rat. *Proc. Soc. Exp. Biol.*, **139**, 66–69

Wrenshall, C. A., Andrus, S. B. and Maver, J. (1955). High levels of pancreatic insulin co-existent with hyperplasia and degranulation of beta-cells in mice with hereditary obese hyperglycemic syndrome. *Endocrinology*, **56**, 335–340

York, D. A. and Bray, G. A. (1972). Dependence of hypothalamic obesity on insulin, the pituitary and the adrenal gland. *Endocrinology*, **90**, 885–894

Zucker, L. M. and Antoniades, H. N. (1972). Insulin and obesity in the Zucker genetically obese rat 'fatty'. *Endocrinology*, **90**, 1320–1330

7

Adipose tissue in
genetically obese rodents

Margaret Ashwell and C. J. Meade* (Division of Clinical Investigation and
Transplantation Biology Section, Clinical Research Centre, Harrow, Middlesex, UK)

SUMMARY

(1)   Genetically determined obesity in rodents is associated with fat cell
hypertrophy (always) and hyperplasia (often).

(2)   Fat cell hypertrophy can be detected in obese rodents in the early post-
natal period.

(3)   Although studies of adipose tissue metabolism from obese and lean animals have produced contradictory results, adipose tissue of obese rodents is generally agreed to be geared towards fat storage and away from fat breakdown. This
is particularly true for young animals in the dynamic phase of obesity.

(4)   Fat transplantation experiments using a wide range of genetically obese
rodents indicate that adipose tissue hypertrophy is readily reversible. Host environment is more important than intrinsic factors in determining the size of fat
cells in grafts.

(5)   Since fat cell size increases so early in obese rodents, yet, from transplantation experiments, is likely to be a secondary defect, the primary defect in genetically obese rodents must therefore be present in the immediate post-natal period,
if not prenatally.

INTRODUCTION

The one abnormality which is common to all species of genetically obese rodents is
their excessive stores of adipose tissue. At 28 days, which is the age at which many
strains of obese mice are visually identifiable from their lean litter mates, the obese
mice (*ob/ob*) have nearly three times as much fat (Thurlby and Trayhurn, 1978).

*Present address: Lilly Research Centre Ltd, Earlwood Manor, Windlesham, Surrey, UK

In this paper we shall discuss some aspects of the morphology and metabolism of this adipose tissue in early and adult life and then devote the rest of the paper to the question of the importance of intrinsic factors within the adipose tissue to the development of the obese syndrome in these animals.

ADIPOSE TISSUE IN EARLY LIFE OF GENETICALLY OBESE RODENTS

Only in the last few years have tests been described which enable genetically obese rodents to be distinguished from their lean litter mates during the first few weeks of life; the introduction of such tests has marked an important stage in the study of adipose tissue of genetically obese rodents.

Animals of the obese genotype can be reliably detected by measuring oxygen consumption (Fried and Antopol, 1963; Kaplan and Leveille, 1973; van der Kroon *et al.*, 1977) or by measuring the cold-induced fall in body temperature (Trayhurn *et al.*, 1977). Obese animals show lower O_2 consumption and a greater fall in body temperature compared with pre-lean animals. Although body weights of *ob/ob* and *+/ob* or *+/+* mice do not begin to diverge significantly until the animals are at least 17 days old, a significant difference in carcass fat has been demonstrated in these mice at 10 days of age (Thurlby and Trayhurn, 1978).

Bergen *et al.* (1975) have suggested that this early increased deposition of fat is accompanied by diminished muscle growth, and Thurlby and Trayhurn (1978) have shown a small reduction in total carcass nitrogen and water content at 17 days of age. These findings would explain the similarity in body weight of pre-obese and pre-lean animals during the first four weeks of life.

Fat cell size in the adipose tissue of very young genetically obese rodents has been studied by several groups, some using a retrospective classification of obesity at a later age, and some using direct tests of pre-obesity (see table 7.1). Joosten and van der Kroon (1974) measured fat cell size in the epididymal fat pads of *ob/ob* mice and their *+/?* litter mates at 12–14 days of age and showed a difference in the frequency distribution and mean cell size of the two groups. Boulangé *et al.* (1977) studied biopsy samples from the inguinal pads of 5–7 day old Zucker rats and reported that *fa/fa* rats showed cellular hypertrophy compared with *+/?* litter mates. Positive identification of genotype of both mice and rats was performed at 6 weeks of age.

Kaplan *et al.* (1976) used the O_2 consumption test to identify homozygous obese mice (C57BL/6J-*ob/ob*) at 3 weeks of age and reported a difference in gonadal adipocyte size frequency distribution between *ob/ob* mice and their litter mates. *ob/ob* mice showed a bimodal distribution of cell size with high proportions of large and small cells, whereas *+/?* animals showed an order distribution of cell size with many small cells and few big cells. This difference in frequency distribution was more apparent in male mice than in female mice. These authors pointed out that the existence of a bimodal distribution makes the calculation of *average* cell size in these animals misleading and in fact the average cell size at 3 weeks was not very different between the two groups. Ashwell and Trayhurn (1978) have

Table 7.1 Adipose tissue cellularity in early life

Animal (fat site)	Age (days)	Test for obesity	Results (obese, cf. lean)	Reference
Obese (ob/ob) mice (epididymal pad)	12–17	observation at 6 weeks	mean cell size increased different frequency distribution	Joosten and van der Kroon (1974)
Zucker (fa/fa) rat (inguinal pad)	5–7	observation at 6 weeks	mean cell size increased	Boulangé et al. (1977)
Obese (ob/ob) mice (gonadal pads)	21	O_2 consumption	different frequency distribution mean cell size same	Kaplan et al. (1976)
Obese (ob/ob) mice (gonadal and subcutaneous)	12–13	cold stress	maximum cell size increased	Ashwell and Trayhurn (1979)

also recognised the difficulty of calculating average fat cell size in adipose tissue of very young genetically obese rodents, and have reported wide heterogeneity in cell size (see figure 7.1). Rather than perform complex frequency distributions of dispersed fixed cells, they examined histological sections of gonadal and subcutaneous adipose tissue from mice ('Aston strain *ob/ob*') identified as *ob/ob* or +/? on the basis of the cold stress test and showed that the *maximum* fat cell size of *ob/ob* mice was significantly greater than that of +/? mice even at 13 days of age (see figure 7.1 a, b).

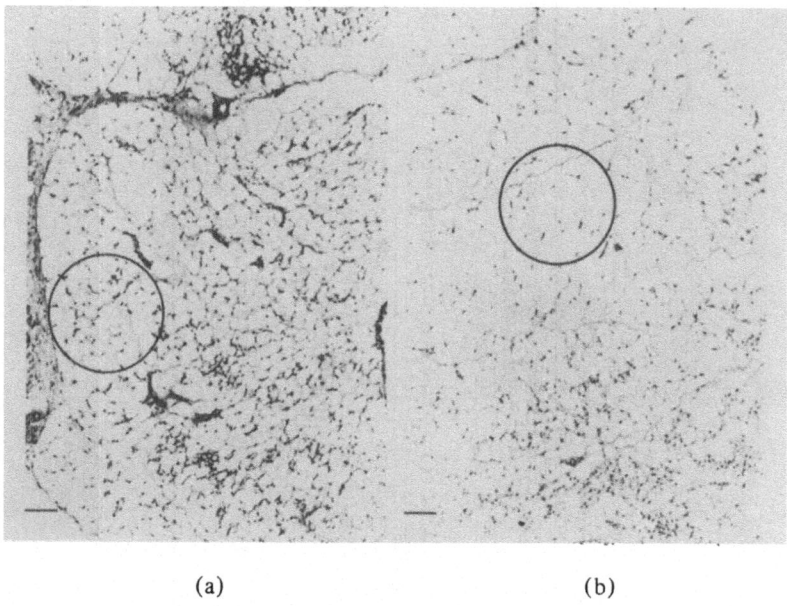

           (a)   (b)

Figure 7.1   (a) Subcutaneous adipose tissue from a 13 day old mouse ('Aston' strain) which was identified as a potential lean (*ob/+* or +/+) animal on the basis of the cold stress test. Body weight of mouse, 10.1 g; maximum fat cell weight (i.e. cells in circle), 0.068 µg. (b) Subcutaneous adipose tissue from a 13 day old mouse ('Aston' strain) which was identified as a potential obese (*ob/ob*) animal on the basis of the cold stress test. Body weight of mouse, 10.3 g; maximum fat cell weight (i.e. cells in circle), 0.135 µg. Scale marker, 100 µm.

There is, therefore, conclusive evidence that very young pre-obese animals from several strains of genetically obese rodents have larger fat cells than their pre-lean litter mates. This difference in cellularity is now so well established that it has been suggested as a possible identification test for pre-obesity (Kaplan *et al.,* 1976; Boulangé *et al.,* 1977). However, the practicability of this suggestion is doubtful, since not only do young mice exhibit large inter-litter differences in cell size, but also mice within a litter can also show quite a variation in cell size (Ashwell and Trayhurn, 1979).

ADIPOSE TISSUE IN ADULT LIFE OF GENETICALLY OBESE RODENTS

Morphology

The cells in adipose tissue can be divided into two broad classes—the mature fat cells and the stromal cells. The fact that some of the stromal cells might represent a pool of adipocyte precursors has not yet been fully investigated in the adipose tissue of genetically obese rodents, but this probability cannot be overlooked, as the precursors have been positively identified in rat and human adipose tissue (Van and Roncari, 1977; Van *et al.*, 1976; Bjorntorp, 1978) and might play important roles in the obese state (Ashwell, 1978).

Stromal cells
Many differences have been shown between genetically obese rodents and their lean litter mates in both categories of adipose tissue cells. An increased number of 'mast cells' has been reported on the basis of histological identification in the adipose tissue of *ob/ob* mice (Hellman *et al.*, 1963a), in yellow obese mice (A^y/-) (Hellman *et al.*, 1963b) and in the gonadal adipose tissue of New Zealand Obese (NZO) mice (Täljedal and Hellman, 1963). Rakow (1974) calculated the total number of stromal cells and fat cells in epididymal fat pads from lean and obese (*ob/ob*) mice (C57BL/6J) by measuring total DNA content and mean fat cell size. He has shown that the fat pads contain equal numbers of identifiable fat cells but that there is a threefold increase in the number of stromal cells in the fat pads of obese mice.

This increase in stromal cells in the adipose tissue of genetically obese rodents explains why the nitrogen content of the adipose tissue (expressed per gram of fat) is similar in lean and obese animals (Hellman *et al.*, 1962; Abraham, 1973).

Further characteristics of the adipose tissue of genetically obese rodents are the pathological changes (increased numbers of mononuclear cells, macrophages and mast cells) which occur with advancing age. These were first reported in *ob/ob* and yellow obese (A^y/−) mice by Hausberger (1966) and have subsequently been shown to occur in five different strains of obese and diabetic mice by Soret *et al.* (1974).

Fat cells
Much more attention has been devoted to the morphology of the mature fat cells in the adipose tissue of genetically obese rodents than to the non-fat cells. Cell size is increased compared with suitable controls in virtually all sites in all species of genetically determined obesity (Täljedal and Hellman, 1963; Johnson *et al.*, 1971; Johnson and Hirsch, 1972; Lemonnier and Alexiu, 1974). The one major exception is the subcutaneous fat depot in NZO mice, particularly females, where the cells remain small (Täljedal and Hellman, 1963; Johnson and Hirsch, 1972).

It is much more difficult to determine total cell number of distinct fat depots, and this has only been done sufficiently often to give definite conclusions for the gonadal fat pads. There is good evidence that cell number increases as well as cell size in all sites in the *ob/ob* mouse. On the other hand, the diabetic mouse (*db/db*) appears to exhibit a pure hypertrophic form of obesity in all depots. (See table 7.2 for details and references.)

Table 7.2   Adipose tissue cellularity in adult life: fat cell number

Animal	Age	Sex	Fat site	Results (obese, cf. lean)	Reference
Obese (*ob/ob*) mouse	adult	M and F	gonadal subcutaneous retroperitoneal }	increased	Hausberger (1957) Herberg *et al.* (1970) Johnson and Hirsch (1972)
Diabetic (*db/db*) mouse	6 months	M and F	all sites	same	Johnson and Hirsch (1972)
Yellow (*A^y/−*) obese mouse	6 months	M	gonadal subcutaneous retroperitoneal }	increased	Johnson and Hirsch (1972)
		F	all sites	slight increase	Johnson and Hirsch (1972)
Zucker (*fa/fa*) rat	12 months	M and F	gonadal subcutaneous	same	Lemonnier and Alexiu (1974)
	12 months	M and F	retroperitoneal }	increased	
NZO mouse	6 months	M and F	all	increased	Johnson and Hirsch (1972)
Desert sand rat	adult	M	gonadal subcutaneous	decreased increased	Robertson *et al.* (1974)
		F	gonadal subcutaneous }	same	Robertson *et al.* (1974)

Table 7.3 Adipose tissue metabolism: fat breakdown

Animal	Character	Results (obese, cf. lean)	Reference
Obese (*ob/ob*) mouse	hormone stimulated lipolysis/mg tissue hormone stimulated lipolysis/cell	decreased increased	Dehaye *et al.* (1977) Steinmetz *et al.* (1969) Herberg and Coleman (1977)
Diabetic (*db/db*) mouse	hormone stimulated lipolysis/mg tissue *in vivo* lipolysis	decreased decreased	Steinmetz *et al.* (1969) Kupiecki and Adams (1974) Allan and Yen (1976)
Yellow (*A^y/−*) mouse	hormone stimulated lipolysis/mg tissue	decreased	Herberg and Coleman (1977) Yen *et al.* (1970)
NZO mouse	hormone stimulated lipolysis/cell	increased	Lovell-Smith and Sneyd (1973)
Zucker (*fa/fa*) rat	hormone stimulated lipolysis/cell	increased	Bray (1977)

Metabolism

Many scientists have compared fat breakdown and fat synthesis in adipose tissue from genetically obese rodents and their lean counterparts. Unfortunately, the results obtained from these studies have been somewhat contradictory. There are many reasons for this. Firstly, the results can vary according to the age and sex of the animals, even if the animals are exactly the same strain (different genetic backgrounds can certainly account for contradictory results). Secondly, the antecedent diet of the animals can affect the results. Thirdly, the weight state of the animal is important—i.e. different results can be obtained from animals in a weight-gaining state compared with those in a steady weight state. Lastly, it is important to note whether the results are expressed on a cell basis or a weight basis, since the obese animals are almost certain to have bigger cells than the lean animals. Results can also vary according to whether the experiment has been done using isolated cells or pieces of tissue, even if the results of both experiments are expressed on the same basis.

Several authors have reported that *in vitro* hormone (usually isoprenaline) stimulated lipolysis is lower on a tissue weight basis but higher on an individual cell basis when adipose tissue from genetically obese rodents is compared with that from lean controls (see table 7.3). *In vivo* hormone stimulated lipolysis has also been investigated in diabetic (*db/db*) mice (Allan and Yen, 1976) by measuring plasma free fatty acid (FFA) levels. The peak response of the diabetic mice was similar to that of the lean controls but the after-peak response was higher in the diabetic mice. This result was interpreted by the authors to be consistent with *in vitro* findings of deficient lipolysis in diabetic mice, since they argue that the triglyceride/blood volume ratio of diabetic mice compared with that of normal mice would predict a lipolytic response of the diabetic mice 6.5 times higher than that of a normal mouse. It should be remembered, however, that any data on FFA levels mean less in the absence of turnover data. Dade and Hems (1972) have shown that the normal plasma FFA levels of *ob/ob* mice are associated with a substantial increase in FFA turnover.

Some of the studies which have compared fat synthesis in the adipose tissue of genetically obese rodents with their lean counterparts are outlined in table 7.4. Here age plays a large role and most authors have shown an increased conversion rate of precursors into glyceride-fatty acids or glyceride-glycerol in young obese animals in the dynamic obese state. Only yellow obese mice do not show this increased capacity for triglyceride synthesis (Hollifield *et al.*, 1960).

The study of individual enzyme activity in adipose tissue from genetically obese rodents and their lean counterparts has borne out the general conclusions reached above—i.e. a general tendency to make fat rather than break it down. Table 7.5 shows some of the enzyme activities which have been studied in three strains of genetically obese rodents. The results of Martin (1974), in which he studied adipose tissue enzymes in pair-fed animals, indicate that some of the enzyme changes might be a consequence of either the obese or the hyperphagic state rather than a cause of the obesity.

Table 7.4 Adipose tissue metabolism: fat synthesis

Animal	Experiment	Results (obese, cf. lean)	Reference
Obese (*ob/ob*) mouse	glucose, acetate → lipids	increased	Renold *et al.* (1960) Hollifield *et al.* (1960)
	glucose → fatty acids	increased (especially in young)	Jansen *et al.* (1967)
Yellow (*A^y/−*) mouse	acetate → fatty acids	increased (young) slight decrease	Zomzely and Mayer (1959) Hollifield *et al.* (1960)
NZO mouse	glucose → fatty acids	increased	Subrahmanyan (1960)
Zucker (*fa/fa*) rat	glucose → fatty acids ± insulin	same (but increased in young)	Bray (1977)

Table 7.5 Adipose tissue metabolism: enzymes

Animal	Enzyme (obese, cf. lean)	Reference
Obese (*ob/ob*) mouse	glycerokinase increased	Koschinsky *et al.* (1970)
	α-glycerol-P-Δ increased	Fried and Antopol (1960)
	glucose-P-Δ increased	Martin *et al.* (1973)
	6-P-gluconate-Δ increased	
	malic enzyme increased	Kaplan and Fried (1973)
	stearic acid desaturase increased	Enser (1975)
	adenyl cyclase decreased	Laudat and Pairault (1975)
Diabetic (*db/db*) mouse	glycerokinase increased	Thenen and Mayer (1975)
Zucker (*fa/fa*) rat (young)	glycerokinase increased	Martin and Lamprey (1975)
	glucose-6-P-Δ increased	Bray (1977)
	malic enzyme increased	Taketomi *et al.* (1975)
	citrate cleavage enzyme increased	Taketomi *et al.* (1975)
	fatty acid desaturase increased	Wahle (1974)
	but in pair-fed animals:	
	malic enzyme decreased	
	hexokinase decreased	
	glucose 6-P-Δ increased	Martin (1974)

Δ = dehydrogenase.

THE IMPORTANCE OF INTRINSIC FACTORS IN ADIPOSE TISSUE IN THE DEVELOPMENT OF OBESITY

Several of the differences reported so far in the adipose tissue of obese and lean rodents would lead one to think that there might be intrinsic factors in the adipose tissue of obese animals which would predispose them to obesity—e.g. the very early appearance of the increased fat cell size, the increased fat storage potential and decreased fat breakdown potential observed in *in vitro* experiments. Other experimental approaches would also suggest this: restriction of food and the subsequent decrease in fat cell size of Zucker rats does not restore lipogenesis or lipolysis to normal (York and Bray, 1973) and streptozotocin-treated *ob/ob* mice have higher body fat contents than lean mice despite similar insulin levels (Boozer and Mayer, 1976).

One of the best direct tests to see whether an abnormality is caused by an intrinsic defect in a certain tissue is to transplant that tissue into a normal environment and see whether the abnormality persists. Fat tissue is easy to transplant, but for meaningful results two conditions must be satisfied. Firstly, the transplant must vascularise properly. The fat cells of a graft that has not vascularised properly decrease in size and extensive necrosis occurs. Secondly, the transplant must not suffer immunological rejection. One way of avoiding rejection is to use a host possessing all the transplantation antigens expressed by the grafted tissue. An F_1 hybrid is thus a suitable host for tissues from either of its inbred parents, because it has both sets of parental transplantation antigens. Recently transplantation has been much simplified by the availability of animals bred by repeated backcrossing into a standard strain, and differing only at a genetic locus determining obesity— e.g. C57BL/6J-*ob/ob*, C57BL/KsJ-*db/db*, C57BL/6J-*A^y/-* and others. However, the earliest attempts at fat transplantation were made before such inbred strains were available. Hausberger (1959) reported that although he had little success in transplanting between very young litter mates by conventional grafting methods, he could transplant subcutaneous fat from obese (*ob/ob*) mice to lean mice and vice versa by a process of parabiosis and separation. The survival rate of these mice was poor, but by examining the surviving fat grafts he showed that fat transplants assumed the typical features of the host tissue and concluded that it was unlikely that the abnormal growth rate of fatty tissue in hereditary obesity was caused by properties inherent in the tissue itself.

Liebelt (1963) reached the opposite conclusion; he transplanted inguinal fat from lean DBA/2 mice and obese NH mice into the ears of host mice which were NH × DBA/2 F_1 hybrids. Examination of the grafts by lipid extraction after 90 days revealed that the ear grafts derived from the NH (obese) parent deposited more lipid than did the grafts of adipose tissue from the DBA/2 (lean) parent. He therefore believed that the source of adipose tissue had more influence than its environment in deciding the character of the adipose tissue in the graft.

In the last few years we have been using a much improved method of fat transplantation in genetically obese rodents to resolve the contradiction between the results of Hausberger and Liebelt and to investigate the role of intrinsic factors in adipose tissue in a wide variety of genetically obese rodents. The technique of fat transplantation is illustrated in figure 7.2. Fat is removed from the epididymal or

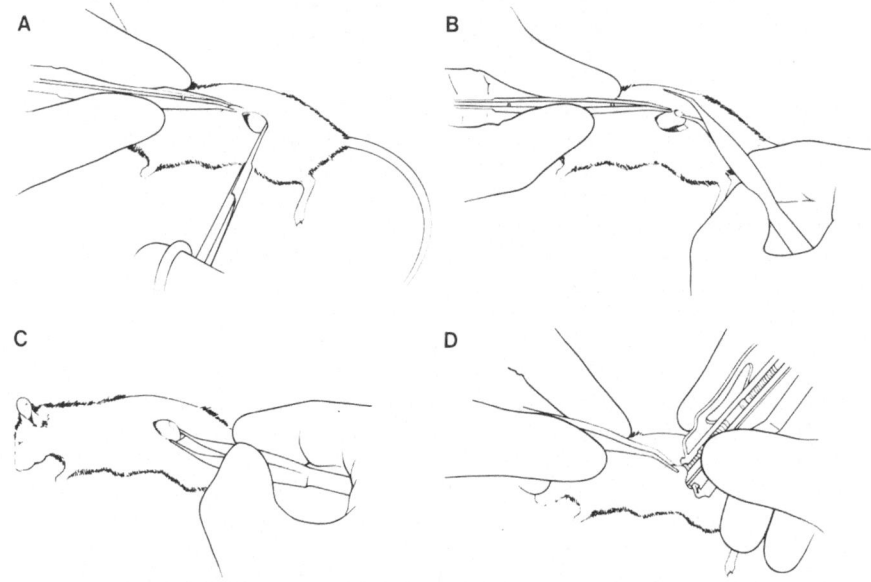

Figure 7.2   The fat transplantation technique. (A) The anaesthetised host animal is prepared for transplantation. (B) The fat from the donor animal is pushed under the kidney capsule. (C) The graft is now placed under the kidney capsule of the host animal. (D) The wound is closed and the animal left to recover from the anaesthetic.

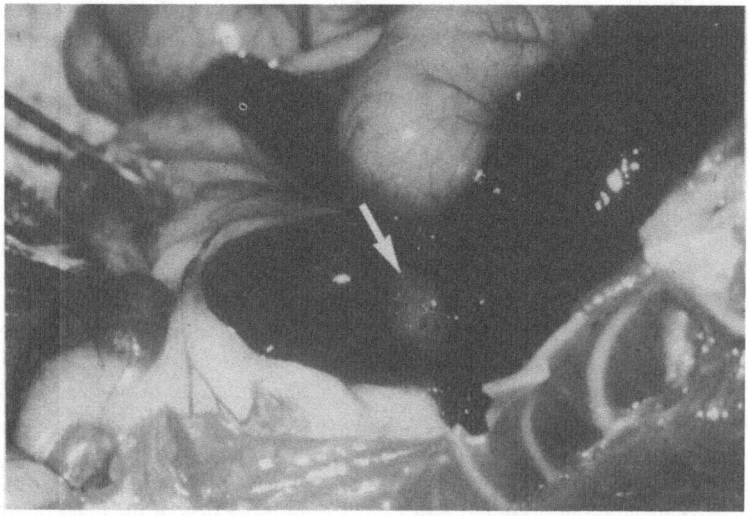

Figure 7.3   The graft *in situ*, i.e. under the kidney capsule of the host mouse. One month after transplantation.

inguinal fat pad of the donor animal, placed into a balanced salt solution and cut into small pieces (5 mg). The anaesthetised host animals are then prepared for transplantation by making a dorsolateral incision in the shaved skin to reveal the kidney (figure 7.2A). A small tear is made in the kidney capsule at the posterior pole and the fat transplant inserted and pushed under the capsule towards the anterior pole of the kidney (figure 7.2B). In 'double kidney' experiments fat transplants (usually from different donors) are inserted under both kidneys. The kidney is replaced (figure 7.2C) and the wound is closed (figure 7.2D). The grafts are left in place for a month, then the animals are killed and the kidney(s) plus graft exposed (see figure 7.3).

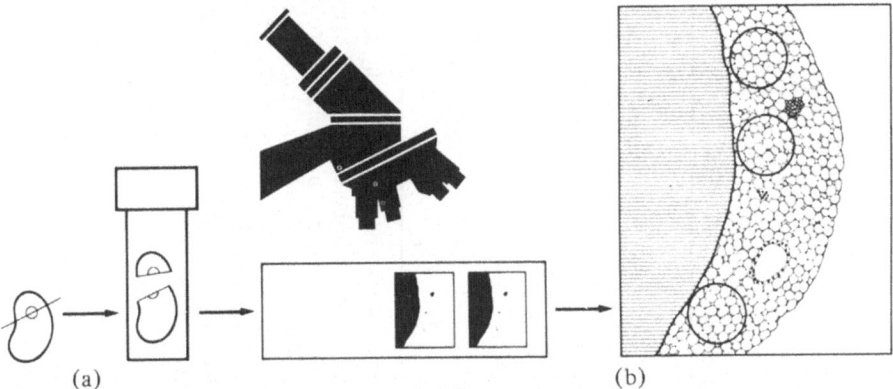

(a)   (b)

Figure 7.4   Measurement of fat cell diameter. (a) The kidney plus graft is sectioned through the plane of the fat graft and put into Bouin's fixative. Thin sections are prepared from the fixed tissue and stained with haematoxylin and eosin. (b) The sections are viewed under a microscope and the cell diameter in the grafts estimated by counting the number of cells in a known area, taking care to choose areas free of vacuoles or pockets of small cells.

The kidney plus graft and also a sample of the host adipose tissue are then put into histological fixative (see figure 7.4a). The kidney is sectioned through the plane of the fat graft and stained with haematoxylin and eosin (figure 7.4b). Fat cell sizing is performed directly on fixed sections using a slight modification of the technique described by Ashwell *et al.* (1976) (see figure 7.4b). The grafts are examined under the microscope and areas chosen which are suitable for cell sizing. Areas containing pockets of small cells are avoided, as are areas which show obvious signs of fat necrosis. In general, the part of the graft nearest the kidney (where vascularisation is best) is used for cell sizing. The first experiments using this fat transplantation technique were performed on C57BL/6J-*ob/ob* mice (Ashwell *et al.*, 1977a, b). The ring skin-grafting procedure of Billingham and Medawar (1951) confirmed that all mice, lean and obese, had similar transplantation antigens. Table 7.6 summarises the results of these experiments. The fat cells of obese donor fat decreased in size in a 'lean environment' to the size typical of 'lean fat', while the cells of lean donor fat transplanted into an 'obese environment' increased to the size typical of 'obese fat'. When both lean and obese fat tissues were transplanted

Table 7.6 Fat transplantation between the obese mouse (C57BL/6J-*ob/ob*) and its lean control

Donor animal	n	Donor fat cell weight after 1 month (μg ± S.D.)	Recipient animal	n	Graft fat cell weight (μg ± S.D.)	Fat cell weight donor vs. graft
Obese	3	0.504 ± 0.207	lean	12	0.100 ± 0.071	P < 0.001
Obese	3	0.504 ± 0.207	obese	7	0.480 ± 0.151	N.S.
Lean	3	0.095 ± 0.039	lean	11	0.102 ± 0.048	N.S.
Lean	3	0.095 ± 0.039	obese	6	0.577 ± 0.437	P < 0.001

Epididymal fat from lean (+/?) or obese (*ob/ob*) mice (about 50 days old) was transplanted to a site underneath the kidney capsule of recipient lean and obese mice (aged between 49 and 64 days). The grafts were left in place for a month and then examined histologically to measure fat cell diameters from which fat cell masses could be estimated (see text). The final column of the table expresses the probability that the fat cell sizes in the grafts belong within the range of masses of the donor fat cells. Where obese fat grafts were transplanted into lean mice and vice versa, it is clear that there is a significant alteration of cell size (P < 0.001). However, transplantation as such (fat → fat or lean → lean) has no significant effect (N.S.) on cell size.

onto opposite kidneys of the same mouse, the tissues acquired cell sizes characteristic of their common host and there was no significant difference in cell size in the
two grafts. These experiments showed that there were no intrinsic defects in the
adipose tissue of *ob/ob* mice which prevented their mobilising fat to decrease
their cell size when placed in a lean environment.

It was still possible, however, that the adipose tissue of obese mice might have
innate capacity for increased fat storage rather than a decreased capacity for fat
mobilisation, so a further series of experiments were designed using the *ob/ob*
mouse (Ashwell and Meade, 1978). In these experiments grafts from lean and
obese donors were put on the kidneys of lean host mice and left for the usual
month. At this point some of the host mice were treated with gold thioglucose
(GTG), which is known to induce obesity, and other host mice were injected with
saline as controls. The mice were then killed at monthly intervals up to three
months and the adipose tissue from the host mice and the grafts were examined in
the normal way. Table 7.7 summarises the results of this experiment. In the control groups, A and B, the obese donor fat cells decreased in size to those typical of
their lean host. The effect of 1, 2 or 3 months' GTG treatment (groups C, D and E,
respectively) was to substantially increase graft and host fat cell size. In no group

Table 7.7 Mean and standard deviations of fat cell masses (μg) of host and
graft fats

Group	n	Host	'Lean graft'	'Obese graft'
A	2	0.108 ± 0.008	0.085 ± 0.004	0.104 ± 0.015
B	4	0.145 ± 0.074	0.121 ± 0.034	0.106 ± 0.028
C	7	0.349 ± 0.057	0.305 ± 0.086	0.328 ± 0.060
D	7	0.284 ± 0.041	0.226 ± 0.024	0.225 ± 0.042
E	7	0.370 ± 0.127	0.338 ± 0.136	0.314 ± 0.130

All mice were male and C57BL/6J strain. Hosts were pure-breeding (+/+)
lean mice aged 60–80 days. Donor obese fat was taken from the epididymal
fat pad of an *ob/ob* mutant aged about 60 days. Donor lean fat was taken
from the same site of a similarly aged lean litter mate (+/?). Transplantation
of fat from lean donor and obese donor under opposite kidneys of the same
host mouse was performed as described in the text.

Group A were killed 1 month after transplantation. Group B were injected
with saline 1 month after transplantation and killed 3 months later. Group C
were injected with GTG 1 month after transplantation and killed 1 month
later. Group D were injected with GTG 1 month after transplantation and
killed 2 months later. Group E were injected with GTG 1 month after transplantation and killed 3 months later.

On sacrifice, the kidneys plus grafts were removed and the fat cells in
the grafts and also the fat cells in the host epididymal fat pad were sized by
the histological method described in the text. Obese donor fat cell mass,
0.520 μg. Lean donor fat cell mass, 0.086 μg.

Paired t-tests between lean and obese graft fat cell masses showed no
significant difference in any group.

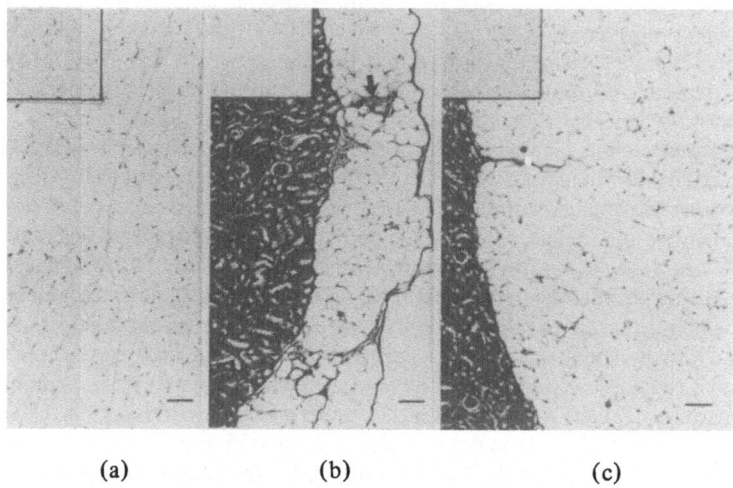

(a)   (b)   (c)

Figure 7.5   Host fat cell size after 3 weeks' GTG treatment. Mean fat cell mass, 0.148 μg. Inset shows fat from lean mouse before GTG treatment (mean fat cell mass, 0.082 μg). (b) 'Obese (*ob/ob*) graft' fat after 3 weeks' GTG treatment of the host mouse. Mean fat cell mass of graft, 0.199 μg. Inset shows the original obese donor fat (mean fat cell mass, 0.520 μg). (c) 'Lean (+/?) graft' fat cell size after 3 weeks' GTG treatment of the host mouse. Mean fat cell mass of graft, 0.180 μg. Inset shows the original lean donor fat (mean fat cell mass, 0.086 μg). Scale marker, 100 μm.

was there a significant difference between the cell size of 'lean' and 'obese' grafts. Thus, both grafts could respond similarly to the new host environment by increasing the size of their cells.

In another experiment the effect of GTG treatment for shorter time intervals was studied. Even after 2 and 3 weeks of GTG treatment, host fat cell size was increased significantly and the fat cell size of 'lean grafts' did not differ significantly from that of the host. Rather than increasing in size faster than the cells of the lean grafts, as anticipated if obesity was associated with an innate capacity for increased fat storage, fat cell size in 'obese grafts' increased significantly less than host or 'lean graft' fat cell size; possibly vascularisation does not take place so easily with 'obese grafts' (see figure 7.5).

The fat transplantation technique has been used in other genetically obese rodents and the results are summarised in table 7.8 (from Meade *et al.*, 1977b). Lean litter mates were the controls for the mouse mutants; for the hamsters, the Bio 4.22 strain was the lean control. In the diabetic, yellow and adipose mouse, 'lean' fat transplanted into the corresponding obese mice or 'obese' fat transplanted into corresponding lean mice underwent a significant change in fat cell mass. Comparison of grafts of the same fat tissue into 'obese' or 'lean' recipient mice showed significant differences ($P < 0.001$) in all cases. Graft cell size altered to that characteristic of the recipient. Figure 7.6 shows some photomicrographs of the grafts in diabetic mice with corresponding host and donor fat.

Table 7.8 Fat transplantation between diabetic mice (C57BL/KsJ-*db/db*), yellow obese mice (C57BL/6J-A^y/*a*) adipose mice (C57BL/6J-*db^{ad}*/*db^{ad}*) and Bio 4.24 hamsters and their corresponding lean control animals (mean values and their standard deviations)

| Obese rodent | No. of mice used for cell sizing | | Body mass of recipient when killed (g) | | Fat cell masses (μg) | | | | | |
| | | | | | Fat from 'lean' donor | | | Fat from 'obese' donor | | |
	lean	obese	lean	obese	Before transplant	After 1 month in lean recipient	After 1 month in obese recipient	Before transplant	After 1 month in lean recipient	After 1 month in obese recipient
Diabetic mouse (C57BL/KsJ-*db/db*)	8	9	26.3 (± 1.6)	38.4 (± 2.1)	0.128	0.101 (± 0.045)	0.765 (± 0.180)	0.688	0.083 (± 0.020)	0.677 (± 0.186)
Yellow obese mouse (C57BL/6J-A^y/*a*)	13	11	27.3 (± 3.2)	40.7 (± 8.7)	0.093	0.118 (± 0.034)	0.449 (± 0.171)	0.575	0.112 (± 0.076)	0.368 (± 0.138)
Adipose mouse (C57BL/6J-*db^{ad}*/*db^{ad}*)	5	6	25.4 (± 2.9)	45.2 (± 3.9)	0.092	0.073 (± 0.022)	0.602 (± 0.134)	0.682	0.069 (± 0.024)	0.519 (± 0.104)
Bio 4.24 hamster	7	7	91.0 (± 8.9)	144.0 (± 14.2)	0.101	0.117 (± 0.025)	0.173 (± 0.043)	0.406	0.098 (± 0.024)	0.145 (± 0.051)

Lean litter mates were the controls for the obese mouse mutants; for the hamsters the Bio 4.22 strain was the lean control. The 'double kidney' transplant technique was used in each case (see legend to table 7.7).

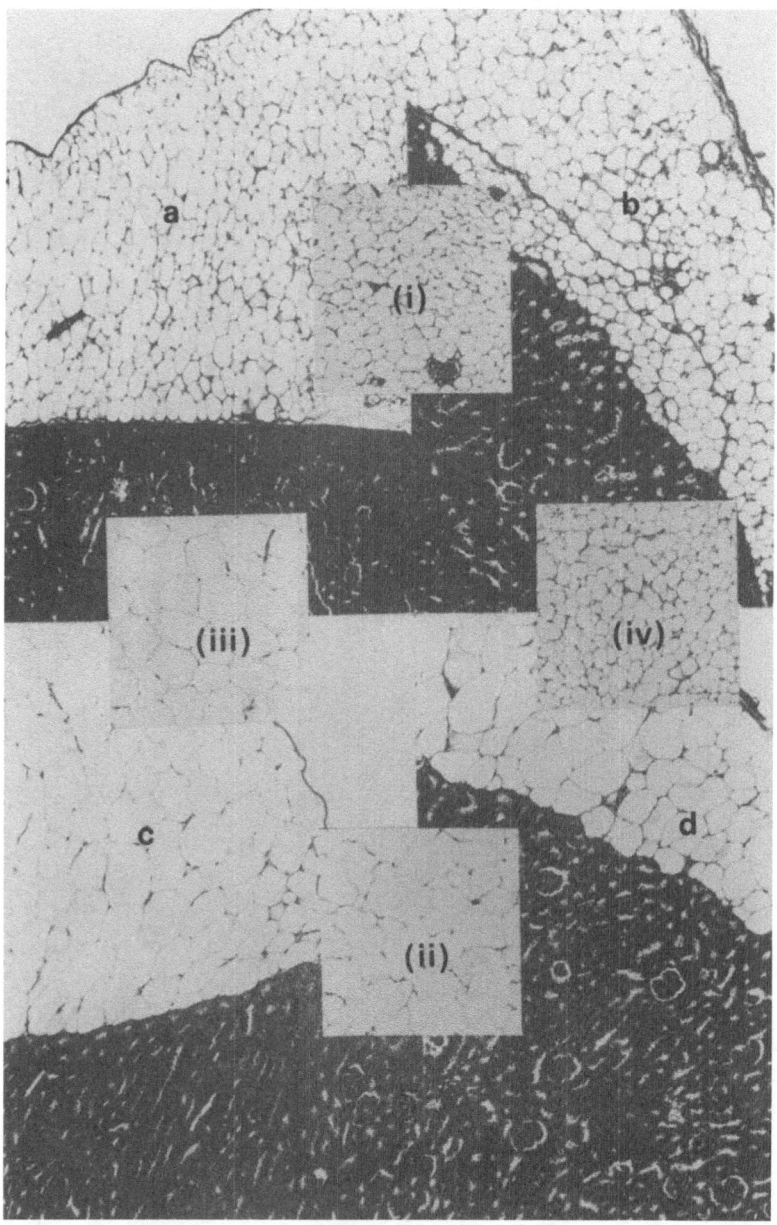

Figure 7.6 Fat transplantation in diabetic (C57BL/KsJ-*db*/*db*) mice. (a) Graft from a diabetic mouse in a lean host. Mean fat cell mass, 0.111 μg. (b) Graft from a lean mouse in a lean host. Mean fat cell mass, 0.069 μg. (c) Graft from diabetic mouse in overweight *db*/*db* host. Mean fat cell mass, 0.829 μg. (d) Graft from a lean mouse in overweight *db*/*db* host. Mean fat cell mass, 0.623 μg.
Insets show the adipose tissue of: (i) lean host (mean fat cell mass, 0.091 μg); (ii) overweight *db*/*db* host (mean fat cell mass, 0.918 μg); (iii) overweight *db*/*db* donor (mean fat cell mass, 0.833 μg); (iv) lean donor (mean fat cell mass, 0.115 μg). Scale marker, 100 μm.

'Lean' hamster fat cells transplanted into obese recipients increased in mass, but rarely reached a mass comparable to that of the recipient fat, perhaps because the hamsters, unlike mice, frequently lost weight after the operation. However, there was no doubt that the 'obese' fat cells shrank in a lean host.

The sixth genetically obese rodent which has been investigated (Meade *et al.*, 1977a) by the technique of transplantation under the kidney capsule is the NH mouse—i.e. the strain used by Liebelt for his ear graft technique. The host mice in these experiments were, as in Liebelt's study, NH × DBA/2 F_1 hybrids; they received grafts of NH (obese) fat and DBA/2 (lean) fat. NH × DBA/2 F_1 hybrids slowly become obese. By transplanting into F_1 recipients of different ages, NH and DBA/2 fat can be placed in environments in which the host fat cells vary in size from animal to animal over a wide range. As figure 7.7 shows, in a common environment NH or DBA/2 cells both adopted the size characteristic of the NH × DBA/2 F_1 host. Thus, the adipose tissue of NH mice is similar to the adipose tissue of the other genetically obese rodents we have investigated, in that environment is of paramount importance in determining cell size.

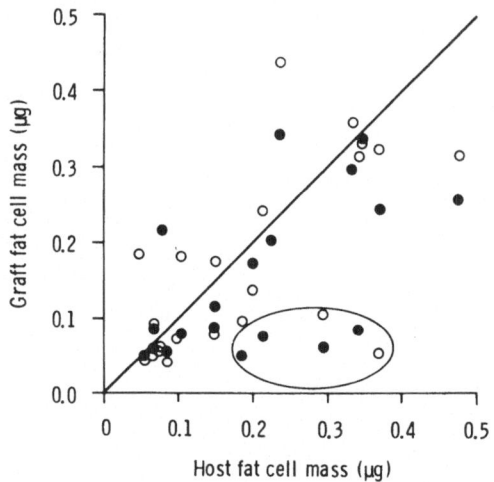

Figure 7.7 Transplantation of NH (obese) and DBA/2 (lean) fat into NH × DBA/2 F_1 hybrids. Plot of graft fat cell mass against host fat cell mass: ●, fat from DBA/2 (lean) donor; ○, fat from NH (obese) donor. Solid line is line of identity.

This is the opposite conclusion from that drawn by Liebelt, who, transplanting fat into older, and therefore obese, NH × DBA/2 F_1 hybrids, found DBA/2 fat tissue failed to accumulate as much lipid as NH tissue. Several differences in technique could explain the discrepancy in results. Liebelt's own data contain a contradiction. He found NH mice, or F_1 mice with an NH parent, showed a significant decrease in body weight or total body lipid content following adrenalectomy, whereas three lean strains showed no such dramatic fall. The NH mouse has a high frequency of adrenal-cortical adenomas (Frantz *et al.*, 1947) and the simplest explanation of Liebelt's adrenalectomy results is that it is differences in adrenal

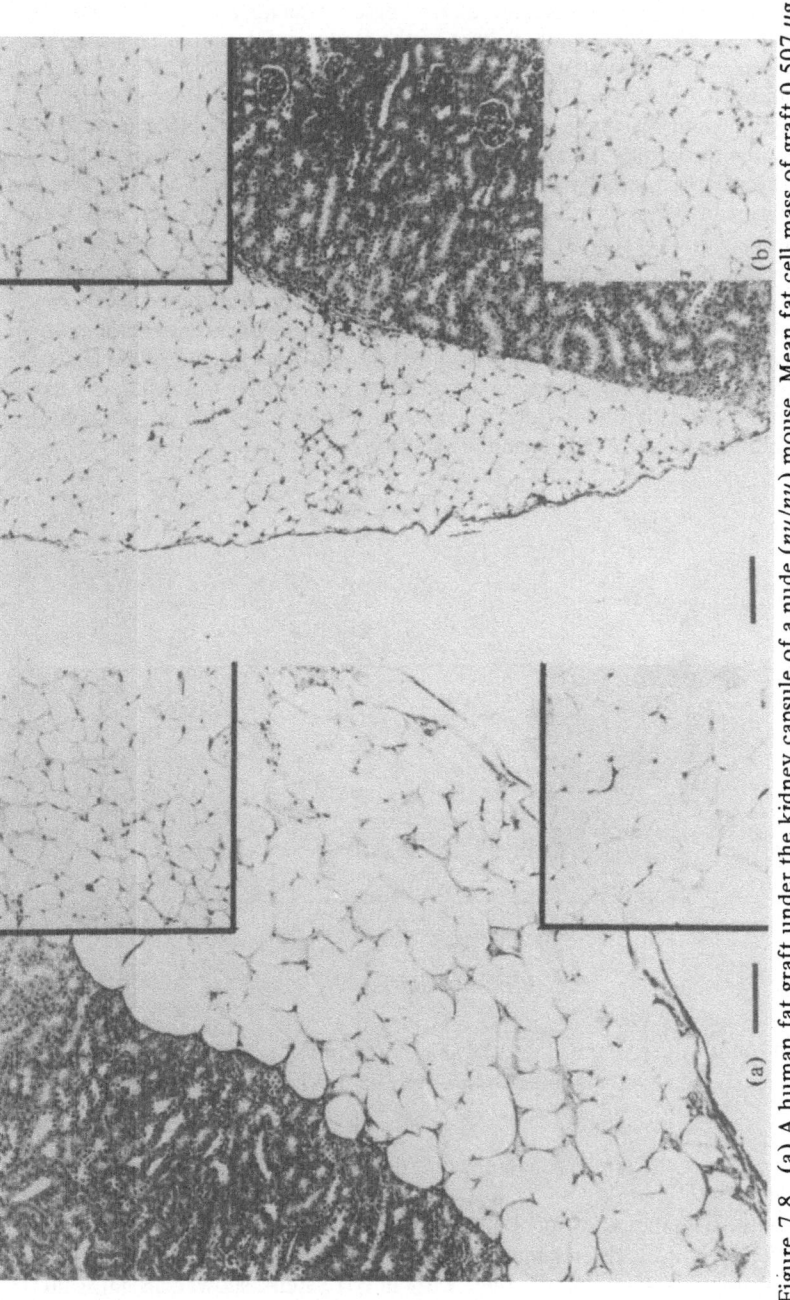

Figure 7.8 (a) A human fat graft under the kidney capsule of a nude (*nu/nu*) mouse. Mean fat cell mass of graft 0.507 µg. Top inset: Host adipose tissue (mean fat cell mass, 0.059 µg). Lower inset: Donor (human) adipose tissue (mean fat cell mass, 0.584 µg). (b) A nude (*nu/nu*) mouse fat graft under the kidney capsule of another nude (*nu/nu*) mouse. Mean fat cell mass of graft, 0.069 µg. Top inset: Host adipose tissue (mean fat cell mass, 0.059 µg). Lower inset: Donor (mouse) adipose tissue (mean fat cell mass, 0.063 µg). Scale marker, 100 µm.

function (i.e. in the hormonal environment of the adipose tissue) which is responsible for obesity. Liebelt himself explains his adrenalectomy results by suggesting
that the lean strains studied failed to lose much weight after adrenalectomy because
of regeneration of adrenal tissue, while the NH strain has a genetically determined
inability to rapidly regenerate such tissue—i.e he accepts his fat transplantation
data but suggests his adrenalectomy data describe an artefact. Liebelt himself, however, points out that his fat grafting results could also be explained by a difference
in the potentiality of NH and DBA/2 tissue to become vascularised. In our studies
we have observed that a few of our DBA/2 grafts, like Liebelt's, failed to assume
the characteristics of the host NH × DBA/2 adipose tissue (ringed points on figure
7.7). These grafts were all poorly vascularised.

Although our studies have provided no evidence for intrinsic differences in the
fat tissue of obese strains of rodents, it must not be concluded that all differences
in fat cell size can be attributed to extrinsic influences. For example, experiments
in our own laboratory have shown that human fat cells remain larger than murine
cells when transplanted into mice bearing the *nu/nu* mutation and therefore unable to reject the human tissue (figure 7.8). This has also been observed by Bach-
Mortensen *et al.* (1976).

To summarise the results of the fat transplantation studies in genetically obese
rodents, it would appear that there is no intrinsic defect in the adipose tissue of
any of the animals studied with respect to the mobilisation of fat from fat cells
and, in the *ob/ob* mouse at least, there is no intrinsic defect in the fat storage
capacity either.

ACKNOWLEDGEMENTS

We wish to thank Mr C. Sowter of the Department of Histopathology for histological expertise and Mr J. Clark of the Division of Electron Microscopy for photo-
micrography. We also wish to thank Dr P. Trayhurn of Cambridge and Dr M. Enser
of Bristol for helpful comments on parts of the manuscript.

REFERENCES

Abraham, R. R. (1973). Some cellular characteristics of the epididymal adipose tissue in lean
and obese-hyperglycaemic mice. *Diabetologia,* 9, 303–306
Allan, J. A. and Yen, T. T. (1976). Lipolytic response of 'diabetic' mice (*db/db*) to isopro-
teronol and propranolol *in vivo. Experientia,* 32, 836–837
Ashwell, M. (1978). The 'fat cell pool' concept. *Int. J. Obesity,* 2, 69–72
Ashwell, M. and Meade, C. J. (1978). Obesity: Do fat cells from genetically obese mice
(C57BL/6J-*ob/ob*) have an innate capacity for increased fat storage? *Diabetologia.* 15,
465–470
Ashwell, M., Meade, C. J., Medawar, P. B. and Sowter, C. (1977a). Adipose tissue: contribu-
tions of nature and nurture to the obesity of an obese mutant mouse (*ob/ob*). *Proc. R.
Soc. London,* B195, 343–353

Ashwell, M., Meade, C. J., Medawar, P. B. and Sowter, C. (1977b). Adipose tissue: contributions of nature and nurture to the obesity of an obese mutant mouse (*ob/ob*). *Proc. Nutr. Soc.*, **36**, 16A

Ashwell, M. A., Priest, P., Bondoux, M., Sowter C. and McPherson, C. K. (1976). Human fat cell sizing–a quick simple method. *J. Lipid Res.*, **17**, 190–192

Ashwell, M. and Trayhurn, P. (1979). In preparation

Bach-Mortensen, N., Romert, P. and Ballegaard, S. (1976). Transplantation of human adipose tissue to nude mice. *Acta Pathol. Microbiol. Scand.*, Section C, **84**, 283–289

Bergen, W. G., Kaplan, M. L., Merkel, R. A. and Leveille, G. A. (1975). Growth of adipose and lean tissue mass in hind limbs of genetically obese mice during preobese and obese phases of development. *Am. J. Clin. Nutr.*, **28**, 157–161

Billingham, R. E. and Medawar, P. B. M. (1951). The technique of free skin grafting in mammals. *J. Exp. Biol.*, **28**, 385–402

Bjorntorp, P. (1978). The fat cell: A clinician's view. In *Recent Advances in Obesity Research*, Vol. II (ed. G. A. Bray), Newman, London, pp. 153–168

Boozer, C. N. and Mayer, J. (1976). Effects of long-term restricted insulin production in obese-hyperglycemic (genotype *ob/ob*) mice. *Diabetologia*, **12**, 181–187

Boulangé, A., Planch, E. and de Gasquet, P. (1977). Excess fat storage and normal energy intake in the newborn Zucker rat (*fa/fa*). Abstract presented at the 2nd International Congress of Obesity, Washington, DC

Bray, G. A. (1977). The Zucker-fatty rat: a review. *Fed. Proc.*, **36**, 148–153

Dade, E. and Hems, D. A. (1972). The turnover of glycerol and free fatty acid in blood of genetically obese mice. *Biochem. J.*, **127**, 41P

Dehaye, J. P., Winand, J. and Christophe, J. (1977). Lipolysis and cyclic AMP levels in epididymal adipose tissue of obese-hyperglycaemic mice. *Diabetologia*, **13**, 553–561

Enser, M. (1975). Desaturation of stearic acid by liver and adipose tissue from obese-hyperglycaemic mice (*ob/ob*). *Biochem. J.*, **148**, 551–555

Frantz, M. K., Kirschbaum, A. and Casas, C. (1947). Endocrine interrelationship and spontaneous tumours of the adrenal cortex in NH mice. *Proc. Soc. Exp. Biol. Med.*, **66**, 645–646

Fried, G. H. and Antopol, W. (1960). Alpha-glycerophosphate oxidation in the tissues of obese hyperglycaemic mice and nonobese controls. *Fed. Proc.*, **19**, 327 (abstract)

Fried, G. H. and Antopol, W. (1963). Oxygen consumption in litters of obese hyperglycaemic mice. *Fed. Proc.*, **22**, 668 (abstract)

Hausberger, F. X. (1957). Composition of adipose tissue in several forms of obesity. *Anat. Rec.*, **127**, 305 (abstract)

Hausberger, F. X. (1959). Behaviour of transplanted adipose tissue of hereditarily obese mice. *Anat. Rec.*, **135**, 109–112

Hausberger, F. X. (1966). Pathological changes in adipose tissue of obese mice. *Anat. Rec.*, **154**, 651–660

Hellman, B., Larsson, S. and Westman, S. (1962). Acetate metabolism in isolated epididymal adipose tissue from obese-hyperglycemic mice of different ages. *Acta Physiol. Scand.*, **56**, 189–198

Hellman, B., Larsson, S. and Westman, S. (1963a). Mast cell content and fatty acid metabolism in the epididymal fat pad of obese mice. *Acta Physiol. Scand.*, **58**, 255–262

Hellman, B., Thelander, L. and Taljedal, I.-B. (1963b). Postnatal growth of the epididymal adipose tissue in yellow obese mice. *Acta Anat.*, **55**, 286–294

Herberg, L. and Coleman, D. L. (1977). Laboratory animals exhibiting obesity and diabetes syndromes. *Metabolism*, **26**, 59–99

Herberg, L., Gries, F. A. and Hesse-Wortmann, Ch. (1970). Effect of weight and cell size on hormone-induced lipolysis in New Zealand obese mice and American obese hyperglycemic mice. *Diabetologia*, **6**, 300–305

Hollifield, G., Parson, W. and Ayers, C. R. (1960). *In vitro* synthesis of lipids from ¹⁴C-acetate by adipose tissue from four types of obese mice. *Am. J. Physiol.*, **198**, 37–38

Jansen, G. R., Zanetti, M. E. and Hutchison, C. P. (1967). Studies on lipogenesis *in vivo*. Fatty acid and cholesterol synthesis in hyperglycemic-obese mice. *Biochem. J.*, **102**, 870–877

Johnson, P. R. and Hirsch, J. (1972). Cellularity of adipose depots in six strains of genetically obese mice. *J. Lipid Res.*, **13**, 2–11

Johnson, P. R., Zucker, L. M., Cruce, J. A. F. and Hirsch, J. (1971). Cellularity of adipose depots in the genetically obese Zucker rat. *J. Lipid Res.*, **12**, 706–714

Joosten, H. F. P. and van der Kroon, P. H. W. (1974). Enlargement of epididymal adipocytes in relation to hyperinsulinemia in obese hyperglycemic mice (*ob/ob*). *Metabolism*, **23**, 59–66

Kaplan, M. L. and Fried, G. H. (1973). Adaptive enzyme response in adipose tissue of obese hyperglycemic mice. *Arch. Biochem. Biophys.*, **158**, 711–719

Kaplan, M. L. and Leveille, G. A. (1973). Obesity: Prediction of preobesity among progeny from crosses of *ob/+* mice. *Proc. Soc. Exp. Biol. Med.*, **143**, 925–928

Kaplan, M. L., Trout, R. and Leveille, G. A. (1976). Adipocyte size distribution in *ob/ob* mice during preobese and obese phases of development. *Proc. Soc. Exp. Biol. Med.*, **153**, 476–482

Koschinsky, Th., Gries, F. A. and Herberg, L. (1970). Glycerol kinase activity in isolated cells of BHob mice. *Horm. Metab. Res.*, **2**, 185–186

Kupiecki, F. P. and Adams, L. D. (1974). The lipolytic system in adipose tissue of Toronto-KK and C57BL/KsJ diabetic mice. *Diabetologia*, **10**, 633–637

Laudat, M.-H. and Pairault, J. (1975). An impaired response of adenylate cyclase to stimulation by epinephrine in adipocyte plasma membranes from genetically obese mice (*ob/ob*). *Eur. J. Biochem.*, **56**, 583–589

Lemonnier, D. and Alexiu, A. (1974). Nutritional, genetic and hormonal aspects of adipose tissue cellularity. In *The Regulation of Adipose Tissue Mass* (ed. J. Vague and J. Boyer), Excerpta Medica, Amsterdam, pp. 158–173

Liebelt, R. A. (1963). Response of adipose tissue in experimental obesity as influenced by genetic, hormonal and neurogenic factors. *Ann. N. Y. Acad. Sci.*, **110**, 723–748

Lovell-Smith, C. J. and Sneyd, J. G. T. (1973). Lipolysis and adenosine 3′,5′-cyclic monophosphate in adipose tissue of the New Zealand obese mouse. *J. Endocrinol.*, **56**, 1–11

Martin, R. J. (1974). *In vivo* lipogenesis and enzyme levels in adipose and liver tissues from pair-fed genetically obese and lean rats. *Life Sci.*, **14**, 1447–1453

Martin, R. J. and Lamprey, P. M. (1975). Early development of adipose cell lipogenesis and glycerol utilisation in Zucker obese rats. *Proc. Soc. Exp. Biol. Med.*, **149**, 35–39

Martin, R. J., Welton, R. F. and Baumgardt, B. R. (1973). Adipose and liver tissue enzyme profiles in obese hyperglycemic mice. *Proc. Soc. Exp. Biol. Med.*, **142**, 241–245

Meade, C. J., Ashwell, M. and Medawar, P. B. (1977a). Adipose tissue: contributions of nature and nurture to obesity in obese rodents. Abstract presented at 2nd International Congress of Obesity, Washington, DC

Meade, C. J., Ashwell, M., Medawar, P. B. and Sowter, C. (1977b). Can genetically transmitted obesity be ascribed to an adipose tissue defect? *Proc. Nutr. Soc.*, **36**, 114A

Rakow, L. (1974). Cellularity of the different cell compartments in white adipose tissue of mice in chronic starvation, refeeding and in two types of obesity. In *The Regulation of the Adipose Tissue Mass* (ed. J. Vague and J. Boyer), Excerpta Medica, Amsterdam, pp. 140–144

Renold, A. E. Christophe, J. and Jeanrenaud, B. (1960). The obese hyperglycemic syndrome in mice. Metabolism of isolated adipose tissue *in vitro*. *Am. J. Clin. Nutr.*, **8**, 719–726

Robertson, R. P., Batchelor, B. R., Johnson, P. R. and Stern, J. S. (1974). Adipocyte cellularity in the desert sand rat (*Psammomys obesus*). *Proc. Soc. Exp. Biol. Med.*, **147**, 134–136

Soret, M. G., Kupiecki, F. P. and Wyse, B. M. (1974). Epididymal fat pad alterations in mice with spontaneous obesity and diabetes and with chemically induced obesity. *Diabetologia*, 10, 639–648

Steinmetz, J., Lowry, L. and Yen, T. T. T. (1969). An analysis of the lipolysis in vitro of obese-hyperglycemic and diabetic mice. *Diabetologia*, 5, 373–378

Subrahmanyan, K. (1960). Metabolism in the New Zealand strain of obese mice. *Biochem. J.*, 76, 548–556

Taketomi, S., Ishikawa, E. and Iwatsuka, H (1975). Lipogenic enzymes in two types of genetically obese animals, fatty rats and yellow KK mice. *Horm. Metab. Res.*, 7, 242–246

Täljedal, I.-B. and Hellman, B. (1963). Morphological characteristics of the epididymal adipose tissue in different types of hereditary obese mice. *Pathol. Microbiol.*, 26, 149–157

Thenen, S. W. and Mayer, J. (1975). Adipose tissue glycerokinase activity in genetic and acquired obesity in rats and mice. *Proc. Soc. Exp. Biol. Med.*, 148, 953–957

Thurlby, P. L. and Trayhurn, P. (1978). The development of obesity in pre-weanling (*ob/ob*) mice. *Br. J. Nutr.*, 39, 397–402

Trayhurn, P., Thurlby, P. L. and James, W. P. T. (1977). Thermogenic defect in pre-obese *ob/ob* mice. *Nature*, 266, 60–61

Van, R. L. R., Bayliss, C. E. and Roncari, D. A. K. (1976). Cytological and enzymological characterisation of adult human adipocyte precursors in culture. *J. Clin. Invest.*, 58, 699–704

Van, R. L. R. and Roncari, D. A. K. (1977). Isolation of fat cell precursors from adult rat adipose tissue. *Cell Tissue Res.*, 181, 197–203

van der Kroon, P. H. N., Van Vroonhoven and Douglas, L. T. (1977). Lowered oxygen consumption and heart rate as early symptoms of the obese, hyperglycemic syndrome in mice (*ob/ob*). *Int. J. Obesity*, 1, 325–330

Wahle, K. W. J. (1974). Fatty acid composition and desaturase activity of tissues of the congenitally obese Zucker rat. *Comp. Biochem. Physiol.*, 48B, 565–574

Yen, T. T. T., Steinmetz, J. and Wolff, G. L. (1970). Lipolysis in genetically obese and diabetes-prone mice. *Horm. Metab. Res.*, 2, 200–203

York, D. A. and Bray, G. A. (1973). Genetic obesity in rats. II: The effect of food restriction on the metabolism of adipose tissue. *Metabolism*, 22, 443–454

Zomzely, G. and Mayer, J. (1959). Fat metabolism in experimental obesities. IX. Lipogenesis and cholesterogenesis in yellow obese mice. *Am. J. Physiol.*, 196, 611–613

8

Non-genetic models
of obesity

D. S. Miller (Queen Elizabeth College, London, UK)

SUMMARY

A description is given of the many methods of producing obesity in experimental animals, that may be used as alternatives to the genetic models. Only a few of these are suitable for providing fat animals in large enough numbers for screening drugs to treat obesity, and these are described in detail in appendices.

A comparison of models, both genetic and non-genetic, shows that there are marked differences between them all. In particular, the relative importance of increased food intake and/or reduced metabolic rate as causative factors of obesity is widely variable. Since it is difficult in the present state of knowledge to say which is the most relevant model of human obesity, it seems better to work with a broad selection of models if the aetiology of obesity and its associated metabolic changes are to be understood.

INTRODUCTION

It may seem strange to present a paper with this title in a volume on genetic obesity but it is right to refer to the alternatives, since non-genetic models may also be relevant to the human condition. It is perhaps also strange that the organisers have not included a paper on genetic leanness, since this is becoming a subject of interest in the breeding of farm animals; but since this is a maverick paper, reference will be given to this model as well. However, the bias of this paper is towards the practical problem of producing obese animals in quantity for the purpose of screening drugs to treat human obesity.

Table 8.1 presents a list of the various techniques described in the literature but many may be rejected on the grounds of cost-effectiveness. In so doing one must be careful not to eliminate a relevant model of obesity, but it can be argued that some are alternatives. The models listed have been arranged according to the method of inducement. Other authors have classified models according to the

131

Table 8.1   Non-genetic models of obesity

I	Dietary-induced
	(1) Early overfeeding: small litters
	(2) Tube-feeding : force feeding
	(3) Gorging : meal eating
	(4) Pampering : cafeteria diets
	(5) Energy-dense diets
II	Chemically induced
	(1) Gold thioglucose
	(2) Monosodium glutamate
	(3) Bipiperidyl mustard
	(4) As : Hg
	(5) Insulin
	(6) Steroid hormones
	(7) Antimetabolites
III	Surgically induced
	(1) Lesions of the hypothalamus
	(2) Castration

reason they believe the obesity develops, but care must be taken in interpreting such classifications, since the aetiology of obesity even in animal models is controversial. The most commonly used dietary method is to feed a high-fat diet simulating those of human affluent societies: such diets are energy-dense and may lead to an increased food consumption, but the prime cause of the obesity is an increased efficiency of food utilisation. The most commonly used chemical methods result in lesions of the hypothalamus, but whether this causes hyper-

Table 8.2   Approximate fat contents and food utilisation of various obese models

			Energetic efficiency (%)	
	% Fat	Fat/N	Gross	Net
Mice				
ob/ob	60	50	15	70
MSG	50	25	10	50
GTG	45	20	7	30
E-D	30	10	5	15
cf. lean	15	5	3	7
Rats				
Zucker (*fa/fa*)	60	40	17	90
E-D	35	15	10	40
cf. lean	15	5	5	20
cf. wild	7	2	2	4

GTG, gold thioglucose; MSG, monosodium glutamate; E-D, the energy-dense diet. Gross energetic efficiency is the energy gain divided by the energy intake: for net efficiency an allowance is subtracted from the intake to allow for the energy cost of maintenance, i.e. it is energy gain divided by energy available for gain.

phagia or an increased efficiency of food utilisation depends upon the interpretation of the data: treated animals eventually eat more per head but less per unit of body weight. Nevertheless, it should be clearly stated that neither dietary-induced nor chemically induced obese models routinely produce animals as fat as the genetic obese models (table 8.2). However, they are significantly fatter than their controls in terms of body weight, the lipid content of their carcasses and their carcass fat : N ratio. As a group they may be regarded as simulating the human overweight condition rather than gross obesity, but such statements are quite misleading when one considers the difference in behaviour of the models in response to drugs or diet restriction. In particular, it is easier to get normal fat : N ratios, an index of obesity independent of body weight, in the non-genetic models of obesity than in genetic obesity, although it is possible in both.

DIETARY-INDUCED OBESITY

Since the classical work of McCance and Widdowson it has been known that manipulation of the quantity of diet given from birth to weaning can produce wide differences in growth rate and body fat. Reducing the number of pups per litter produces higher weaning weights which persist throughout the rest of life. It is a matter of value judgement as to whether animals from small or large litters are regarded as normal, but the difference between the two is considerable. The phenomenon has received attention in recent years because of the fat cell theory, which claims that overfed human infants have an induced greater number of adipocytes which predisposes them to obesity later in life. Animal data support this view but have been vigorously criticised on technical grounds. The practical problem of using this model is that the smaller the size of the litter the greater the effect, and, hence, one needs a very large number of litters to produce a reasonable supply of animals.

Manipulation of the quantity of food consumed after weaning is perhaps a more sensible approach, but it could be that a critical period of cellular growth is missed. Overfeeding laboratory animals, however, presents a different set of problems. Their food intake may be reduced by restricted feeding, or varied by altering the concentration of nutrients, especially protein, but although such animals may have very different growth rates, their carcass composition remains remarkably constant, depending only upon age. Adult animals voluntarily eat about 140 kcal/$W^{0.75}$ (W = weight in kg) and it is difficult to persuade them to eat more. Attempts to overcome this limit by flavouring laboratory diets have not been successful – although it has recently been claimed that pigs will overeat diets flavoured with apple. Some species are known to eat more than others (e.g. the polar bear), but these are not normal laboratory animals. The pig, the spiny mouse (*Acomys cahirinus*) and some breeds of dog are possible exceptions: at least they appear to be naturally fat species capable of overeating. It should be noted that in normal laboratory animals the limit to consumption is not one of bulk, since diluting the diet with an inert filler results in a compensatory increase in food consumption to maintain the intake at 140 kcal/$W^{0.75}$. To produce obesity by overfeeding it is necessary either to pamper the animals with a varied diet –

the cafeteria system (Stock and Rothwell, this Symposium) — or to force-feed. Strangely it is not necessary to overfeed by force-feeding to produce obesity: force-feeding alone will do it. This surprising result is associated with adaptations to the diurnal pattern of eating. Meal-fed animals (gorgers) are fatter than continuous eaters (nibblers) even if they consume the same amount of food. Thus one can achieve effects similar to force-feeding by training animals simply to eat all their food in 2 h/day. Overfeeding by force-feeding has severe physical limits. The normal technique is to tube-feed a liquid diet but the volumes necessary are excessive because of the required low dry weight of the food, which is determined by the need for it to pass a narrow cannula. Both meal-feeding and force-feeding are tedious techniques and not practical methods for producing obese animals in large numbers. The observation that obesity may be so produced is, however, of considerable theoretical interest and the metabolic consequences are adequately reviewed by Fabry (1969).

Producing obese animals by increasing the quantity of food consumed has its limitations, but varying the quality of the diet has been more productive. As mentioned earlier, diluting diets with an inert filler does not reduce energy intake, but increasing the energy density of the diet may raise energy intake marginally. This is achieved by increasing the fat content of the diet, and there is some evidence that fat calories are used more efficiently than carbohydrate calories, at least in farm animals. But fat should not simply replace starch weight for weight in such diets, since the ratio of protein to energy will fall below the requirements for maximum growth. In practice a diet containing 60 per cent fat and 30 per cent protein is used. The original diet consisted of a mixture of dried egg and butter, but is prohibitive in cost. Lard and casein are suitable substitutes, but the original diet is slightly superior, provided it is supplemented with biotin (0.1 g/kg) to avoid the effects of the antimetabolite avidin in the egg. Saturated fat is more effective than unsaturated oils, but it is advisable to add 1 per cent maize oil to avoid a deficiency of essential fatty acids. Casein does not have the same biological value as egg protein, but if 0.2 per cent methionine is added, the value is nearly the same. (See Appendix 1.) Such a diet has been variously called the high-fat diet, the high-fat–high-protein diet and the carbohydrate-free diet, but it is more appropriate to call it the energy-dense diet, since its effects are primarily a result of this function. In a recent study we constructed diets in such a way that those containing a high proportion of metabolisable energy (ME) from fat were not necessarily the most energy-dense: this was done by varying the proportions of fat, starch and cellulose (containing 9, 4, 0 ME kcal/g, respectively). When fed to rats the partial correlation of body fat against dietary energy density was 0.96 ($P < 0.001$) but that against dietary fat was only 0.18 (N.S). In fact, animals fed diets with more metabolisable energy per unit weight use this food more efficiently irrespective of the sources of that metabolisable energy. The correlations of dietary energy density with both gross energetic efficiency ($r = 0.94$) and net energetic efficiency ($r = 0.84$) were both highly significant. But if this increased efficiency is not due to fat, it is difficult to provide a metabolic explanation based on energy density alone. Digestibility is of course accounted for in determining the metabolisable energy of the diets, but it could be that we have underestimated the energy cost of moving food along the gut: certainly gut weights are negatively correlated with energy density ($r = 0.90$). Rats consuming the energy-dense diets also had a higher energy intake per head, but when these data are expressed per unit body weight, per metabolic body size ($W^{0.75}$) or per lean body mass, there were no correlations. Thus, the prime cause of obesity with this model is that dietary energy density improves the utilisation of metabolisable energy.

The energy-dense diet produces obesity in all the strains of laboratory animals examined so far. However, it does not raise the fat content of the already obese Zucker rat. Nor does it have any effect on the lean wild rat. Nevertheless, in a recent experiment with seven strains of rat fed three different diets, one of which was the energy-dense diet, an analysis of variance showed that the genetic influences were far greater than the dietary influence (table 8.3). But if the Zucker and wild rats are excluded from the analysis, the dietary effects are considerable,

Table 8.3 Fraction of total variation due to genetic and dietary factors of 7 (4*) strains of rat

	Genetics	Diet
Body weight	81 (36)	8 (38)
Body length	73 (61)	6 (12)
Body fat	83 (13)	7 (50)
Food intake	80 (47)	11 (42)
Heat production	75 (55)	12 (31)
Energetic efficiency	76 (15)	10 (46)
Adipose tissue		
cell no.	94 (−)	− (21)
cell size	− (−)	26 (27)

*Excluding Zucker and wild rats, i.e. laboratory strains only (hooded, Sprague-Dawley, Wistar and WAG).
The figures represent the percentage of total variation of various parameters measured on 7 (4*) strains of rat fed three different diets for 1 year. Results are only presented if they have reached a statistical significance of $P < 0.01$ by analysis of variance. (Boroumand-naini, 1977.)

having a greater positive influence than genetics on body fat, energetic efficiency, fat cell number and fat cell size. The resistance of the wild rat to dietary influences is quite remarkable and we have been unable to increase their carcass fat to more than 10 per cent. They, like other wild animals, including the wild boar and feral sheep, must carry a lean gene. Alternatively, it is open to argument that the normal strains of animal we all use have been selected by generations of animal breeders to produce what we ought to call genetically obese strains.

CHEMICALLY AND SURGICALLY INDUCED OBESITY

It has been known for at least 100 years that lesions of the hypothalamus can produce obesity, and 50 years ago physiologists started to produce obesity in experimental animals using stereotactic techniques selecting the ventromedial nuclei. It is often claimed that animals so treated are hyperphagic, but this conclusion depends on the method of expression of food intake; it has also been suggested that the rapid weight gain may be due to an alteration in feeding habits such that the treated animals become gorgers rather than nibblers. The technique requires skill, the mortality rate is not low, and one cannot always be certain that the ventromedial part of the hypothalamus and *only* that part has been destroyed (Morrison, 1977). It is therefore not a suitable method for producing large numbers of obese animals. However, lesions may be induced chemically, at least in mice, and the technique forms the basis of a number of models. But the relevance to human obesity is not clear.

In all of these methods there is the general problem that in order to produce obesity the dose required is close to the toxic dose. A further complication is that both successful treatment and toxicity depend upon the strain of mouse used. The percentage of obese animals among the survivors increases with dose but so does the death rate. A compromise is therefore necessary where the number of obese survivors is maximised for a given strain of animal. In such circumstances there are inevitably a number of treated animals that do not become obese, and only time will tell which is which. In the case of the gold thioglucose (GTG)-treated animals the obese may be selected by appearance, but when monosodium glutamate (MSG) is used, the obese animals can only be identified early by carcass analysis, since they increase their fat content without a marked change in weight. Both agents are used on young animals, often by a single injection: however, recently it has been shown that by giving MSG in six injections during the first week of life virtually all animals become obese. (See Appendix 2.) Bipiperidyl mustard (BPM) has also been used, but the compound is difficult to obtain and is carcinogenic. All three agents have been shown to cause lesions in the ventromedial part of the hypothalamus, although their histology is not identical. The animals treated with GTG eat more food per head in the early stages of life, but expressing their energy intake per unit of body weight, metabolic body size or lean body mass does not indicate hyperphagia. This intellectual problem is not apparent in the case of the MSG-treated animals, since they tend to have similar body weights and similar food intakes to their controls, and clearly hyperphagia does not exist; even when their body weights increase later in life, their food intakes are the same as those of the controls. Both models produce animals with a marked enhanced efficiency of energy utilisation, from which one can conclude that the hypothalamus controls feed efficiency. This is not unreasonable, since it is known to be involved in temperature regulation and, hence, possibly the control of heat production. If the latter were reduced, obesity would be a consequence.

There are a number of other chemicals which have been claimed in the old literature to induce obesity, possibly as a result of hypothalamic lesions. Of these arsenic and mercury are of interest in view of their involvement in pollution, but no recent clinical or experimental findings have confirmed the claim. Mercury was used in the treatment of venereal disease, but the obesity associated with its use may have had other sociological causes.

Various hormones have a marked effect on metabolic rate and lipogenesis, and some have been used to increase the rate of fat deposition. Insulin lowers blood glucose and by way of compensation increases voluntary food intake; without food the animals die. Metabolic rate is decreased concomitant with increased lipogenesis, and the positive energy balance that results is due both to the increase in food intake and to an increased efficiency of food utilisation, although the former is more important. As a method for producing large numbers of obese animals, however, the technique is too tedious, since daily injections are required. Oral hypoglycaemic agents incorporated into the diet, sadly, do not have the same effect as insulin. Both the sulphonylureas and the diguanides increase energy utilisation slightly, but this is compensated for by a slight reduction in food intake, thus producing no weight change. Perhaps this is as well, since they are used in controlling mild maturity-onset diabetes in the obese.

It is well known that steroid hormones have a marked effect on energy balance, but the literature is vast and the effects of any one compound show marked specific differences. Compounds that cause weight gain in one species reduce the carcass fat in another. Stilboestrol is used commercially in poultry husbandry for caponisation, a technique to produce rapid weight gains in cockerels, but it reduces the body fat content of obese mice. Generally, in rodents progesterones tend to increase body fat, whereas oestrogens reduce it, but in women it is the high-oestrogen-containing contraceptive pills that have the worst reputation for weight gain. No one has proposed an obese animal model based on the use of steroid hormones (or castration) nor surprisingly have they proposed an animal model based on thyroid antimetabolites (e.g. thiouracil), although these are known to lower metabolic rate. Nevertheless, the use of hormones for manipulating energy balance is a useful tool in understanding the aetiology of obesity. An antimetabolite to adrenalin would be particularly useful, but β blockers do not seem to produce obesity.

DIFFERENCES BETWEEN MODELS

In providing the above catalogue, emphasis has been given to the importance of raising the efficiency of energy utilisation to compensate for a general but erroneous view that animal models become obese only by overfeeding. Looking at the list as a whole, one could propose a universal theory based on the control of glucose utilisation for lipogenesis by substrate flow and hormonal levels, but supporting evidence is sparse. Certainly, all the models produce animals fatter than their controls, but some are easily reversible (Stock and Rothwell, this Symposium), others reverse only with difficulty, while others become lean on treatment but return to their obese state when treatment is withdrawn. We are primarily interested in producing obese animals for testing thermogenic drugs — that is, drugs for the treatment of human obesity that increase metabolic rate rather than reduce appetite. We take the view that since some fortunate lean individuals can overeat without suffering the consequences, in contrast to those with an obese disposition, it should be possible to provide a treatment which does not involve a frugal diet with its poor prognosis. Drugs of this type reveal differences between animal models and it is of some importance to know which models closely resemble the human conditions(s).

To illustrate this problem data are presented in table 8.4 on the effects of triiodothyronine (T_3) and ephedrine (E) on six animal models of obesity and a lean

D. S. Miller

Table 8.4 Effect of two drugs on various obese models

| | Ratios to control animals without drugs | | | | | | | | | |
| | Food intake | | Oxygen consumption | | Body weight | | Carcass protein | | Carcass fat | |
	T_3	E	T_3	E	T_3	E	T_3	E	T_3	E
Mice										
ob/ob	*	0.79	*	1.16	*	0.80	*	0.86	*	0.72
GTG	1.53	0.66	1.88	1.13	0.78	0.64	0.85	0.83	0.43	0.29
MSG	1.33	0.82	1.43	1.22	0.96	0.69	1.08	0.91	0.78	0.43
E-D	1.57	1.02	1.44	1.22	1.00	0.82	1.10	0.98	0.69	0.32
lean	1.49	1.00	1.85	1.22	1.00	0.95	1.13	0.94	0.83	0.71
Rats										
Zucker	*	0.55	*	1.25	*	0.59	*	0.88	*	0.34
E-D	1.13	0.72	1.48	1.08	0.77	0.81	0.78	0.84	0.24	0.37

GTG, gold thioglucose; MSG, monosodium glutamate; E-D, the energy-dense diet; T_3, triiodothyronine 400 μg/kg diet; E, ephedrine 1 g/kg diet (Massoudi, 1978).
*Animals died.

control. The drugs were incorporated into the diet and the same diet fed to all models (T_3, 400 μg/kg; E, 1 g/kg). Food intake, metabolic rate, and body weight and composition are expressed as ratios relative to control animals that did not receive drugs, in order to make the changes in these parameters more apparent. Generally, T_3 increased oxygen consumption but also increased food intake, the balance between the two producing weight losses in some models but not in others. Carcass fat was reduced in all models, but the effect was greatest with the GTG-treated mice and the rats fed the energy-dense diet. However, the same diet containing the same level of T_3 killed all the genetically obese animals, both rats and mice. The effect of ephedrine was more consistent, since all models lost body weight and body fat without increasing food intake and had elevated oxygen consumptions. Nevertheless, it was least effective in the genetically obese mouse and most effective in the genetically obese rat. It should be noted that both drugs reduced body fat without substantial changes in body protein, and, hence, the fat : N ratio is reduced, in contrast to reducing body weight by food restriction i.e. the animals have become more lean irrespective of their body weights. Both drugs raise the metabolic rate in man, but ephedrine is much less toxic and, indeed, may be bought across the counter without prescription. Its relative safety may be judged by the fact that we currently maintain our breeding colony of genetically obese mice on the ephedrine diet simply because it makes the normally sterile males fertile. When they are used on heterozygous females the litters contain an equal number of obese and lean animals, which is very convenient experimentally.

In conclusion, non-genetic models of obesity are not, in practice, as grossly obese as the genetic variety but they may have relevance to the human condition. Five or six models are suitable for the production of reasonably large numbers of animals for experimental purposes. If further selection is necessary, one would choose the CFLP stock of mice and treat one-third with MSG, give one-third the energy-dense diet and retain one-third as controls: add to these three groups the *ob/ob* and possibly wild mice and one would have five groups covering a very wide range indeed. Our own view is that no one model simulates human obesity. The non-genetic models differ both between themselves and from the genetic models and even the genetic models are not identical. The aetiology of the obesity of all models resides in changes of both food intake and the efficiency of energy utilisation. The latter is the most important factor and its various biochemical explanations are the most intriguing. In the present state of knowledge it seems better to work with a broad selection of obese animals if only because they *are* different.

REFERENCES

Most of the work in this paper is unpublished but available in thesis form from London University.

Boroumand-naini, M., PhD, 1977. Nutrition and genetics
Djazayery, A., PhD, 1973. Energy metabolism with particular reference to obesity
Massoudi, M., PhD, 1978. Obesity and thermogenic drugs
Parsonage, S. R., PhD, 1971. Some factors affecting energy utilisation

Other references are:

Bunyan, D., Merrell, E. A. and Shah, P. D. (1976). The induction of obesity in rodents by means of monosodium glutamate. *Br. J. Nutr.*, 35, 25–39
Fabry, P. (1969). *Feeding Patterns and Nutritional Adaptations*, Butterworths, London
Morrison, S. D. (1977). The hypothalamic syndrome in rats. *Fed. Proc.*, 36, 139–142

Appendix 1   Energy-dense diets (parts by weight)

	Egg/butter	Lard/casein
Egg powder	48	–
Butter	44	–
Lard	–	61
Casein	–	30
Salts	5	5
Vitaminised cellulose powder	3	3
Maize oil	–	1
Biotin	0.01	–
Methionine	–	0.2

Appendix 2   Chemically induced obesity

Monosodium glutamate

Mice of the CFLP strain* (Anglia) are injected on the following days after birth: 1, 2, 3, 6, 7, 8. The solution contains 300 g/l normal saline. The dose rate is 0.01 ml/g body weight. The expected death rate is 10% : all survivors become obese. Sources : Bunyan *et al.* (1976); Massoudi (1978).

Gold thioglucose

Mice of the CBA strain (Bantin and Kingman) are injected at weaning with a single dose. The solution contains 50 mg/ml oil (solganol:Schering). The dose rate is 0.012 ml/g body weight. The expected death rate is 15% : 90% of the survivors become obese. Source : Massoudi (1978).

NOTE: The number of injected mice that become obese is much less with other strains — e.g. CBA, 75%; CS-1, 14%; TO, 18%; LACA, 14%; and CSD, 9%. These figures cannot be improved by changing the dose rate.

* Now discontinued but genetically similar to the ICI Alderley Park stock of mice.

9

Energy balance
in reversible obesity

M. J. Stock and Nancy J. Rothwell (Department of Physiology,
Queen Elizabeth College, London, UK)

SUMMARY

Various methods of producing reversible obesity in animals are reviewed, and
include high-fat feeding, insulin-induced hyperphagia, force-feeding and feeding
a mixed and varied diet (the cafeteria system). Of these, the most successful
seem to be force-feeding and the cafeteria feeding system, both of which have
been utilised to produce obesity from which animals can spontaneously recover
when allowed free access to the stock diet alone.

Our own results suggest that the obesity induced by tube-feeding varying
proportions of normal daily energy intake, allowing voluntary feeding to continue,
results entirely from an increase in metabolic efficiency, since the animals consume the same amount of energy as their free-feeding controls. When tube-feeding
is stopped, the weight loss which follows is accompanied by a marked hypophagia,
while energy expenditure remains normal. Conversely, animals fed the cafeteria
diet become obese entirely as a result of hyperphagia without any changes in
resting oxygen consumption. When returned to the stock diet, these obese rats
rapidly return to control body weight as a result of a decrease in energy intake and
a simultaneous increase in resting oxygen consumption.

It is concluded that these two examples of experimental obesity provide further
evidence for the existence of lipostasis in the rat and imply that controls acting on
both intake and expenditure operate to maintain energy balance.

INTRODUCTION

Although this Symposium is concerned with obesity, and genetic models in particular, we presume that this reflects an underlying interest in the normal regulation of energy balance, and therefore feel that a discussion on mechanisms of

141

weight gain and loss in essentially normal rats is not out of context. There seems little doubt that the normal adult laboratory rat regulates its energy balance, since although body fat tends to increase with age, this trend is remarkably consistent in all normal rats. Conventionally this regulation is ascribed to the precise control of energy intake exhibited by the rat, but more recently the possible involvement of controls operating on expenditure has become apparent*. The study of genetically and hypothalamically obese animals certainly suggests that derangement of energy balance can occur as a result of changes in both energy intake and metabolic efficiency (Cox and Powley, 1977; Djazayery *et al.*, 1979). The description of these two types of obesity will obviously increase our understanding of the possible origins and metabolic consequences of obesity, but both could be considered as pathological conditions, and as such can only provide limited information on the normal physiological processes involved in energy balance regulation. An alternative approach is to study the regulation of energy balance in animals where the regulatory mechanisms are intact and actively operating rather than being absent or inactive, as in the permanently obese animal.

In taking this alternative approach we have based our experiments on the assumption that the long-term regulation of energy balance operates to maintain body fat stores constant. This assumption forms the basis of Kennedy's 'lipostatic' theory for the control of energy intake (Kennedy, 1953). If body fat is indeed the regulated variable, it follows that experimentally induced perturbations in body fat content should activate compensatory mechanisms in order to restore the system to its 'set-point'. The first experimental requirement, therefore, is to find a means by which body fat can be altered in a reversible manner. It would also be desirable if the chosen method only produced changes in fat and did not influence lean body mass, thus avoiding complications when interpreting results from a lipostatic viewpoint. Fat depletion is easily achieved by food restriction , but the production of excessive fat deposition, from which the animals can spontaneously recover, has rarely been described. A successful search for such a technique would enable us to study not only how animals could become obese, but also, more importantly, to what extent they have control over their body fat content and the means by which they achieve this control. This is obviously not possible with genetically or hypothalamically obese animals, where spontaneous recovery cannot occur.

We will now review various methods of inducing obesity in order to compare:

(1)   The time required to produce significant increases in body fat (and the effect on other components of body composition).

(2)   To what extent these gains are reversible — i.e. do the animals exhibit 'lipostasis'?

(3)   The means by which gains and losses are achieved.

Some of these methods have been described in the previous chapter, but not in this context, and we will include in our review some of our own more recent findings.

---

*In this paper, changes in energy expenditure refer to changes in heat production resulting from alterations in metabolic efficiency and therefore include changes in dietary-induced thermogenesis.

It is worth remembering that Kennedy's original lipostatic theory and the more recent modifications of it (Le Magnen *et al.*, 1973; Booth, 1975) are concerned solely with the control of energy intake and would therefore predict that compensation for excessive fat gains will be achieved by depression of intake. We hope to demonstrate that this is not always the case and that changes in heat production (i.e. dietary-induced thermogenesis) are involved in both the development of the obese state and its reversal. In this respect we see no reason why control of thermogenesis should not be included, along with the control of energy intake, in the lipostatic hypothesis.

HIGH-FAT DIETS

One of the simplest ways of producing obese animals is to feed high-fat diets (Peckham *et al.*, 1962; Schemmel and Mickelsen, 1973), although most workers have found this method only to be reliable if feeding is commenced at weaning and maintained for several months. Furthermore, the degree of obesity which results depends to a large extent on hereditary factors, since not all strains respond to high-fat feeding (Schemmel and Mickelsen, 1973; Miller, 1978).

Recovery from the obese state has rarely been studied, although Schemmel's group have reported that if the high-fat diet is withdrawn during the post-weaning period, obesity is not maintained. Peckham *et al.* (1962) and Macdonald (1977) report fat losses in adult animals when the high-fat diet is withdrawn, but the degree of adiposity was either very small (Peckham *et al.*, 1962), or could not be compared with that of a control group (Macdonald, 1977). In most cases the greater rate of energy retention in animals on high-fat diets has been ascribed to an increase in energy intake (Schemmel and Mickelsen, 1973), but there is evidence for a change in metabolic efficiency. Both Lemonnier (1972) and Herberg's group (Herberg *et al.*, 1974) have reported greater body fat gains in rats and mice, respectively, without any increase in energy intake, and Miller (1978) argues that the increase in efficiency has a more prominent role in the development of obesity than the changes in energy intake.

This type of obesity has been described in more detail in the previous paper (Miller, 1978), but we can summarise by saying that high-fat feeding does not appear to be suitable to meet our experimental requirements because: (a) the time required to produce significant changes in body fat is fairly lengthy, (b) when very young animals are used, changes in lean body mass may occur, and (c) it is uncertain whether spontaneous reversal can occur.

INSULIN INJECTIONS

Daily injections of insulin will induce hypoglycaemic hyperphagia and increased fat deposition, eventually leading to a larger body fat content (Hausberger and Hausberger, 1958; Hoebel and Teitelbaum, 1966; Macdonald *et al.*, 1976). These workers have reported excess weight gains of between 0.7 and 3.0 g/day in animals treated with 8–24 units of protamine insulin per rat per day. Withdrawal of insulin treatment results in an immediate fall in food intake, and body weight soon reaches that of controls. Hausberger and Hausberger (1958) have shown that the increased weight gain occurring in insulin-treated rats is due solely to an increased body fat content, with no changes in lean body mass. There are few data on the body composition of animals allowed to recover from treatment, but we have noted that the increase in epididymal fat pad weight and adipocyte diameter in animals made obese with insulin is completely reversed by the time body weight has returned to normal (Macdonald *et al.*, 1976).

Although there can be no doubt that hyperphagia is the primary cause of the excessive weight gain, there is some evidence for changes in metabolic efficiency during treatment. Calculations based on the data of Hoebel and Teitelbaum (1966) and Macdonald (1977) suggest that the greater weight gains are due in part to an increased feed efficiency (i.e. body weight gain in grams per gram of food eaten). There is also a suggestion that the hypophagia following withdrawal of insulin injections is accompanied by a reduced metabolic efficiency, since the rate of weight loss is more rapid than one would predict from the depression of intake alone. Thus, we have a situation where it is possible to demonstrate that changes in energy intake and dietary thermogenesis are involved in the weight gain and the weight loss.

From this description it would appear that the use of insulin should provide a suitable means of studying reversible obesity, but one important drawback is that the effective dose of insulin is very variable and unpredictable. Some workers have used exceptionally high doses and have not been able to induce significant increases in body weight (Mozes *et al.*, 1977), while others have reported deaths resulting from quite low doses (Macdonald, 1977). A more fundamental reservation about the use of insulin is that it has many profound metabolic effects and the interpretation of these types of experiments has to take account of the effects of the hormone *per se* as well as the consequences of a greater fat mass.

TUBE-FEEDING

Force-feeding by gastric intubation can be used to produce excessive fat deposition in two ways: firstly, by overfeeding such that experimental animals are forced to consume more food than controls, and secondly, by delivering the food in discrete meals, since meal-feeding is well known to increase metabolic efficiency (Fabry, 1969).

Cohn and Joseph (1959, 1960, 1962) were able to induce extremely large weight gains in rats by intragastric feeding, and found that when tube-feeding was removed, the obese animals became hypophagic and lost weight rapidly until they had reached control weight, by which time food intake had returned to normal. The same workers (1959) found that when animals were fed 100 or 80 per cent of control intake by stomach tube the rats gained similar amounts of weight to free-feeding controls. However, in spite of identical or lower energy intakes, the tube-fed animals still gained considerably more fat than controls, thereby demonstrating the marked increase in metabolic efficiency that accompanies meal-feeding. Quatermain *et al.* (1971) fed rats 100 per cent of their normal daily

Table 9.1 Effect of tube-feeding on body weight gain and energy intake (Rothwell and Stock, 1978b)

		No. of animals	Duration of experiment (days)	Body weight gain (g)	Energy intake (kJ/rat)	Energy intake delivered by tube (% control intake)	Total energy intake (% control intake)
Expt. 1:	Control	6	30	44 ± 5	11 340 ± 110	34.3	103.0
	Tube-fed	5		71 ± 10*	11 650 ± 140 (N. S.)		
Expt. 2:	Control	12	21	67 ± 5	8500 ± 110	47.5	99.7
	Tube-fed	8		100 ± 9†	8400 ± 130 (N. S.)		
Expt. 3:	Control	6	23	58 ± 5	8860 ± 90	68.2	95.3
	Tube-fed	6		95 ± 5†	8420 ± 80 (N. S.)		
Expt. 4:	Control	12	21	40 ± 5	7700 ± 30	74.7	100.0
	Tube-fed	9		89 ± 5‡	7710 ± 50 (N. S.)		

Tube-fed animals (adult, male Sprague–Dawley rats) received one (Expt. 1), two (Expt. 2) or three (Expts. 3 and 4) daily intra-gastric loads of Complan (Glaxo) mixed into a slurry (dry weight of slurry varied from 30 to 60 per cent w/v). All animals had free access to a semisynthetic powdered diet with a nutrient composition similar to that of Complan.
Mean ± S. E. M.
Significantly different from controls: * $P < 0.05$;
                                       † $P < 0.01$;
                                       ‡ $P < 0.001$.

N.S. = not significant.

food intake by intragastric infusion with the same diet available for consumption by mouth. They found that animals continued to eat voluntarily, whether the load was infused continuously or in discrete meals, and this resulted in excessive weight gains which were completely reversed after cessation of gastric feeding.

We have also given rats a portion of their daily food intake by stomach tube and allowed free access to the normal diet, in order to observe the effects on residual voluntary intake and body weight gain (Rothwell and Stock, 1978b). It was found that when anything from 34 to 75 per cent of *ad libitum* intake was delivered intragastrically, voluntary intake was depressed to the extent that the total energy intake of the animals was exactly the same as that of controls throughout the experiment (table 9.1). This finding is by itself a remarkable demonstration of the precision to which the rat can control energy intake in spite of the stresses imposed by gastric intubation and the disruption of its normal meal pattern. However, as a consequence of the increased metabolic efficiency induced by the meal-feeding, this perfect compensation of voluntary intake was accompanied by excessive weight gains in all experimental animals (table 9.1), the rate of energy retention being dependent on the fraction of daily intake delivered by tube. Since the animals became obese, it would appear that in this situation the control of energy intake becomes completely dissociated from the regulation of energy balance. Tube-feeding may be unique in this respect, but these findings have raised serious questions about existing theories of intake control (Rothwell and Stock, 1978b).

We found that after 20–30 days of tube-feeding, rats had gained significantly more weight than controls, and when tube-feeding was stopped and the rats were allowed to eat freely, their food intake dropped and weight loss occurred. In many cases the body weights of obese rats did not completely return to those of controls, and were still elevated some 20 days after treatment had ended. The body water content of tube-fed and control animals was measured at several stages of the experiment by an *in vivo* tritium dilution method, and by assuming a constant value for the water content of fat-free mass (Pace and Rathbun, 1945) body fat could be estimated. These measurements revealed that most (80 per cent) of the excess weight gain of the tube-fed animals was due to fat, although some small increases in fat-free mass were apparent. The weight loss after treatment was also comprised mainly of fat, and it seems that the absence of full recovery of body weight might be due to permanent changes in fat-free mass induced by the tube-feeding. From the data on body fat and food intake we were able to obtain an approximate estimate of energy expenditure during the period of weight loss by assuming that changes in body fat accounted for all the changes in body energy. This we found to be similar for obese and control animals, suggesting that changes in energy intake were solely responsible for the weight loss.

Unfortunately, it is difficult to calculate the relative contributions of intake and expenditure to weight gains and losses in the experiments of other workers, because in most cases the data given on these parameters are rather limited. However, from our experiments we can say that the obesity resulting from tube-feeding is brought about by increases in metabolic efficiency, while energy intake remains normal. The reversal of obesity, on the other hand, is brought about by a reduction of intake and there is no evidence for a change in efficiency. The only obvious disadvantages of this method of inducing reversible obesity for our purposes are the alterations in lean body mass reported by us and Cohn and Joseph, and the possible long-term metabolic effects of meal-feeding.

CAFETERIA FEEDING

It is the contention of many human nutritionists that, at least as regards the control of energy intake, man is not a rat — in other words, social, psychological and economic factors determine man's food intake as much as hypothalamic and metabolic influences. While we would agree with the line of this argument, we feel it is rather unfair to suppose that the rat is indifferent to all of these external influences. Because the normal laboratory animal is maintained under constant environmental conditions, deprived of social interaction and fed a bland and monotonous diet, it is not surprising that its food intake should be so constant. However, recent behavioural studies indicate that 'environmental enrichment' can profoundly influence the food intake of laboratory rats. In an attempt to influence the voluntary intake of rats in this way we added flavours (e.g. vanilla) or sweeteners (e.g. saccharine) to the conventional rat diet. Unfortunately, these additions were insufficient to induce any marked changes in food consumption or body weight gain, and any increase in intake was only transient. Sclafani and Springer (1976), however, have found that feeding a varied diet, mainly comprised of foods normally consumed by man, results in obesity. The rat is offered a diet varying not only in flavour, but also in appearance, texture and composition, and because the choice of foods is changed daily, this feeding regimen has been called the cafeteria system.

Sclafani and Springer (1976) were able to produce excess weight gains of 80 g in 55 days in male rats on cafeteria feeding, but since food intake was not determined, it is not known whether this obesity resulted entirely from an increased energy intake or whether changes in metabolic efficiency were also contributory. These workers also found that the excess weight could be lost when rats were returned to stock diet only, and during this period obese animals showed a reduced motivation for feeding as assessed by their willingness to eat quinine-adulterated food or lever-press for food

Rolls and Rowe (1977) have utilised the cafeteria system to produce obesity in male and female rats. Feeding a mixed diet for 13 weeks, they obtained excess weight gains of 102 g but found that the obese animals did not lose any of the extra weight when returned to stock diet and would defend their elevated weight even after starvation. Once again food intake was not measured either during mixed diet feeding or afterwards and neither Rolls' nor Sclafani's group report any data on the body composition of cafeteria-fed rats. The differences regarding the reversibility of cafeteria-induced obesity between these two groups is difficult to explain, although it is possible that hereditary factors may be involved, as with high-fat feeding.

Our own work (Rothwell and Stock, 1978a) shows that very large weight gains resulting from cafeteria feeding can be spontaneously reversed when rats are returned to the normal stock diet. Figure 9.1 shows that significant increases in body weight can be produced in cafeteria animals after only 17 days, and when treatment was withdrawn, the body weight of these rats had returned to control levels within 12 days. An unexpected finding was that during this period, after withdrawing the cafeteria diet, the energy intake was the same for both control and

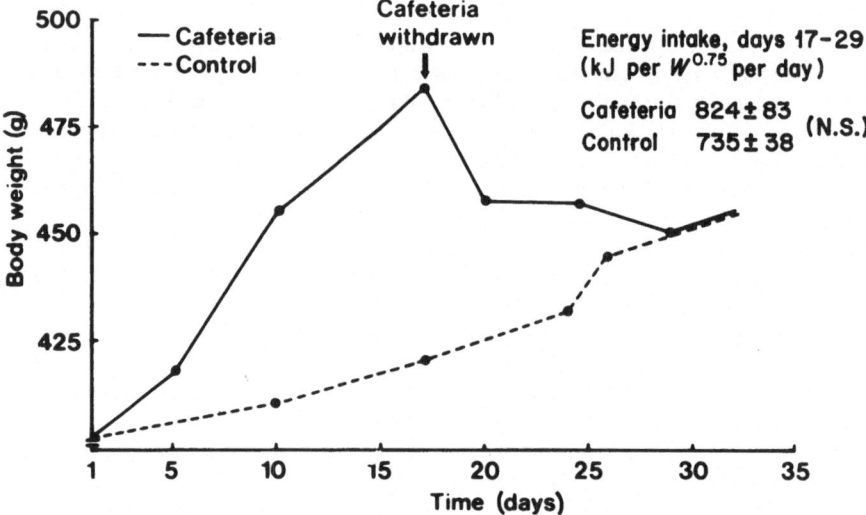

Figure 9.1 Body weight of control and experiment rats (adult, male Sprague–Dawley) during and after cafeteria feeding. Control animals received stock diet throughout; cafeteria animals received stock diet plus four different food items each day until day 17, after which they were allowed stock diet only. Energy intake was determined for both groups on days 17 – 29 and values are given as mean ± S.E.M.

experimental animals, even though the cafeteria group was rapidly losing weight and the control group slightly gaining. We therefore deduce that there must have been a considerable increase in the energy expenditure of the cafeteria rats to account for this weight loss. In a second experiment energy intake was also measured during the period of cafeteria feeding, and it was found that experimental animals ate between 25 and 70 per cent more than controls. There was no evidence for any changes in metabolic efficiency during this period (body weight gain/kJ eaten), and resting oxygen consumption, measured in a closed-circuit respirometer (Stock, 1975), was the same for both groups. We conclude that the weight gain was entirely due to hyperphagia and it is worth noting that the cafeteria rats in fact consumed not only more energy than controls, but also a greater weight of food; all of which demonstrates that the precise control of food intake normally exhibited by the laboratory rat can be overridden by psychological and environmental factors.

In our second experiment body fat was measured by the *in vivo* method described earlier, and it was found that all the weight gains and losses of cafeteria rats could be accounted for by changes in body fat, with fat-free mass remaining un-altered throughout treatment and recovery. Once again, cafeteria animals lose weight when returned to the stock diet alone, but this time there was a reduction in food intake, with the obese animals consuming significantly less than controls during the period of weight loss. In order to determine whether this decreased in-take was sufficient to account for the weight loss or whether there were simultan-eous changes in metabolic efficiency, we estimated energy expenditure from the change in body energy stores (i.e. fat) and food intake and found that energy expenditure was elevated in cafeteria rats by some 44 per cent over the 4 days immediately following removal of the cafeteria diet. Thus, in both experiments we have evidence for an increased energy expenditure contributing to the weight loss of the obese rats. More direct confirmation of this comes from the measure-ments of oxygen consumption made on the days following withdrawal of the cafeteria diet. The resting oxygen uptake of cafeteria rats was significantly (24 per cent) higher than that of controls, even when corrected for body size. Since these values involve no activity and cannot be explained by a greater energy intake in the cafeteria animals (they were hypophagic at this time), it is reasonable to postu-late that the elevated oxygen consumption is due to increased thermogenesis.

The cafeteria method of inducing obesity seems to provide an excellent experi-mental model for the study of mechanisms involved in the regulation of energy balance. Obesity is quickly and easily achieved and involves no alterations in lean body mass. When treatment is withdrawn, all the excess fat is lost, as a result of both a reduction in intake and a relatively large increase in heat production; this model could therefore prove particularly useful in the study of dietary-induced thermogenesis and its metabolic origins.

A comparison of the tube-feeding and cafeteria systems reveals several salient features of energy balance regulation and clearly contrasts the varied means by which body fat can be gained and lost. The excessive weight gains of the tube-fed rats are due to an increase in metabolic efficiency, while in the cafeteria rats the weight gain is a consequence of hyperphagia. Interestingly, both examples show a conspicuous lack of appetite control — the simplest case being the cafeteria feeding, where the rat apparently eats 'for pleasure' rather than 'for calories', while the tube-fed rats voluntarily adjust their intake to match that of their free-feeding controls but fail to adjust it in response to the increased rate of energy retention. Although the tube-fed animals gain weight as a result of an increase in metabolic efficiency, the subsequent loss of weight is due to a depressed food intake. The cafeteria rats, on the other hand, having previously overeaten, lose most of the excess weight by decreasing metabolic efficiency (i.e. increasing heat production). These contrasting models of reversible obesity therefore provide a cogent demonstration of the need to identify changes in both intake and expen-diture if a complete understanding of energy balance is desired.

It is apparent that the tube-fed and cafeteria-fed rats exhibit lipostasis, since both types of obesity are readily reversed. The cafeteria system probably provides the best experimental test of lipostasis so far devised, since it permits the induction of excessive weight gains with the minimum of interference to the animals and the alterations in body weight are entirely due to changes in body fat. At the same time, it has to be admitted that this lipostatic control is not rigidly enforced, particularly as it is easily overcome merely by offering the rat a varied and palatable diet. It is also clear that the rat is not so different from man in this respect and it may be that man could also control his body fat just as well as the normal laboratory rat if he were to eat a pelleted stock diet! Such a suggestion may not be so far-fetched, because in many parts of the world the bulk of man's intake comes from one staple food. Together with low fat intakes, this constitutes a fairly bland and monotonous diet, and even when supplies are ample, obesity is rarely seen. Perhaps we can think of the normal laboratory rat as an example of this human condition and the cafeteria rat as an analogy of man in more affluent societies.

REFERENCES

Booth, D. A. (1975). Normal metabolic control of hunger. In *Hunger: Basic Mechanisms and Clinical Implications* (ed. D. Novin, W. Wyrwicka and G. A. Bray), New York, Raven Press, pp. 127-143

Cohn, C. and Joseph, D. (1959). Changes in body composition with force-feeding. *Am. J. Physiol.*, **196**, 965-968

Cohn, C. and Joseph, D. (1960). The effects of feeding frequency in intermediary metabolism. *Metabolism*, **9**, 492-501

Cohn, C. and Joseph, D. (1962). Influence of body weight and body fat on the appetite of normal lean and obese rats. *Yale J. Biol. Med.*, **34**, 598-607

Cox, J. E. and Powley, T. L. (1977). Development of obesity in diabetic mice pair-fed with lean siblings. *J. Comp. Physiol. Psychol.*, **91**, 347-358

Djazayery, A., Miller, D. S. and Stock, M. J. (1979). Energy balance in obese mice. *Nutr. Metab.* (in press)

Fabry, P. (1969). *Feeding Patterns and Nutritional Adaptations*, Butterworths, London

Hausberger, F. X. and Hausberger, B. C. (1958). Effect of insulin and cortisone in weight gain, protein and fat content of rats. *Am. J. Physiol.*, **193**, 455-460

Herberg, L., Doppen, W., Major, E. and Gries, F. A. (1974). Dietary induced hypertrophic hyperplastic obesity in mice. *J. Lipid Res.*, **15**, 580-585

Hoebel, B. G. and Teitelbaum, P. (1966). Weight regulation in normal and hypothalamic hyperphagic rats. *J. Comp. Physiol. Psychol.*, **61**, 180-193

Kennedy, G. C. (1953). The role of depot fat in the hypothalamic control of food intake in the rat. *Proc. R. Soc. (London)*, **B140**, 578-592

Le Magnen, J., Devos, M., Gaudilliere, J., Louis-Sylvestre, J. and Tallon, S. (1973). Role of a lipostatic mechanism in regulation by feeding of energy balance in rats. *J. Comp. Physiol. Psychol.*, **81**, 1-23

Lemonnier, D. (1972). Effect of age, sex and site on the cellularity of the adipose tissue in mice and rats rendered obese by a high fat diet. *J. Clin. Invest.*, **51**, 2907-2915

Macdonald, I. A. (1977). Metabolic effects of variations in total body lipid. PhD thesis, London

Macdonald, I. A., Rothwell, N. J. and Stock, M. J. (1976). Lipolytic and lipogenic activities of adipose tissue during spontaneous fat depletion and repletion. *Proc. Nutr. Soc.*, **35**, 129A

Miller, D. S. (1978). Non-genetic models of obesity in laboratory animals. This volume

Mozes, S., Kuchar, S. and Bodak, K. (1977). Hypoglycaemia and food intake in rats given graduated doses of insulin. *Physiol. Bohemoslov.*, **26**, 159-164

Pace, N. and Rathbun, E. N. (1945). Studies on body composition. III. The body water and chemically combined nitrogen content in relation to fat content. *J. Biol. Chem.*, **158**, 689-691

Peckham, S. C., Centerman, H. W. and Carroll, J. (1962). The influence of hypercaloric diet on gross body and adipose tissue composition in the rat. *J. Nutr.*, 77, 187–197

Quatermain, D., Kissileff, H., Shapiro, R. and Miller, N. E. (1971). Suppression of food intake with intragastric loading: relation to the natural feeding cycle. *Science*, 173, 941–943

Rolls, B. J. and Rowe, E. A. (1977). Dietary obesity: permanent changes in body weight. *J. Physiol. (London)*, 272, 2P

Rothwell, N. J. and Stock, M. J. (1979a). Mechanisms of weight gain and loss in reversible obesity in the rat. *J. Physiol. (London)* (in press)

Rothwell, N. J. and Stock, M. J. (1979b). A paradox in the control of energy intake. *Nature* (in press)

Schemmel, R. and Mickelsen, O. (1973). Influence of diet, strain, age and sex in fat depot mass and body composition of the nutritionally obese rat. In *The Regulation of Adipose Tissue Mass* (ed. J. Vague and J. Boyer), Excerpta Medica, Amsterdam, pp. 238–253

Sclafani, A. and Springer, D. (1976). Dietary obesity in adult rats: similarities to hypothalamic and human obesity syndromes. *Physiol. Behav.*, 17, 461–471

Stock, M. J. (1975). An automatic, closed-circuit oxygen consumption apparatus for small animals. *J. Appl. Physiol.*, 39, 849–850

10

Lipogenesis and hormone resistance in liver and adipose tissue of genetically obese mice

D. A. Hems (Department of Biochemistry, St George's Hospital Medical School, London, UK)

SUMMARY

Genetically obese rodents provide a good model of human obesity and late-onset diabetes. In the *ob/ob* mouse obesity can develop on fat or carbohydrate diets. On starch-based diets obesity is sustained by enhanced fatty acid synthesis in both liver and adipose tissue. The main precursors for newly synthesised fatty acid are glucose in adipose tissue and glycogen and lactate in liver, all ultimately derived from dietary glucose. Lipogenesis in the liver of normal mice is rapidly inhibited by hormones which act via cyclic-AMP (glucagon) and hormones which do not (vasopressin, angiotensin II). The liver of *ob/ob* mice exhibits resistance to these inhibitory effects, even after pair-feeding to maintain normal weight, whereas glycogenolysis is normally stimulated. Thus, *ob/ob* mice could have a selective defect among cellular membrane response mechanisms which do not involve cyclic-AMP. In contrast, lipogenesis in adipose tissue of pair-fed *ob/ob* mice exhibits a near-normal inhibitory response to adrenaline. Insulin-resistance in tissues of obese (*ob/ob*) mice, demonstrable in liver with respect to the anti-glucagon action of insulin, is likely to be a secondary consequence of obesity. The pathogenesis of obesity could involve a failure of cellular responses to extracellular agents—e.g. of those catabolic regulatory mechanisms which normally restrain insulin action in tissues.

153

INTRODUCTION

The major nutritional problem in the materially developed countries is obesity. This condition is associated with an increased risk of vascular disease, diabetes, hypertension and a variety of less serious ailments.

There are obvious limitations to the investigation of fundamental aspects of obesity in human subjects. Hence, increasing attention is being directed towards obesity in animals. Genetically obese mice and rats provide some of the most clear-cut models of obesity. The most widely studied obese rodent is the homozygote for the autosomal recessive gene *obese* (*ob*). Such mice are hyperglycaemic and hyperinsulinaemic (Genuth, 1969; Abraham *et al.,* 1971)—i.e. there is relative inadequacy of insulin action (Stauffacher and Renold, 1969; Abraham and Beloff-Chain, 1971) and they also exhibit alterations in lipid metabolism (to be elaborated in this article). The characteristics of genetically obese, hyperglycaemic rodents are described in excellent review articles (Bray and York, 1971; Loten *et al.,* 1974a; Herberg and Coleman, 1977).

The aim of this article is to review aspects of lipid metabolism in obese rodents, and present some of our own relevant experiments. Unless otherwise stated, the data presented in this paper refer to obese (*ob/ob*) mice with an undefined but close-bred genetic background, maintained by controlled random mating at the Biochemistry Department, Imperial College, London (see Abraham and Beloff-Chain, 1971; Abraham *et al.,* 1971; Beloff-Chain *et al.,* 1975).

GENERAL ASPECTS OF LIPID METABOLISM IN OBESE MICE

There is a range of alterations in lipid metabolism in genetically obese (*ob/ob*) mice (table 10.1), which may be categorised as a general increase in synthesis and turnover of fatty acids and glycerides, and which particularly involves alterations in liver and adipose tissue.

The changes in lipid metabolism in obese rodents are more intractable to starvation than are changes in carbohydrate metabolism. For example, the capacity for hepatic glycogenolysis is reversibly increased in obese mice (Elliott *et al.,* 1971), whereas their increased rates of hepatic lipogenesis (Hems *et al.,* 1975) or of turnover of plasma triglyceride (Salmon and Hems, 1973) are much less reversible on starvation.

For this reason, and in view of the fact that lipid metabolism is by definition implicated in the pathogenesis of obesity, great interest attaches to the elucidation of the events of lipid biochemistry in obesity (table 10.1).

Fat can be laid down in adipose tissue as a result of uptake of either triglyceride or carbohydrate from blood.

Obesity sustained by a fat diet involves the uptake of plasma (chylomicron) triglyceride by the lipoprotein lipase process, and this enzyme shows increased activity in adipose tissue of obese mice (table 10.1; Rath *et al.,* 1974).

On the other hand, in obesity sustained by a starch-based diet, fatty acid synthesis must obviously contribute to the obesity. This process is considered at length in later sections. However, one general point merits emphasis here. This is

Table 10.1 Lipid metabolism in blood and tissues of genetically obese (*ob/ob*) mice

	Lean mice	Obese mice
Total fatty acid synthesis (μmol per g per h):		
liver, daytime	2	5
adipose tissue, daytime	1	2
liver, night-time	4	12
adipose tissue, night-time	3	3
Turnover in blood (μmol per mouse per min)		
triglyceride-fatty acid	0.25	1.0
glycerol	0.31	1.13
free fatty acid	0.9	1.8
Plasma concentrations (μmol/ml):		
glycerol	0.16	0.36
free fatty acid	0.44	0.53
free fatty acid after noradrenaline	0.82	1.06
triglyceride	0.68	0.82
total cholesterol	2.02	2.64
ketone bodies (total)	0.23	0.20
Lipoprotein lipase activity (U/mg tissue powder):		
skeletal muscle	0.3	0.4
epididymal fat	5	10

Mean values of rates of metabolic processes in intact mice are given. The age or sex of mice was comparable for each pair of measurements. In most cases age was 8–10 weeks. Data, with permission, are from the papers cited in table 10.2, Rath *et al.* (1974) and Abraham *et al.* (1971). All differences (*ob/ob* versus lean) are significant, except the plasma lipid concentrations (see original papers).

the apparent paradox that circulating glucose is not a favoured precursor for lipogenesis in liver (see later section; Salmon *et al.*, 1974; Hems *et al.*, 1975), whereas glycogen and lactate are preferred carbon sources (Salmon *et al.*, 1974). Blood glucose may, however, still convert to fatty acids in liver, via glycogen or recycled lactate produced in muscle, intestine or red cells. Thus, in carbohydrate-sustained obesity, fatty acid synthesis *de novo* (ultimately from dietary glucose) is crucial in the laying down of fat, but liver and adipose tissue are different circulating precursors for this process.

Turnover in plasma of free fatty acids and glycerol (Elliott *et al.*, 1974) and also triglyceride (Salmon and Hems, 1973) is increased in freely fed *ob/ob* mice (table 10.1), and these changes are not significantly reversed by starvation. Similar increases in triglyceride turnover are observed in other obese syndromes—e.g. in obese sand rats (*Psammomys obesus*) (Robertson *et al.*, 1973). These increased rates of turnover are associated in *ob/ob* mice with increased plasma levels of glycerol (Elliott *et al.*, 1974) but not of free fatty acid (Abraham *et al.*, 1971) or triglyceride (Salmon and Hems, 1973). Liver and adipose tissue are both hypertrophied in

obese rodents, so these tissues clearly have the capacity to assimilate plasma free
fatty acid or triglyceride (respectively) at a sufficient rate to keep plasma concentrations normal, despite the increased turnover (table 10.1).

Adipose tissue lipolysis does not appear to be impaired *in vivo* in *ob/ob* mice,
since plasma free fatty acids exhibit increased turnover (Elliott *et al.*, 1974) and a
normal rise in response to noradrenaline or starvation (Abraham *et al.*, 1971). The
same is true of the *fa/fa* rat (Zucker, 1972).

Genetically obese rodents usually exhibit hypercholesterolaemia (e.g. *ob/ob*
mice: table 1, Salmon and Hems, 1973; Enser, 1972). Nevertheless they do not
develop vascular disease. This may be connected with the fact that the plasma cholesterol in mice is mainly located in the high-density (HD or α) lipoprotein class
(Salmon and Hems, 1973), whereas that in human plasma is in low-density (LD or
β) lipoprotein. LD-lipoprotein (LDL) may be more prone to bind to vascular walls
and cause fatty vascular disease.

RELATIVE ROLES OF LIVER AND ADIPOSE TISSUE IN FATTY ACID SYNTHESIS

In animals whose obesity is sustained by a carbohydrate-based diet the process of
synthesis *de novo* of fatty acids is clearly important. There is a marked diurnal
rhythm in the total rate of fatty acid synthesis, in liver or adipose tissue of mice
allowed free access to a starch-based diet (Hems *et al.*, 1975; figure 10.1). Both
lean and obese mice ingest most of their food during the dark period; the increased
synthesis of fatty acid in this period is dependent on the availability of food, and
so may be due at least partly to alterations in the pattern of circulating substrates
and hormones, in response to feeding. In obese mice the diurnal rhythms in lipogenesis in both liver and adipose tissue show the same feeding-dependent pattern
as in lean mice (Hems *et al.*, 1975).

In mice with free access to a carbohydrate diet, fatty acid synthesis, expressed
per whole animal, is about twice as rapid in liver as in free adipose tissue (i.e in the
discrete adipose organ), as described in detail by Hems *et al.* (1975). This conclusion obtains at all times. Recycling of newly synthesised fatty acid between liver
and adipose tissue (or other organs) is not sufficiently rapid to invalidate this conclusion, or $^{14}C/^3H$ ratios in liver and adipose tissue of mice receiving [^{14}C]-glucose
and ^3H_2O (figure 10.1) would resemble each other more closely. Measurements of
^{14}C and ^3H in plasma fatty acid, in similar experiments, also suggest that there is
not extensive release of hepatic newly synthesised fatty acid within periods as
short as 1–2 h.

Previous experiments with ^{14}C-labelled precursors have tended to underestimate the significance of hepatic fatty acid synthesis, perhaps because 'isotope dilution' by other sources of acetyl-CoA (e.g. glycogen) is more extensive than in adipose tissue. For example, experiments with [^{14}C]-glucose do not provide a useful
measure of hepatic fatty acid synthesis, as blood glucose is of minor significance as
a carbon source in mice (Salmon *et al.*, 1974) or rats (Clark *et al.*, 1974).

The occurrence of rapid fatty acid synthesis in the liver need not imply that the
liver has a significant role in the production of lipids stored in adipose tissue, which

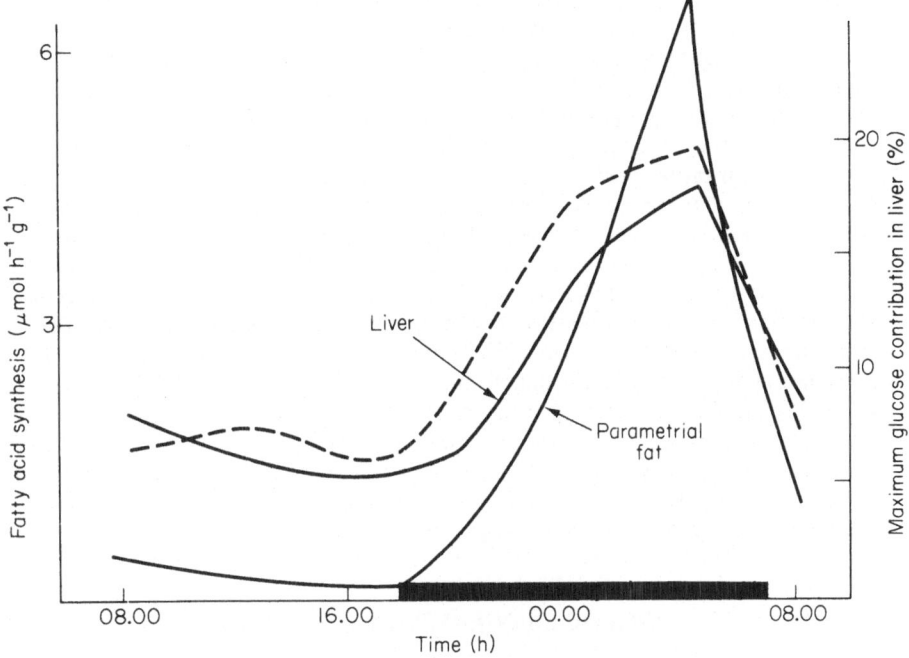

Figure 10.1 Fatty acid synthesis in normal mice. Mice were injected with ^3H_2O and [^{14}C]-glucose at times throughout the 24 h cycle. The rate of fatty acid synthesis (h^{-1}) (———) was calculated from ^3H values after 1 h, and the maximal contribution of glucose (– – –) to lipogenesis in liver from the ratio ($^{14}C/^3H$ in adipose tissue)/($^{14}C/^3H$ in liver). The shaded area indicates the dark periods during which most feeding occurred. Modified from Hems *et al.* (1975).

will depend on the proportion of newly synthesised fatty acid secreted by the liver and taken up by adipose tissue. Definitive results on this point are not available. Nevertheless, the synthesis of fatty acids *de novo* in the liver of obese mice could contribute to adipose-tissue fat storage, since the turnover of plasma triglyceride is enhanced (Salmon and Hems, 1973), and so is the lipoprotein lipase activity of adipose tissue (Rath *et al.*, 1974; Enser, 1972).

In genetically obese (*ob/ob*) mice adipose tissue is the main site of extra lipogenesis, as shown by results with ^3H_2O (Hems *et al.*, 1975), despite the clear enhancement in the liver (compared with lean mice). This applies both in young animals before much fat has accumulated and in older mice, as the rates of fatty acid synthesis (per whole mouse) in adipose tissue (or liver) do not change markedly between the ages of 1 and 3 months, in lean or obese mice (see also Yen *et al.*, 1976). Thus, in genetically obese mice the massive accumulation of fat that occurs during this period takes place on a relatively unchanged total complement of the metabolic apparatus involved in fat synthesis. This explains the dramatic decline, with age, in the rate of lipogenesis per gram of wet adipose tissue in obese mice.

The above considerations suggest that 'insulin-resistance' of adipose tissue in obese mice does not preclude enhancement of fatty acid synthesis.

During the light (feeding) period, the $^{14}C/^3H$ ratio (in experiments such as those described in figure 10.1) in lean mice is lower in parametrial than brown scapular fat, implying that a significant contribution of precursors other than glucose is present in parametrial fat (Hems *et al.*, 1975). In obese mice this is true for brown scapular rather than parametrial fat. These precursors remain to be identified; they could include lactate (Rath *et al.*, 1975) or plasma triglyceride fatty acid synthesised *de novo* in the liver. During the dark period, the $^{14}C/^3H$ ratios in these two regions of adipose tissue becomes equal; thus, glucose can contribute equally to fatty acid synthesis in both tissues, and may be the sole carbon precursor for adipose tissue lipogenesis in the eating phase, in either lean or obese mice.

From experiments *in vivo* with ^3H_2O, approximate 'integrated' total rates of lipogenesis per 24 h can be calculated. In obese mice (aged 1 month) these are about 350 μmol of fatty acid (per mouse) in the liver and 1200 in adipose tissue. These rates are commensurate with the rate of fat deposition, even in younger obese mice, which may be as much as 0.5 g of fat/day–i.e. about 1500 μmol of fatty acid. Thus, the synthesis of fatty acids in adipose tissue, in particular, is fast enough to account for the gain in weight in *ob/ob* mice.

HEPATIC LIPOGENESIS IN OBESE MICE

Liver metabolism is altered in genetically obese (*ob/ob*), hyperglycaemic mice (table 10.2). For example, the organ is hypertrophied, by a factor of about 2. The role of the liver in fat synthesis must be considered. Young obese mice may gain as much as 0.5 g/day of adipose tissue triglyceride on a starch-based diet (see above). The rate of synthesis *de novo* of fatty acid from dietary non-fat precursors such as glucose may exceed 1 g/day, since some fatty acid is degraded by the mouse. Measurements of fatty acid synthesis have shown that some of this extra lipogenesis in genetically obese mice occurs in liver (Loten *et al.*, 1974a, b; Chan and Exton, 1977; Hems *et al.*, 1975; Yen *et al.*, 1976), as is also true of the *fa/fa* rat (Godbole and York, 1978). As would be expected from the role of glycogen in fatty acid synthesis (Salmon *et al.*, 1974), their hepatic glycogen synthesis and breakdown are both enhanced compared with matched lean controls (Abraham *et al.*, 1971; Elliott *et al.*, 1971; Das and Hems, 1974; Chan *et al.*, 1975; Elliott *et al.*, 1974), whereas gluconeogenesis in liver is not significantly altered (Elliott *et al.*, 1971), except that there may be an enhancement from glycerol (Elliott *et al.*, 1974).

The question arises of the nature of the blood-borne or tissue precursors of fatty acids synthesised in liver. Plasma glucose is not a major carbon source for fatty acid synthesised in the liver, as shown by the observation that the $^{14}C/^3H$ ratio (in mice that receive [^{14}C]-glucose and ^3H_2O simultaneously) is consistently higher in adipose tissue than in liver (Hems *et al.*, 1975; figure 10.1). In normal mice during the light period, this ratio is higher in brown scapular fat (than in liver) by a factor of about 19. Thus, the carbon contribution of glucose to hepatic lipogenesis at this time is 5 per cent at most, in general agreement with results in the perfused

Table 10.2   Lipid metabolism in liver of genetically obese (*ob*/*ob*) mice

	Lean mice	Obese mice
Liver composition: (μmol/g)		
neutral glyceride	31	72*
lipid phosphorus	37	39
total cholesterol	11	11
Metabolic processes (μmol min^{-1} g^{-1}):		
ketogenesis from oleate (2 mM)	0.24	0.16*
conversion of free fatty acid to glycerolipid	0.2	1.3*
total fatty acid synthesis (night-time)	4.0	12.0*
conversion of glucose to fatty acid		
(% of total rate) (night-time)	< 20	< 20
Metabolite concentrations (μmol/g):		
acetyl-CoA	0.05	0.05
α-glycerophosphate	0.18	0.13
citrate	0.74	0.45
cytoplasmic $\dfrac{[\text{free NADP}^+]}{[\text{free NADPH}]}$:	0.006	0.013*

Mean values for parameters in livers of fed mice are given. All measurements were in intact mice, except for the rate of ketogenesis, which was in perfused liver. Data, with permission, are from Elliott *et al.* (1974), Salmon and Hems (1973) and Hems *et al.* (1975).

*$P < 0.05$ (vs. lean mouse value).

mouse liver (Salmon *et al.*, 1974). In *ob*/*ob* mice the corresponding value is about 8 per cent. During the daylight period, glycogen is likely to provide significant carbon for hepatic fatty acid synthesis (Salmon *et al.*, 1974), rather than precursors such as lactate or alanine, which do not appear to be rapidly converted into hepatic fatty acids in mice during daylight hours (Elliott *et al.*, 1974). In this 'post-absorptive' situation the release of glucose by the liver could occur in association with lipogenesis. In *ob*/*ob* mice both these processes are enhanced (Elliott *et al.*, 1971; Hems *et al.*, 1975).

Despite the relative lack of importance of plasma glucose as a carbon source of fatty acids synthesised in liver, some blood glucose may enter the liver for fat synthesis, especially during periods of net glycogen synthesis (e.g. in obese mice feeding at night on a starch-based diet). Experiments with simultaneous [^{14}C]-glucose and ^{3}H$_2$O permit calculation of maximal rates for this process. Thus, during the period of enhanced lipogenesis (20:00–04:00 h), the ^{14}C/^{3}H ratio in adipose tissue exceeded that in liver by a factor of only 9 in lean mice and 5 in obese mice. The approximate rate of fatty acid synthesis from blood-borne glucose in the liver can be taken to be (at most) 20 per cent of the total rate (see previous section), i.e. about 70 μmol of fatty acid per mouse per day for the whole liver, in *ob*/*ob* mice.

During feeding and glycogen deposition in obese mice, the major plasma source of carbon for hepatic lipogenesis could consist of substrates such as lactate, which can easily generate pyruvate or acetyl residues. Lactate itself can fulfil this role in isolated liver preparations, as well as markedly enhancing fatty acid synthesis (Salmon *et al.*, 1974; Clark *et al.*, 1974). *In vivo* such lactate would presumably be derived from the degradation of glucose by muscle or intestine.

It is not clear why the liver should carry out lipogenesis from blood-borne lactate rather than glucose, since, of course, the original source of all plasma lactate is dietary or endogenous glucose. At all events, there seems to be no qualitative alteration in this cycle of events in obese animals, although the rate of hepatic lipogenesis is enhanced.

In the liver of normal mice fatty acid synthesis is accelerated by either glucose or lactate, independently of their role as precursors (Salmon *et al.*, 1974). In the perfused liver of genetically obese (*ob/ob*) mice these regulatory responses were less marked (table 10.3), perhaps because the basal rate of synthesis of fatty acids was faster than that in control lean mice. This suggests that glycogen has assumed the role of predominant precursor of fatty acid in obese mouse liver, in accord with the fact that in obese mice circulating glucose provides an even smaller proportion of synthesised fatty acids than in lean mice (table 10.3; Salmon and Hems, 1974). Similar observations have been made in hepatocytes from obese rats (Bloxham *et al.*, 1977; Katz *et al.*, 1976).

Table 10.3 Fatty acid synthesis in perfused liver of genetically obese (*ob/ob*) mice

Substrates (mM)	No. of mice	Newly synthesised fatty acid	
		Total rate (μmol C_2-units/g)	% from ^{14}C-substrate
[14C]-Glucose (5)	3	68 ± 12	6 ± 1
[14C]-Glucose (11)	4	76 ± 16	15 ± 2
[14C]-Glucose (25)	4	124 ± 17	27 ± 3
Glucose (15), [14C]-lactate (11)	5	81 ± 14	32 ± 5

Livers of fed female *ob/ob* mice were perfused, and fatty acid synthesis was measured over 3 h, by the incorporation of ^3H from ^3H_2O, and of ^{14}C from [U^{14}C]-substrates added to perfusate at zero time, into total liver fatty acids (Salmon *et al.*, 1974). Results are means ± S.E.M.

The liver is a major user of circulating free fatty acids. In genetically obese (*ob/ob*) mice (table 10.2) the conversion of free fatty acids to glycerolipid in liver is enhanced (Salmon and Hems, 1973), whereas the conversion to ketone bodies is diminished (Elliott *et al.*, 1974). This is explicable by an alteration in the disposition of fatty acyl-CoA within the liver.

Hence, the major route for hepatic triglyceride or phospholipid synthesis in obesity, even in starch-sustained obesity, is apparently via uptake of preformed fatty acids (non-esterified) in plasma. Yet the excessive turnover in these fatty acid pools in liver must ultimately depend on the enhanced rate of net synthesis *de novo* of fatty acid, which maintains the content of fatty acid in the pools.

HORMONAL CONTROL OF HEPATIC FATTY ACID SYNTHESIS
IN OBESE MICE

Increased hepatic fatty acid synthesis in obese rodents may be demonstrated *in vitro*—e.g. in mice (Salmon and Hems, 1974; Assimacopoulos-Jeannet *et al.*, 1974) or rats (in hepatocyte suspensions: Katz *et al.*, 1976; Bloxham *et al.*, 1977). This allows the possibility of investigation of the control of lipogenesis by hormones which act directly on liver.

If we first consider the role of insulin, it may be shown from experiments with intact *ob/ob* mice that hyperinsulinaemia is implicated in their excessive hepatic lipogenesis (Assimacopoulos-Jeannet *et al.*, 1974; Loten *et al.*, 1974a, b; Winand *et al.*, 1968). Yet insulin stimulation of hepatic fatty acid synthesis *in vitro* is notoriously hard to reproduce (in contrast to the situation in adipose tissue). So far, it is not possible to say whether an enhanced direct action of insulin on liver (Winand *et al.*, 1968, 1973; Loten *et al.*, 1974a, b) is responsible for the increased lipogenesis of genetically obese rodents. Nevertheless, it is a reasonable presumption that insulin action is direct on liver—e.g. to bring about extra induction of regulatory enzymes of lipogenesis. Thyroid hormones may also be implicated in this type of control, and obese mice show resistance to this effect (Volpe and Marasa, 1975).

Adrenal corticosteroids can also promote hepatic fatty acid synthesis in intact animals, perhaps via increases in plasma insulin concentration (Kirk *et al.*, 1976). Obese (*ob/ob*) mice exhibit enhanced adrenocortical function (Herberg and Kley, 1975), so this factor could be relevant (albeit indirectly) to their enhanced hepatic lipogenesis.

Catabolic effects of hormones on hepatic fatty acid synthesis have been easier to document experimentally. There is a group of hormones able to exert short-term catabolic effects on liver, which are not mediated by purine nucleoside cyclic monophosphates (for review, see Hems, 1977). This group of hormones comprises α-adrenergic agonists, vasopressin, oxytocin and angiotensin II (so far).

One hormone which can rapidly inhibit hepatic fatty acid synthesis, at least in the mouse, is vasopressin (Ma and Hems, 1975; table 10.4). In perfused liver of genetically obese (*ob/ob*) mice vasopressin does not inhibit fatty acid synthesis, i.e. there is 'vasopressin-resistance' in respect of this process (Hems and Ma, 1976; table 10.4). Glycogen breakdown, on the other hand, occurs in response to vasopressin in the liver of *ob/ob* mice (table 10.4; Hems and Ma, 1976), as it does in the normal liver of the mouse (Ma and Hems, 1975) or rat (Hems and Whitton, 1973; Hems *et al.*, 1978b).

Further hormones which can inhibit fatty acid synthesis in the mouse liver include adrenaline and angiotensin II (Hems, 1977; Ma *et al.*, 1977). Again, the liver of the genetically obese (*ob/ob*) mouse does not respond to these hormones with a decrease in fatty acid synthesis (table 10.5).

The failure of response to vasopressin, angiotensin II or adrenaline, in respect of the process of lipogenesis in the liver of *ob/ob* mice, is not reversed by severe food deprivation (tables 10.4 and 10.5), despite the maintenance of normal weight in these *ob/ob* mice.

Table 10.4 Responses to vasopressin in perfused liver of genetically obese
(ob/ob) mice

Mice	Hormone	No. of mice	Fatty acid synthesis (μmol C_2-units per g in 2 h)	Glucose output (μmol $\times$ g $\times$ min^{-1})
Lean	none	4	79	0.6
Lean	vasopressin (10^{-8} M)	6	37*	4.0*
Obese	none	7	46	0.6
Obese	vasopressin (10^{-8} M)	3	40	3.6*

Livers of fed female mice were perfused with glucose (15 mM) and lactate (10 mM).
Perfusate containing ^3H_2O was recycled for 2 h. Obese mice were diet-restricted,
receiving 4 g food/day; they weighed 24–30 g–i.e. the same as lean mice. Hormones
were present from 45 min, being added every 15 min. Data, with permission, are
from Hems and Ma (1976).
*$P < 0.05$ (vs. hormone-free value).

Table 10.5 Effect of adrenaline and angiotensin II in perfused liver of genetically
obese (ob/ob) mice

Mice	Hormone	No. of mice	Fatty acid synthesis (μmol C_2-units per g/h)	Glucose output (μmol g^{-1} min^{-1})
Lean	none	10	22 ± 2	0.9 ± 0.2
Lean	adrenaline (10^{-7} M)	5	12 ± 2*	8.6 ± 0.7*
Lean	angiotensin II (10^{-8} M)	4	8 ± 1*	3.1 ± 0.2*
Obese	none	4	25 ± 4	0.6 ± 0.3
Obese	adrenaline (10^{-7} M)	4	21 ± 3	5.2 ± 0.4*
Obese	angiotensin II (10^{-8} M)	4	24 ± 3	3.2 ± 0.5*

Livers of fed female mice were perfused (Ma and Hems, 1975; Hems and Ma, 1976)
with glucose (15 mM) and lactate (10 mM). Perfusate containing ^3H_2O and hormones was passed through liver in a non-recirculating system for 40–50 min. Lean
mice were +/+. Obese mice were diet-restricted ob/ob (see table 10.4). Results are
unpublished data (G. Y. Ma, M. Cawthorne and D. A. Hems), and are means of the
number of perfusions indicated.
Results are means ± S.E.M.
*$P < 0.05$; cf. with hormone-free value.

Resistance to catabolic-acting effectors could be relevant in genetically obese rodents to the consequences of their hyperinsulinaemia (which itself could be directly implicated in the development of the obesity: Assimacopoulos-Jeannet *et al.*, 1974; Loten *et al.*, 1974a, b). There could be loss of catabolic regulatory effects which (in normal mice) override insulin stimulation of tissues. Therefore, the influence of insulin on the effects of vasopressin on glycogenolysis and fatty acid synthesis in perfused livers of mice was investigated. Insulin did not counter-act the catabolic effects of vasopressin on these processes in normal mice (table 10.6).

Glucagon can also inhibit fatty acid synthesis (table 10.7; Ma *et al.*, 1978), but there may be a difference between this inhibition and that exerted by other hormones. The inhibitory effect of glucagon in the perfused liver is not exerted at concentrations which occur in plasma (i.e. is less potent than that on glycogen breakdown), where-as there is inhibition in preparations such as slices where functional viability is not as good, when glucagon is tested in the same conditions as in perfused liver experiments (Raskin *et al.*, 1974). The inhibitory action of glucagon on fatty acid synthesis may be secondary in time to the primary metabolic effects of the hor-mone on carbohydrate metabolism (Ma *et al.*, 1978). In particular, glucagon depletes the liver of glycogen, a favoured source of acetyl units for fatty acid synthesis (Salmon *et al.*, 1974).

In the perfused livers of genetically obese (*ob/ob*) mice glucagon is less effec-tive in inhibiting fatty acid synthesis (table 10.7). Again, glycogenolysis responded normally to the hormone. This finding constitutes yet another case of hormone-resistance, shown by the process of fatty acid synthesis in obese mouse liver (Ma *et al.*, 1978).

Glucagon also inhibits cholesterol synthesis in perfused liver of normal mice, and the *ob/ob* mouse liver shows resistance to this effect (table 10.7). The relationship between the regulation of cholesterol and fatty acid synthesis in liver is obscure,

Table 10.6   Effect of insulin on vasopressin action on glycogen content and fatty acid synthesis in the perfused liver

Treatment	No. of perfusions	Fatty acid synthesised (μmol C_2-units per 2 h per g)		Final glycogen content (μmol glucose/g of fresh liver)
		total	from lactate	
Control	4	38 ± 4	14 ± 2	177 ± 25
Vasopressin	3	18 ± 4	7 ± 2	50 ± 12
Vasopressin plus insulin	3	23 ± 5	6 ± 1	45 ± 19

Livers were perfused for 3 h with recirculating medium containing ^3H_2O, [U-^{14}C]-lactate (initially about 12 mM) and glucose (initially about 15 mM). Radioisotope-labelled water and lactate were added after 60 min. Vasopressin (4 mU/ml) was added after 45 min. In one group insulin (300 mU) was added from the start of the perfusion, and subsequently every 60 min until the end of the perfusion. Livers were analysed after 3 h, and results are means $\pm$ S.E.M. of the numbers of perfusions indicated (Ma *et al.*, 1978). Values in the groups with vasopressin are not significantly different from each other. but are different from control values.

Table 10.7   Responses to glucagon in perfused liver of genetically obese
             (ob/ob) mice

| Mice | Addition | No. of mice | Lipid synthesis (μmol C_2-units per g per 2h) | | Glucose output (μmol g^{-1} min^{-1}) |
			Fatty acid	Cholesterol	
Lean	none	5	44.4 ± 3.7	2.3 ± 0.4	0.44 ± 0.10
Lean	glucagon	5	9.2 ± 2.0*	0.6 ± 0.1*	5.01 ± 0.65*
Obese	none	4	49.7 ± 4.1	2.2 ± 0.4	0.59 ± 0.13
Obese	glucagon	4	36.8 ± 5.6	2.2 ± 0.3	4.78 ± 0.57*

Livers were perfused as in table 10.4. Glucagon, when present, was added to a
concentration of 10^{-9} M, every 15 min during 2 h. Obese mice were fed 4 g/day
(i.e. diet-restricted).
Results are means ± S.E.M. (Ma *et al.*, 1978).
*$P < 0.05$, vs. hormone-free value.

and the role in obesity of impaired control of cholesterol synthesis cannot yet be
assessed.

If glucagon is relevant to the control of hepatic fatty acid synthesis *in vivo*
(albeit in a secondary manner), then the fact that the content of cyclic-AMP in
liver of obese rodents does not increase on starvation (Lavine *et al.*, 1975) could
be relevant to the slower adaptive response of hepatic lipogenesis on starvation
(Hems *et al.*, 1975).

ARE THERE ABNORMALITIES OF INTRACELLULAR (NON-HORMONAL) CONTROL OF FATTY ACID SYNTHESIS IN LIVER OF *ob/ob* MICE?

Many factors combine to regulate the rate of hepatic fatty acid synthesis. Control
by hormones and substrates has been covered in preceding sections.

Another type of control involves the intracellular ('feedback') devices, whereby
the availability of acetyl residues for lipogenesis is geared to the ATP status of the
cell (e.g. in regard to consumption by energy-consuming processes, or the state
of oxygenation of the liver). For example, α-glycerophosphate, citrate and acetyl-
CoA have been proposed as key regulatory metabolites. Their role may be evaluated from measurements of the hepatic content of metabolites. Such measurements do not suggest that the increased lipogenesis in obese mice is due to increased
hepatic content of α-glycerophosphate, acetyl-CoA or citrate (table 10.2).

When there is increased hepatic fatty acid synthesis in liver, there may be a decreased ratio of [free NADPH]/[free $NADP^+$] in cytoplasm (Veech *et al.*, 1969),
showing that the cytoplasmic reducing equivalents are 'pulled' into the fatty acid
product, rather than 'pushing' the increased synthesis. This general finding applies
in the liver of genetically obese *(ob/ob)* mice (table 10.2; Elliott *et al.*, 1974).

Fatty acid synthesis would be expected to be inhibited by anoxia. The key
step in lipogenesis is that catalysed by acetyl-CoA carboxylase, which shows an

inhibitory response in ischaemic liver, i.e. the proportion of active ('initial') enzyme decreases (table 10.8). This response must involve depolymerisation (or phosphorylation perhaps) of the active enzyme; details are not yet to hand. In the ischaemic liver of *ob/ob* mice the inactivation of the enzyme appeared to develop at the same rate as in lean mice (table 10.8).

Table 10.8   Acetyl-CoA carboxylase activity in aerobic and ischaemic livers of genetically obese (*ob/ob*) mice

Mouse	Time of ischaemia (min)	No. of observations	Acetyl-CoA carboxylase activity (μmol min^{-1} g^{-1} wet tissue)		
			Initial	Total	% Initial
Lean	0.2	11	4.1 ± 0.4	7.4 ± 0.6	55
Lean	0.5	5	1.8 ± 0.3	6.7 ± 1.0	27
Lean	1.0	3	1.2 ± 0.2	5.6 ± 0.9	21
Lean	2.0	3	2.0 ± 0.3	8.4 ± 0.8	24
Obese	0.2	12	7.5 ± 0.6	12.5 ± 0.9	60
Obese	0.5	3	5.1 ± 0.8	13.2 ± 2.4	39
Obese	1.0	3	4.3 ± 0.5	15.8 ± 1.6	27
Obese	2.0	3	4.4 ± 0.4	14.8 ± 1.9	30

Acetyl-CoA carboxylase activity was assayed in livers from intact (unanaesthetised) female mice aged 10 weeks. One piece of liver was rapidly frozen, while the remainder was cut into pieces which became ischaemic, for various periods following cessation of blood flow (Hems and Brosnan, 1970). The enzyme was assayed (Gove and Hems, 1978) in a supernatant fraction, either directly (initial activity) or after treatment with citrate (total activity).
Results, means ± S.E.M., are unpublished data (C. D. Gove and D. A. Hems).

Taken together, the above considerations suggest that intracellular factors (i.e. metabolites and inorganic solutes) which control fatty acid synthesis are able to exert their normal influence in the liver of *ob/ob* mice.

HORMONAL CONTROL OF ADIPOSE TISSUE METABOLISM

Obesity, by definition, is a condition of adipose tissue; therefore, a key question is whether in the genetic obese syndromes there is a primary alteration in the biochemistry of this tissue.

There is not strong evidence, so far, for a primary inborn lesion in adipose tissue in any genetically obese rodent. Many alterations have been elucidated in adipose tissue metabolism of mature freely fed obese animals, but they are usually less marked in young or starved animals, which suggests that the changes concerned are secondary consequences of florid obesity.

There are major difficulties in studying adipose tissue of obese rodents *in vitro*. One is that viable adipocytes cannot be isolated from the entire spectrum of sizes of cell in the tissue from the obese animal (Jamdar, 1978). Therefore, experiments are best carried out with pieces of tissue ('pads', 'slices'), although they do not allow normal diffusion of hormones and substrates to and from cells. Also, the inert mass of the abnormal tissue could act as a store for hormones, which would require to be leached out before a true picture could be obtained about the metabolic capacities of the adipose tissue. Finally, although the large abnormal cells in the adipose tissue of genetically obese rodents may not all be viable in isolated preparations, they still contribute inert mass to the total tissue. Hence, the problem of assigning a reference parameter to data (e.g. per mass of DNA or protein, or per cell, etc.), which is not easy to resolve for normal tissue, is greatly compounded when working with tissue from obese animals.

These difficulties are relevant, for example, to evidence bearing on the action of insulin in adipose tissue of obese mice. The hyperinsulinaemia of obese rodents would be expected to be relevant to their obesity, in view of the well-known action of insulin to promote lipogenesis in adipose tissue. Experiments with streptozotocin in intact mice suggest that this is the case (Loten *et al.*, 1974a, b). Experiments *in vitro*, however, suggest that the adipose tissue of obese rodents is resistant to stimulation by insulin of lipogenesis (Loten *et al.*, 1976) or of glucose oxidation (Abraham and Beloff-Chain, 1971). Part at least of this resistance may be artefactual, and due merely to sequestered insulin, which can be leached out before incubation (Loten *et al.*, 1976).

Longer-term insulin action on adipose tissue in obese states probably contributes to the excess capacity for lipogenesis *de novo* (Loten *et al.*, 1974a, b; 1976) or from plasma triglyceride (Rath *et al.*, 1974).

Resistance to catabolic-acting control factors (such as hormones) can occur in genetically obese mouse tissues, and was described for liver in a previous section. If such resistance occurred in adipose tissue, this could lead to a predisposition to obesity.

Catecholamines are major catabolic regulators of adipose tissue metabolism, which can stimulate fat degradation ('lipolysis'), in particular. In mature obese animals there may be resistance to this effect (Enser, 1970; Yen and Steinmetz, 1972; Otto *et al.*, 1976). However, contrary results have been reported (Shepherd *et al.*, 1977); furthermore, this resistance is significantly reversed by starvation, and is not manifest in intact *ob/ob* mice (Abraham *et al.*, 1971). The question of catecholamine action on adipose tissue lipolysis in obesity is not yet resolved.

Adrenaline can inhibit fatty acid synthesis in adipose tissue, so that it is relevant to ask whether this control is operating normally in mice whose obesity is sustained by a starch-based diet—i.e. where fatty acid synthesis *de novo* is relevant to the fat deposition.

Although adipose tissue pieces from freely fed *ob/ob* mice show resistance to inhibition of fatty acid synthesis by adrenaline, this defect is significantly ameliorated in diet-restricted mice (figure 10.2). This suggests that such catecholamine-resistance is not of primary significance in the development of obesity, and contrasts with the resistance to vasopressin action shown by liver, which was not reversed by starvation.

There is an impaired response of adipose tissue adenylate cyclase to adrenaline in *ob/ob* mice (Laudat and Pairault, 1975). This impairment is manifest as a decrease in extent of response, but not in the concentration-dependence of the hormone effects. The authors inferred that there could be a defect in the coupling between the hormone receptor and the cyclase system. This is possible, as the study involved a subcellular 'membrane' fraction (although a simpler explanation could be merely that the cyclase appeared less 'active' in absolute terms, as a result of inert tissue decreasing the 'slope' of the response curve). A similar conclusion was drawn by Shepherd *et al.* (1977) from a study of cyclic-AMP responses to noradrenaline in intact *ob/ob* mouse adipocytes.

Vasopressin has been reported to exert catabolic effects on the adipose tissue of some species. Hence, this hormone was tested on adipose tissue pieces from mice. No effect of the hormone on fatty acid synthesis *de novo* was detected, in lean or genetically obese (*ob/ob*) mice (table 10.9).

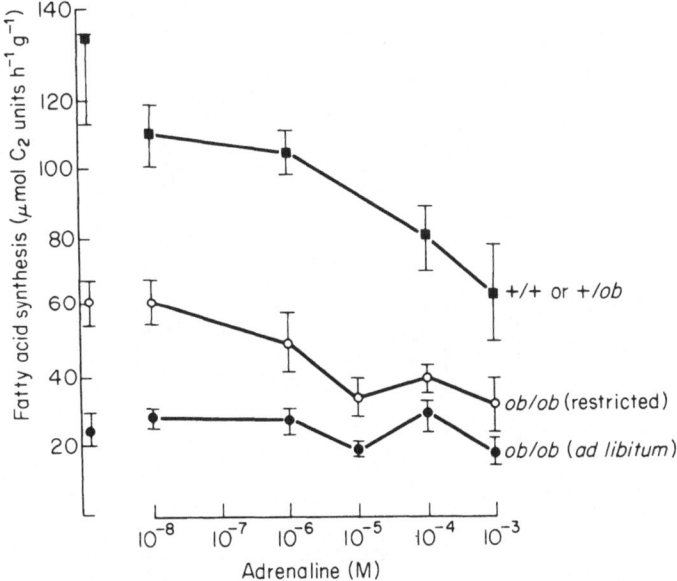

Figure 10.2 Reversible adrenaline-resistance in adipose tissue of genetically obese (*ob/ob*) mice. Pieces (about 30 mg wet weight) of mesenteric adipose tissue from female mice were incubated in 2 ml of Krebs–Ringer bicarbonate saline (Rath, 1977). Fatty acid synthesis was measured with ^3H_2O (Jungas, 1968) over 1 h, in the presence of adrenaline at various initial concentrations. Tissue was obtained from lean mice (■), *ob/ob* mice fed *ad libitum* (●), or diet-restricted (*ob/ob*) mice (○). Results are means ± S.E.M. of 3–6 measurements. Modified (with permission) from Rath (1977).

Table 10.9 Effects of hormones on adipose tissue lipogenesis

Region of tissue	Hormones (M, in addition to insulin)	Total fatty acid synthesis (μmol C_2-units h^{-1} g^{-1})		Fatty acid synthesis (% from glucose)	
		Lean	Obese	Lean	Obese
Parametrial	none	51 ± 9 (9)	30 ± 4 (9)	83 ± 12 (9)	70 ± 10 (9)
Mesenteric	none	102 ± 7 (9)	38 ± 5 (9)	93 ± 13 (9)	75 ± 9 (9)
Parametrial	vasopressin (5×10^{-9})	52 ± 9 (9)	28 ± 4 (9)	79 ± 9 (9)	63 ± 6 (9)
Mesenteric	vasopressin (5×10^{-9})	91 ± 11 (9)	38 ± 9 (9)	75 ± 9 (9)	65 ± 10 (9)

Pieces of adipose tissue from lean or genetically obese mice aged 8 weeks were incubated in Krebs–Ringer bicarbonate saline at 37° for 1 h in the presence of [^{14}C]-glucose (15 mM) and insulin (25 mU/ml). Total fatty acid synthesis was measured with ^3H_2O (Rath, 1977). No vasopressin effects are statistically significant.

INSULIN-RESISTANCE IN OBESE RODENTS

Obese rodents (or people) are almost invariably hyperinsulinaemic and hyper-glycaemic, at least when allowed free access to food. This combination of circum-stances implies that 'insulin-resistance' (in respect of processes of glucose assimila-tion at least) exists in the major insulin-responsive tissues—i.e. muscle and adipose tissue (see Czech *et al.*, 1977, for review).

Insulin-resistance may be demonstrated *in vivo* (Stauffacher and Renold, 1969) or *in vitro*, in incubations of either adipose tissue or muscle (Abraham and Beloff-Chain, 1971). The resistance is partially reversed by starvation (Abraham and Beloff-Chain, 1971), which leads to the general inference that such resistance is secondary to the events of obesity, rather than of primary pathogenic significance.

'Insulin-resistance' cannot be appraised unless one has identified a clear-cut direct action of insulin on normal tissue. The best-documented such effect in liver is counteraction by insulin of glucagon-induced glycogenolysis. Therefore, this insulin effect has been investigated in the perfused liver of *ob/ob* mice (figure 10.3). The liver of normal mice showed the expected antagonism by insulin to the glyco-genolytic effect of glucagon. Freely fed obese mice did not show this insulin effect. However, when *ob/ob* mice were deprived of food so that their weight re-mained steady at approximately that of lean mice, the anti-glucagon effect of in-sulin was present (figure 10.3). Therefore, this hepatic insulin-resistance cannot be of primary significance in the *ob/ob* syndrome. Also, this reversibility of the alter-ation in control of glycogenolysis in *ob/ob* mice is in keeping with the general tendency of carbohydrate metabolism to exhibit reversible changes in obesity, whereas changes in lipid metabolism are more intractable to starvation.

Insulin acts on tissues by first binding to a receptor in the plasma membrane. Insulin-resistance can sometimes reflect alterations in their number of properties (for review, see Czech *et al.*, 1977). There are fewer insulin receptors in the liver of genetically obese rodents, compared with the number in matched lean animals (Soll *et al.*, 1975a, b; Baxter and Lazarus, 1975). These receptors are functionally normal (Soll *et al.*, 1975b), and the diminution in the number of insulin receptors in liver of obese mice is reversible by food privation (Soll *et al.*, 1975a; Le Marchand *et al.*, 1977).

In summary, several points may be made about insulin-resistance in obese rodents:

(1) Insulin-resistance is difficult to document quantitatively in the strict sense—i.e. in terms of a relative or absolute failure of insulin action *in vitro*—in the light of difficulties which include the possible sequestration of insulin from plasma in tissues isolated from freely fed obese animals. This is especially true for adipose tissue.

(2) Insulin-resistance (in respect of anabolic effects of the hormone) tends to be reversible (e.g. by starvation), and less marked in young obese animals.

(3) Insulin-resistance in muscle, adipose tissue or liver, to anabolic effects of the hormone, is likely on general grounds to be a secondary consequence of obesity, as primary insulin resistance would not be expected to cause a predispo-sition to obesity or to excessive fatty acid synthesis in liver.

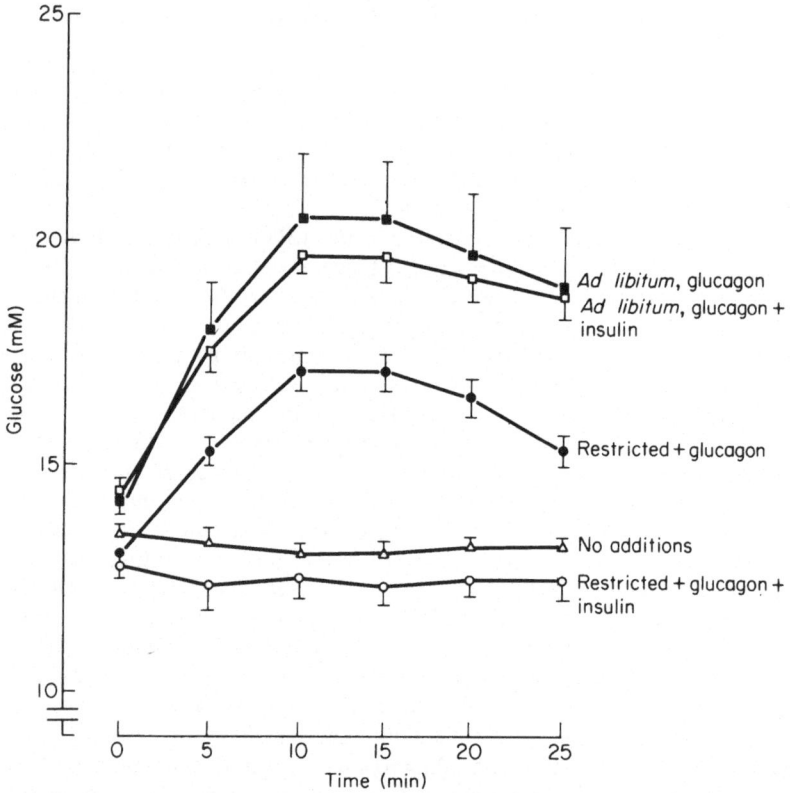

Figure 10.3 Reversible insulin-resistance in liver of genetically obese mice. Livers of fed female *ob/ob* mice were perfused with recirculating perfusate (Ma and Hems, 1975) containing 10 mM glucose, for 30 min. Then the perfusion was changed to non-recirculating, hormones were added and glucose was measured in effluent perfusate, over the next 30 min. When present, glucagon was 10^{-10} M, and insulin 10^{-8} M. Perfusions contained no additions (Δ), glucagon ($\blacksquare$, $\bullet$), or glucagon plus insulin ($\square$, Δ), in obese mice fed *ad libitum* ($\blacksquare$, $\square$, Δ) or obese mice on restricted diet, i.e. 4 g/day ($\bullet$, $\circ$). Results are means ± S.E.M. (bars) of three perfusions. From Ma *et al.*, 1978.

GENERAL SIGNIFICANCE OF RESISTANCE TO CATABOLIC-ACTING HORMONES IN OBESITY

The failure of vasopressin, angiotensin II, adrenaline and glucagon to inhibit fatty acid synthesis in the liver of *ob/ob* mice constitutes the first evidence in genetically obese mice of a failure of the process of fatty acid synthesis to respond normally to inhibitory hormones. This impairment is likely to reflect closely the inborn error in these animals, for three reasons: (1) it was not reversible by relatively severe food privation; (2) a lesion in the inhibitory control of fatty acid synthesis would be of crucial pathogenic importance, as obesity is, by definition, a state in which there is excess deposition of fat; (3) the increase in fatty acid synthesis is intractable to starvation in intact obese mice (Hems *et al.*, 1975), as are other processes of lipid metabolism (see earlier section). Such irreversibility suggests that the lesion in *ob/ob* mice could reside in the control properties of lipid metabolism.

The likely importance of impairments in the response mechanisms to extra-cellular stimuli is highlighted by the fact that the effect of intracellular anoxia on hepatic acetyl-CoA carboxylase was not markedly defective in liver from *ob/ob* mice.

Therefore, the genetic obesity in the *ob/ob* mouse could be a consequence of a defective response in cells, whereby extracellular effectors fail to exert normal constraints—on fatty acid synthesis, in particular. Since cholesterol synthesis is also not inhibited by vasopressin (as it is from lactate, in particular, in normal mouse liver: Ma and Hems, 1975), it is possible that the lesion in obese mice tissues is located within the general mechanisms of response of lipid metabolism to effectors, rather than being specific to the pathways of fatty acid synthesis. The observed defect does not involve the adenylate cyclase system, as vasopressin and angiotensin II do not act through this system in the liver (Hems *et al.*, 1978a).

It must be emphasised that the 'resistance' to these specific hormone effects in liver is here being interpreted as a clue to a regulatory derangement in cells. It is not likely that the lack of hepatic response to the catabolic hormones *specifically* leads to obesity.

The above inference, that genetic obesity could reflect an impaired response in cells, whereby extracellular effectors fail to exert normal effects on cells (including catabolic effects on lipogenesis), is reminiscent of a proposal by Chang *et al.* (1975) based on experiments with lectins, that *ob/ob* mice exhibit a generalised receptor defect in cells, which does not include impairments of the adenylate cyclase system or of responses to $3',5'$-cyclic-AMP. However, the defect revealed by our studies cannot be a generalised receptor failure (even of receptor systems other than adenylate cyclase), as glucose output is normally stimulated by vaso-pressin, in diet-restricted obese mice, at concentrations that act on the liver in normal mice (Ma and Hems, 1975). Also, the defect is not total, as very high con-centrations of vasopressin inhibit lipogenesis, even in obese mice fed *ad libitum* (Hems and Ma, 1976). Since the mechanism of vasopressin action on liver lipid metabolism has not been elucidated, the nature of the defect in obese mice cannot be described in specific terms.

The particular interest of the vasopressin-resistance in the liver of obese mice arises from its persistence in diet-restricted animals. In the experiments described, these mice were fed as a routine at 09.00 h, whereupon they rapidly consumed their food. Thus their regimen was of a meal-feeding (as opposed to nibbling) type. They were severely deprived of food (they received 4 g/day), so that they gained no weight (above 30–35 g) over several weeks. The rate of fatty acid synthesis in their livers was lower than that in lean mice, which suggests that insulin action on their liver was not excessive. The return of the full response of glycogen breakdown to vasopressin, after diet restriction, could also carry this implication.

Nevertheless, even intractable vasopressin resistance could perhaps be second-ary to an aspect of the meal-fed state, or to excessive insulin action on the liver.

Such difficulties have dogged attempts to elucidate the inborn error in genetically obese rodents.

Resistance to catabolic-acting hormones may be viewed in the light of the hyperinsulinaemia of obese rodents (Genuth, 1969), which is likely to be implicated in their tendency to store excess fat (Winand *et al.*, 1968; Loten *et al.*, 1974a, b; Assimacopoulos-Jeannet *et al.*, 1974). One may ask: Why do catabolic-acting agents not counteract this continuous insulin stimulation of muscle or adipose tissue? Since the adenyl cyclase system sometimes responds normally to hormones in tissues from obese rodents (at least if they are young or fasted), then the key question may be: Are there catabolic (or related) effects in cells which do not involve purine nucleoside cyclic monophosphates, and which are defective in tissues from obese rodents?

Given this approach, it is relevant that insulin does not prevent the rapid action of vasopressin in the perfused liver of normal mice. Thus, resistance to vasopressin action on the liver could be a direct manifestation of the genetic lesion in *ob/ob* mice—i.e. the lesion could reside within mechanisms (other than adenylate cyclase) that mediate responses to extracellular effectors, and normally offset insulin action. In the case of vasopressin action on the liver, this mechanism may involve Ca^{2+} ions, at least in respect of its glycogenolytic effect (Hems *et al.*, 1978b). However, it would be speculative to state at the present stage that the lesion in obese rodents involves cation permease or pump systems.

A defect of inhibitory or other control responses in cells in genetically obese rodents could explain many of the manifestations of these syndromes (reviewed by Bray and York, 1971; Loten *et al.*, 1974a). In particular, a failure of catecholamine action could contribute to their altered adipose tissue metabolism (Enser, 1970; Laudat and Pairault, 1975), although the evidence for a primary pathogenic role here is not yet overwhelming (see earlier section). Also, failure of adrenaline action on the pancreas could lead to excessive insulin secretion, since catecholamines can inhibit insulin secretion.

It is widely felt that the inborn defect in genetically obese rodents resides at least partly in the hypothalamus. The present work bears on this aspect in two ways. Firstly, the intractability of the defect in the hepatic response to vasopressin (on starvation) suggests that the inborn lesion in the *ob/ob* mice is manifested in non-nervous tissues, such as liver and adipose tissue. Secondly, if the present finding is taken to be representative of the inborn error in *ob/ob* mice (as we believe to be warranted), then it follows that their impaired nervous tissue function will be located in post-synaptic functions (i.e. in responses to transmitters). This remains to be tested, as do many aspects of any proposed pathogenic significance of 'hormone-resistance' in obesity.

It remains probable, however, that insights gained into the genetic obese syndromes in rodents will provide useful clues about the nature of obesity in people.

ACKNOWLEDGEMENTS

I thank Sir Ernst Chain and Dr Beloff-Chain for encouragement with our experiments, which were carried out at Imperial College, London. Also, I am grateful to

Drs E. Rath, M. Salmon and G. Ma, and Mr C. Gove, who carried out some of the experiments, and kindly gave permission, where relevant, for unpublished data to be presented.

REFERENCES

Abraham, R. R. and Beloff-Chain, A. (1971). Hormonal control of intermediary metabolism in obese hyperglycaemic mice. I. Sensitivity and response to insulin in adipose tissue and muscle *in vitro*. *Diabetes,* **20,** 522–534

Abraham, R. R., Dade, E., Elliott, J. and Hems, D. A. (1971). Hormonal control of intermediary metabolism in obese hyperglycaemic mice. II. Levels of plasma free fatty acid and immunoreactive insulin and liver glycogen. *Diabetes,* **20,** 535–541

Assimacopoulos-Jeannet, F., Singh, A., le Marchand, Y., Loten, E. G. and Jeanrenaud, B. (1974). Abnormalities in lipogenesis and triglyceride secretion by perfused livers of obese-hyperglycaemic (*ob/ob*) mice: relationship with hyperinsulinaemia. *Diabetologia,* **10,** 155–162

Baxter, D. and Lazarus, N. R. (1975). The control of insulin receptors in the New Zealand obese mouse. *Diabetologia,* **11,** 261–267

Beloff-Chain, A., Hawthorn, J. and Green, D. (1975). Influence of the pituitary gland from the homozygote (+/+) and heterozygote (*ob/+*) lean mouse on insulin secretion *in vitro*. *FEBS Letters,* **55,** 72–74

Bloxham, D. P., Fitzsimmons, J. T. R. and York, D. A. (1977). Lipogenesis in hepatocytes of genetically obese rats. *Horm. Metab. Res.,* **9,** 304–309

Bray, G. A. and York, D. A. (1971). Genetically transmitted obesity in rodents. *Physiol. Rev.,* **51,** 598–646

Chan, T. M. and Exton, J. H. (1977). Hepatic metabolism of the genetically diabetic (*db/db*) mouse. II. Lipid metabolism. *Biochim. Biophys. Acta,* **489,** 1–14

Chan, T. M., Young, K. M., Hutson, N. J., Brumley, F. T. and Exton, J. H. (1975). Hepatic metabolism of genetically diabetic (*db/db*) mice. I. Carbohydrate metabolism. *Am. J. Physiol.,* **229,** 1702–1712

Chang, K.-J., Huang, D. and Cuatrecasas, P. (1975). The defect in insulin receptors in obese-hyperglycaemic mice: a probable accompaniment of more generalized alterations in membrane glycoproteins. *Biochim. Biophys. Res. Comm.,* **64,** 566–573

Clark, D. G., Rognstad, R. and Katz, J. (1974). Lipogenesis in rat hepatocytes. *J. Biol. Chem.,* **249,** 2028–2036

Czech, M. P., Richardson, D. K. and Smith, C. J. (1977). Biochemical basis of fat cell insulin resistance in obese rodents and man. *Metab. Clin. Exp.,* **26,** 1057–1080

Das, I. and Hems, D. A. (1974). Glycogen synthetase and phosphorylase activities in different tissues of genetically obese mice. *Horm. Metab. Res.,* **6,** 40–44

Elliott, J. A., Beloff-Chain, A. and Hems, D. A. (1971). Carbohydrate metabolism of the isolated perfused liver of normal and genetically obese hyperglycaemic (*ob/ob*) mice. *Biochem. J.,* **125,** 773–780

Elliott, J. A., Dade, E., Salmon, D. M. W. and Hems, D. A. (1974). Hepatic metabolism in normal and genetically obese mice. *Biochim. Biophys. Acta,* **343,** 307–323

Enser, M. (1970). Fatty acid mobilization in obese mice. *Nature,* **226,** 175–177

Enser, M. (1972). Clearing-factor lipase in obese hyperglycaemic mice (*ob/ob*). *Biochem. J.,* **129,** 447–453

Genuth, S. (1969). Hyperinsulinism in mice with genetically determined obesity. *Endocrinology,* **84,** 386–391

Godbole, V. and York, D. A. (1978). Lipogenesis *in situ* in the genetically obese Zucker fatty rat (*fa/fa*): role of hyperphagia and hyperinsulinaemia. *Diabetologia,* **14,** 191–197

Gove, C. D. and Hems, D. A. (1978). Fatty acid synthesis in the regenerating liver of the rat. *Biochem. J.,* **170**, 1–8

Hems, D. A. (1977). Short-term hormonal control of hepatic carbohydrate and lipid catabolism. *FEBS Letters,* **80**, 237–245

Hems, D. A. and Brosnan, J. T. (1970). Effects of ischaemia on content of metabolites in rat liver and kidney *in vivo. Biochem. J.,* **120**, 105–111

Hems, D. A., Davies, C. J. and Siddle, K. (1978a). Effects of hormones on content of purine nucleoside cyclic monophosphates in perfused rat liver. *FEBS Letters,* **87**, 196–201

Hems, D. A. and Ma, G. Y. (1976). Resistance to hepatic action of vasopressin in genetically obese (*ob/ob*) mice. *Biochem. J.,* **160**, 23–28

Hems, D. A., Rath, E. A. and Verrinder, T. R. (1975). Fatty acid synthesis in liver and adipose tissue of normal and genetically obese (*ob/ob*) mice during the 24 hr cycle. *Biochem. J.,* 150, 167–173

Hems, D. A., Rodrigues, L. M. and Whitton, P. D. (1978b). Rapid stimulation by vasopressin, oxytocin and angiotensin II of glycogen degradation in hepatocyte suspensions. *Biochem. J.,* **172**, 311–317

Hems, D. A. and Whitton, P. D. (1973). Stimulation by vasopressin of glycogen breakdown and gluconeogenesis in the perfused rat liver. *Biochem. J.,* **136**, 705–709

Herberg, L. and Coleman, D. (1977). Laboratory animals exhibiting obesity and diabetes syndromes. *Metab. Clin. Exp.,* **26**, 59–99

Herberg, L. and Kley, H. K. (1975). Adrenal function and the effect of a high-fat diet on C57BL/6J and C57BL/6J-*ob/ob* mice. *Horm. Metab. Res.,* **7**, 410–415

Jamdar, S. C. (1978). Glycerolipid biosynthesis in rat adipose tissue. Influence of adipose cell size and site on triacylglycerol formation in lean and obese rats. *Biochem. J.,* **170**, 153–160

Jungas, R. L. (1968). Fatty acid synthesis in adipose tissue incubated in tritiated water. *Biochem.,* **7**, 3708–3717

Katz, J., Wals, P., Golden, S., Goldman, J. K. and Bernardis, L. L. (1976). Lipogenesis by hepatocytes of rats with hypothalamic obesity. *Horm. Metab. Res.,* **9**, 59–63

Kirk, C. J., Verrinder, T. R. and Hems, D. A. (1976). Fatty acid synthesis in the perfused liver of adrenalectomized rats. *Biochem. J.,* **156**, 593–602

Laudat, M. H. and Pairault, J. (1975). An impaired response of adenylate cyclase to stimulation by epinephrine in adipocyte plasma membranes from genetically obese mice (*ob/ob*). *Eur. J. Biochem.,* **56**, 583–590

Lavine, R. L., Voyles, N., Perrino, P. V. and Recant, L. (1975). The effect of fasting on tissue cyclic AMP and plasma glucagon in the obese hyperglycaemic mouse. *Endocrinology,* **97**, 615–620

Le Marchand, Y., Loten, E. G., Assimacopoulos-Jeannet, F., Forgue, M.-E., Freychet, P. and Jeanrenaud, B. (1977). Effect of fasting and streptozotocin in the obese-hyperglycaemic (*ob/ob*) mouse. Apparent lack of a direct relationship between insulin binding and insulin effects. *Diabetes,* **26**, 582–590

Loten, E. G., Assimacopoulos-Jeannet, F., Le Marchand, Y., Singh, A. and Jeanrenaud, B. (1974a). Regulation of carbohydrate and lipid metabolism in the liver and adipose tissue of normal and diabetic mice. *Adv. Enz. Reg.,* **12**, 45–71

Loten, E. G., Le Marchand, Y., Assimacopoulos-Jeannet, F., Denton, R. M. and Jeanrenaud, B. (1976). Does hyperinsulinaemia in *ob/ob* mice cause an insulin-stimulated adipose tissue? *Am. J. Physiol.,* **230**, 602–607

Loten, E. G., Rabinowitch, A. and Jeanrenaud, B. (1974b). *In vivo* studies on lipogenesis in obese hyperglycaemic (*ob/ob*) mice. Possible role of hyperinsulinaemia. *Diabetologia,* **10**, 45–52

Ma, G. Y., Gove, C. D. and Hems, D. A. (1977). Inhibition of fatty acid synthesis and stimulation of glucose release by adrenaline and angiotensin II in the perfused mouse liver. *Biochem. Soc. Trans.* **5**, 986–989

Ma, G. Y., Gove, C. D. and Hems, D. A. (1979). Effects of glucagon and insulin on fatty acid synthesis and glycogen degradation in the prfused liver of normal and genetically obese (*ob/ob*) mice. *Biochem. J.* (in press)

Ma, G. Y. and Hems, D. A. (1975). Inhibition of fatty acid synthesis and stimulation of glycogen breakdown by vasopressin in the perfused mouse liver. *Biochem. J.*, **152**, 389–392

Otto, W., Taylor, T. G. and York, D. A. (1976). Glycerol release *in vitro* from adipose tissue of obese (*ob/ob*) mice treated with thyroid hormones. *J. Endocrinol.*, **71**, 143–155

Raskin, P., McGarry, J. G. and Foster, D. W. (1974). Independence of cholesterol and fatty acid biosynthesis from cyclic adenosine monophosphate concentration in the perfused rat liver. *J. Biol. Chem.*, **249**, 6029–6032

Rath, E. (1977). Metabolism of fatty acids in adipose tissue of normal and genetically obese (*ob/ob*) mice. PhD Thesis, London University

Rath, E., Beloff-Chain, A. and Hems, D. A. (1975). Contribution of lactate carbon to fatty acid synthesis in adipose tissue of normal and genetically obese (*ob/ob*) mice. *Biochem. Soc. Trans.*, **3**, 513–515

Rath, E., Hems, D. A. and Beloff-Chain, A. (1974). Lipoprotein lipase activities in tissues of normal and genetically obese (*ob/ob*) mice. *Diabetologia*, **10**, 261–266

Robertson, R. P., Gavareski, D. J., Henderson, J. D., Porte, D. Jr. and Bierman, E. L. (1973). Accelerated triglyceride secretion: a metabolic consequence of obesity. *J. Clin. Invest.*, **52**, 1620–1626

Salmon, D. M. W., Bowen, N. L. and Hems, D. A. (1974). Synthesis of fatty acids in the perfused mouse liver. *Biochem. J.*, **142**, 611–618

Salmon, D. M. W. and Hems, D. A. (1973). Plasma lipoproteins and the synthesis and turnover of plasma triglyceride in normal and genetically obese mice. *Biochem. J.*, **136**, 551–563

Salmon, D. M. W. and Hems, D. A. (1974). Control of fatty acid synthesis by substrate supply in the perfused mouse liver. *Biochem. Soc. Trans.*, **2**, 1011–1014

Shepherd, R. E., Malbon, C. C., Smith, C. J. and Fain, J. N. (1977). Lipolysis and adenosine 3′,5′-phosphate metabolism in isolated white fat cells from genetically obese-hyperglycaemic mice (*ob/ob*). *J. Biol. Chem.*, **252**, 7243–7248

Soll, A. H., Kahn, C. R. and Neville, D. M. Jr. (1975a). Insulin binding to liver plasma membranes in the obese hyperglycaemic (*ob/ob*) mouse. *J. Biol. Chem.*, **250**, 4702–4704

Soll, A. H., Kahn, C. R., Neville, D. M. Jr. and Roth, J. (1975b). Insulin receptor deficiency in genetic and acquired obesity. *J. Clin. Invest.*, **56**, 769–780

Stauffacher, W. and Renold, A. E. (1969). Effect of insulin *in vivo* on diaphragm and adipose tissue of obese mice. *Am. J. Physiol.*, **216**, 98–105

Veech, R. L., Eggleston, L. V. and Krebs, H. A. (1969). The redox state of free nicotinamide adenine dinucleotide phosphate in the cytoplasm of rat liver. *Biochem. J.*, **115**, 609–616

Volpe, J. J. and Marasa, J. C. (1975). Regulation of hepatic fatty acid synthetase in the obese-hyperglycaemic mutant mouse. *Biochim. Biophys. Acta*, **409**, 235–248

Winand, J., Furnelle, J. and Christophe, J. (1968). Le metabolisme lipidique du foie chez la souris normal et la souris obese-hyperglycaemique. *Biochim. Biophys. Acta.*, **152**, 280–292

Winand, J., Furnelle, J., Wodon, C., Hebbelinch, M. and Christophe, J. (1973). 7-day time study of lipid metabolism in normal and obese-hyperglycaemic Bar Harbor mice. Qualitative and quantitative aspects. *Biochimie*, **55**, 63–73

Yen, T. T. T., Allan, J. A., Yu, P. L., Acton, M. A. and Pearson, D. V. (1976). Triacylglycerol contents and *in vivo* lipogenesis of *ob/ob, db/db* and A^{vy}/a mice. *Biochim. Biophys. Acta*, **441**, 213–220

Yen, T. T. T. and Steinmetz, J. (1972). Lipolysis of genetically obese and/or hyperglycaemic mice with reference to insulin response of adipose tissue. *Horm. Metab. Res.*, **4**, 331–337

Zucker, L. M. (1972). Fat mobilization *in vitro* and *in vivo* in the genetically obese Zucker rat 'fatty'. *J. Lipid Res.*, **13**, 234–243

11

The hypothalamo – pituitary system in obesity

J. A. Edwardson and Amanda Donaldson (Department of Physiology,
St. George's Hospital Medical School, London, UK)

SUMMARY

The widespread metabolic and endocrine abnormalities of obesity involve changes in pituitary function. In the genetically obese (*ob/ob*) mouse there are disturbances in secretion of all the known pituitary anterior lobe and intermediate lobe hormones.

There is increased secretion of corticotrophin (ACTH) by the anterior pituitary of *ob/ob* animals, and also of two related peptides, corticotrophin-like intermediate lobe peptide (CLIP) and melanophore-stimulating hormone (MSH) by the pars intermedia. Increased release of ACTH presumably accounts for the adrenal hypertrophy and elevated glucocorticoid levels observed in the *ob/ob* mouse. Evidence is described showing that CLIP, the free 18-39 COOH terminal fragment of ACTH, has insulin-releasing properties. The increased secretion of CLIP in the *ob/ob* mouse may be implicated in the development of hyperinsulinaemia.

In contrast to the corticotrophin-related peptides, there is evidence suggesting diminished secretion of gonadotrophins (FSH and LH), thyrotrophin (TSH), growth hormone (GH) and prolactin (PRL). In some cases this is associated with changes in levels of the appropriate hypothalamic releasing or release-inhibiting hormones.

These widespread changes indicate some biochemical defect at hypothalamic or higher level, perhaps in the neurotransmitter systems which regulate release of the hypophysiotropic peptides. Dopamine and noradrenaline levels are raised in the *ob/ob* mouse, but there is, as yet, no indication of the way in which impaired monoaminergic mechanisms could produce the characteristic abnormalities of pituitary secretion in this strain.

INTRODUCTION

The hypothalamo–pituitary system plays a fundamental role in the regulation of metabolic and chemical homeostasis. Through this axis the behavioural mechanisms of feeding and satiety are integrated with appropriate autonomic and endocrine responses. Thus it is not surprising that the widespread metabolic and endocrine abnormalities of severe obesity in man and experimental animals involve changes in pituitary function. It is clearly important to determine whether the altered pattern of hypothalamo–pituitary activity represents (1) an essential component in the expression of the obese condition, (2) a secondary change which may reflect the nature and locus of the primary defect or (3) a compensatory response serving to mitigate some of the metabolic consequences of obesity. In this paper we present evidence that the conspicuous abnormalities of adrenal, pancreatic and gonadal function in the genetically obese (*ob/ob*) mouse may all reflect some disturbance of the hypothalamo–pituitary system, and attempt to consider some of the above questions.

The hypothalamus is phylogenetically old and functionally complex. Within the hypophysiotropic region most of the known neurotransmitters and neuropeptides occur, often in high concentrations. Each of the anterior pituitary hormones is controlled independently by one or more hypothalamic hormones released from neurosecretory nerve terminals in the median eminence. These neurons discharge their secretory products into portal blood vessels which provide a direct humoral route of access to the anterior pituitary. Release of the hypothalamic hormones is regulated through multiple neuronal projections from elsewhere in the brain and also by local mechanisms concerned with the feedback control of target gland hormones, blood nutrients and other humoral stimuli. All of these activities are compressed within a relatively minute region of enormous biochemical and structural complexity, and recent advances in understanding the mechanisms of hypothalamic neurosecretion constitute a major landmark in endocrinology.

Recent developments in this field include a battery of techniques which are relevant to the study of the hypothalamo–pituitary system in obesity. Immunohistochemical procedures permit the precise cellular localisation of specific neurotransmitters, neuropeptides and pituitary hormones, and the minute tissue concentrations of these substances can be measured by radioimmunoassay and other sensitive procedures. A variety of *in vitro* preparations have been developed, including incubated or superfused tissue fragments, slices and isolated nerve endings (synaptosomes), which permit the investigation of biochemical aspects of hypothalamo–pituitary function at a level of specificity unobtainable with the whole animal. The present report indicates the potential value of some of these procedures for the study of neuroendocrine mechanisms in obesity.

CONTROL OF ADRENOCORTICAL FUNCTION IN THE *ob/ob* MOUSE

Hypertrophy and hyperplasia of the adrenal cortex are characteristic of the *ob/ob* mouse and most other forms of genetically transmitted rodent obesity (for references, see Bray and York, 1971). There is some evidence that increased secretion of glucocorticoids may be important for the development of the obese syndrome, since adrenalectomy of *ob/ob* mice at 2 months of age causes significant lowering of blood glucose and body weight, abolishes insulin resistance and restores to normal the responses to fasting or a glucose load (Naeser, 1973; Solomon and Mayer, 1973). Interest in the hypothalamic regulation of corticotrophin (ACTH) secretion led us to investigate possible reasons for the increased adrenocortical function of *ob/ob* mice, and it was found that the pituitary ACTH content of these animals was 14 times greater than that of lean litter mate controls (Edwardson and Hough, 1975). This increase in ACTH content was paralleled in the extent to which the isolated pituitary gland of adult *ob/ob* mice released corticotrophin in an *in vitro* superfusion system, the mean rate of release (577 ± 57 pg ACTH per mg pituitary per min; ± S.E.M.; *n* = 7) being considerably greater

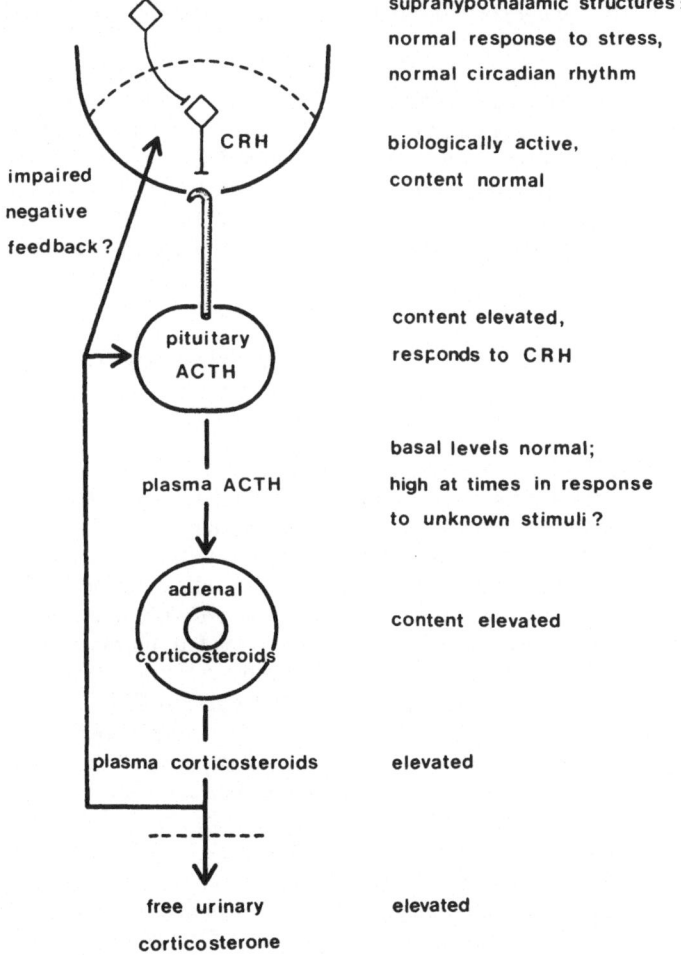

Figure 11.1 Hypothalamo–pituitary–adrenocortical function in the *ob/ob* mouse.

than that for lean glands (86 ± 13 pg mg^{-1} min^{-1}; $n = 5$). A group of adult *ob/ob*
mice which had been maintained on a restricted diet from 5 weeks of age and
whose body weights were only slightly above normal was also studied and it was
found that levels of ACTH released during superfusion (457 ± 73 pg mg^{-1} min^{-1};
$n = 5$) were comparable with those of the *ob/ob* group fed *ad libitum*. This finding
indicates that the hyperactivity of the hypothalamo-pituitary-adrenocortical axis
in *ob/ob* animals is independent of hyperphagia or the metabolic consequences
of adiposity.

Acid extracts of the hypothalamus were prepared from lean and *ob/ob* mice
and tested for ACTH-releasing activity. While no difference in the hypothalamic
content of corticotrophin-releasing hormone (CRH) was found, this does not exclude the possibility of a substantial increase in the turnover and release of CRH
in *ob/ob* animals. This seems likely in view of the increased pituitary ACTH content, increased ACTH release from the superfused pituitary, and elevated plasma,
adrenal and free urinary corticosteroids observed in the *ob/ob* mouse (Edwardson
and Hough, 1975; Kley *et al.*, 1976).

The main features of the hypothalamo-pituitary-adrenocortical system of the
ob/ob mouse are summarised in figure 11.1 and there are some interesting parallels
with findings obtained in human obesity. Several reports indicate that in human
obesity the rate of cortisol production is increased (Migeon *et al.*, 1973; Garces
et al., 1968), although paradoxically the plasma cortisol concentration appears to
remain unchanged. As with the *ob/ob* mouse, an increase in ACTH secretion is
implied by these findings, although this remains to be established by direct
measurement.

THE HYPOTHALAMO-PITUITARY SYSTEM IN THE REGULATION
OF INSULIN SECRETION

Elsewhere in this volume (Beloff-Chain, 1979) the crucial role of insulin in obesity
has been described together with evidence for the existence of a hypophysial
factor which may regulate insulin secretion. In summary, the isolated neurointermediate lobe of the *ob/ob* mouse has been shown to release an insulin secretagogue which is absent from the anterior pituitary. Studies with purified and synthetic peptides indicate that the insulin-releasing activity is due to corticotrophin-
like intermediate lobe peptide (CLIP) or a closely related substance, and we have
shown recently (Beevor *et al.*, 1979) that the insulin-releasing activity is abolished
by incubation with an antiserum which cross-reacts against CLIP.

If the insulin-releasing factor is CLIP, then the levels of this peptide in the *ob/ob*
pituitary intermediate lobe should be elevated in comparison with lean glands,
which do not show stimulatory activity in the superfusion system. Extracts of
ob/ob mouse intermediate lobes were prepared and the content of immunoreactive
CLIP, ACTH and α-melanophore-stimulating hormone (α-MSH) was determined by
radioimmunoassay. Figure 11.2 shows that the content of CLIP in the intermediate
lobe of *ob/ob* animals was 36-fold greater than in lean litter mate controls. Chromatographic separation of these extracts on a column of Bio-Gel P6 using the method
described by Lowry and Chadwick (1970) showed that the immunoreactive 'CLIP'

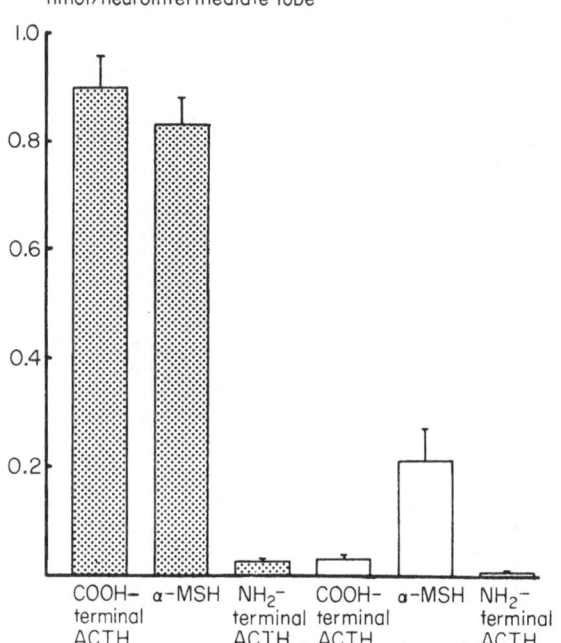

nmol/neurointermediate lobe

Figure 11.2   Content of 'immunoreactive CLIP' and α-MSH in 0.1 M HCl extracts
of the neurointermediate lobe from *ob/ob* mice (stippled) and lean (+/+) controls
(blank). The content of ACTH determined by an immunoassay against the NH_2
terminal sequence is small compared with values measured by the COOH terminal
immunoassay. This difference (COOH value minus NH_2 value) represents a free
COOH terminal fragment with chromatographic properties similar to those of CLIP.
Values show mean (± S./E.) of measurements from 6 glands in each group.

was due to a substance of the appropriate molecular size. The content of α-MSH
in obese gland extracts was similar to that of CLIP, whereas in the lean gland extracts the content of α-MSH was considerably greater than that of CLIP.

When neurointermediate lobes from *ob/ob* mice were superfused *in vitro*, there
was a fivefold increase in the rate of CLIP release compared with lean controls
(figure 11.3). Also, the molar ratios of CLIP : α-MSH released from lean neuro-
intermediate lobes was 1:1, whereas with the *ob/ob* preparation this ratio was 2:1.
These results are significant in a number of respects. Firstly, they provide further
evidence that the insulin-releasing factor detected in superfusates of *ob/ob* pituitary glands is CLIP. Secondly, they provide insight into the mechanisms of
biosynthesis of corticotrophin-related peptides in the pituitary intermediate lobe.
Also, they raise the possibility that hyperactivity of the intermediate lobe may be
a contributory or even essential factor in the development of hyperinsulinaemia
in *ob/ob* mice.

Many extra-adrenal effects of ACTH have been reported (Lebovitz, 1973), but
these have usually been ascribed to the NH_2-terminal steroidogenic fragment. The
insulin-releasing effects which we have described (Beloff-Chain, 1979) provide

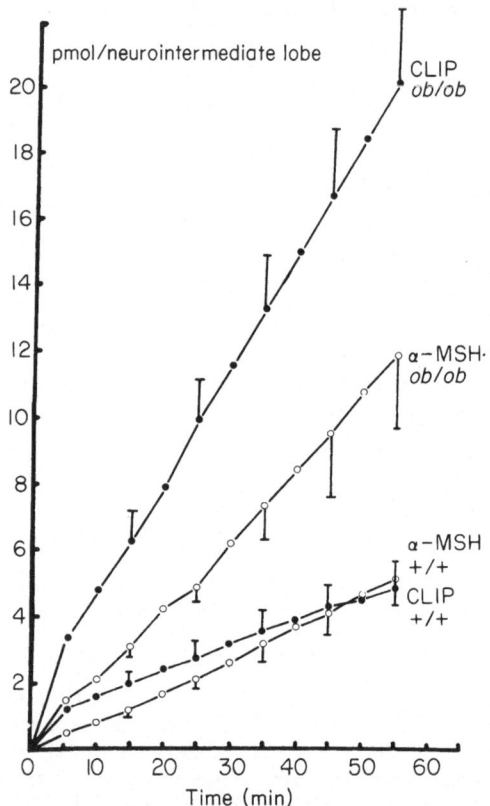

Figure 11.3   Release of immunoreactive CLIP (●) and α-MSH (○) from superfused neurointermediate lobes of *ob/ob* and lean (+/+) mice. Glands were superfused with Krebs–Ringer bicarbonate medium containing 5mM glucose at 37°C,
using a flow rate of 340 µl/min. Results expressed as a cumulative plot during 1 h.
NH_2-terminal ACTH-like activity was not detectable. Mean (± S.E.) of six super-
fusions in each group, except for α-MSH in +/+, where *n* = 5.

the first demonstration of biological activity for the COOH-terminal amino acid
sequence of ACTH. All of the experimental evidence obtained to date is in favour
of the insulin-releasing factor being corticotrophin-like intermediate lobe peptide
rather than ACTH itself. CLIP was first described by Scott *et al.* (1973), who
showed it to be the free 18-39 COOH-terminal fragment of ACTH, localised within the intermediate lobe of the pituitary, and postulated that it may be a hormonal
peptide in its own right. Our studies have demonstrated that (1) the insulin-releasing factor originates from the superfused neurointermediate lobe and is not
released by the pars anterior, (2) intact ACTH and synthetic 17-39 ACTH have
insulin-releasing activity, whereas the 1-24 steroidogenic fragment of ACTH is
without effect, (3) activity in pituitary superfusate medium is abolished by incubation with an antiserum against 17-39 ACTH but not with control serum, and
(4), as shown above, the intermediate lobe of the *ob/ob* mouse contains high

levels of a peptide with immunological and chromatographic properties resembling those of CLIP. Finally, it remains to be shown that purified and characterised extracts of CLIP from the intermediate lobe have full insulin-releasing activity.

In relation to an insulin-releasing role for CLIP, it is of interest to note that Larsson (1977, 1978) has shown by immunofluorescence techniques that a COOH-terminal fragment of ACTH is present in endocrine cells of the pancreas and gastrointestinal tract. Thus CLIP may be involved in the regulation of insulin secretion via the entero—pancreatic axis, as well as by neuroendocrine reflexes involving the hypothalamo–hypophysial system. If CLIP is a true hormone, then the mechanisms involved in control of its secretion by the hypothalamus must involve three important stages: biosynthesis of the macromolecular precursor to CLIP; enzymic processing of this precursor to produce CLIP; release of the active peptide from its site of storage. Mains *et al.*(1977) have shown that both ACTH and lipotrophin are derived from a common macromolecular precursor which presumably gives rise to the related smaller peptides, including endorphins, CLIP and α-MSH (figure 11.4). In the *ob/ob* mouse there is a concomitant increase in ACTH in the anterior lobe together with α-MSH and CLIP in the intermediate lobe, suggesting that there is some common element in the hypothalamic control of their biosynthesis. However, since CLIP is released from the superfused neuro-intermediate lobe in a 2:1 ratio with α-MSH, there are clearly independent mechanisms of secretion. Recently we have shown (Edwardson and Donaldson, 1979) that short exposure (5 min) to hypothalamic extracts causes a sustained (20 min) increase in CLIP release from the superfused neurointermediate lobe, whereas the

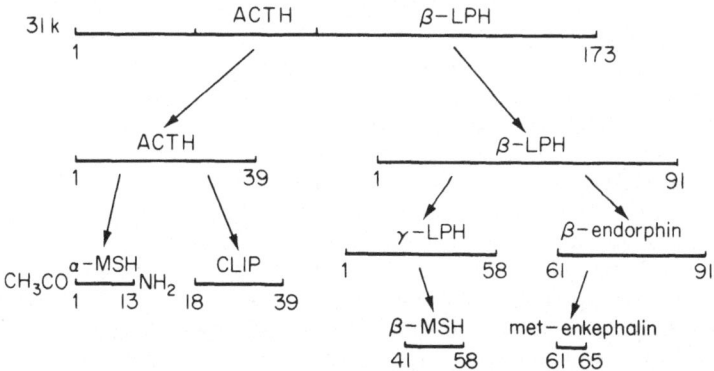

Figure 11.4   Diagram showing structural relationships between corticotrophin- and lipotrophin-related peptides, all of which originate from a common glycoprotein precursor with a molecular weight of approximately 31 000. Routes for the enzymic processing of the precursor into these smaller fragments are not known and arrows indicate only homologous sequences (not drawn to scale). 31 k = precursor; ACTH = corticotrophin (steroidogenic activity); LPH = lipotrophin (functions unknown); α-MSH = α-melanophore-stimulating hormone (actions on skin pigmentation and hair follicles); CLIP = corticotrophin-like intermediate lobe peptide (insulin-releasing activity); γ-LPH = γ-lipotrophin (contains 'β-MSH', functions unknown); β-endorphin and methionine enkephalin are both endogenous agonists at the opiate receptors.

effect on α-MSH release is frequently absent or even inhibitory. Tilders and Smelik (1977) have presented evidence that α-MSH secretion may be regulated through dopaminergic secretomotor neurons which originate in the arcuate nucleus of the hypothalamus. Dopamine was without effect on CLIP release in our super-fusion system and the response obtained with hypothalamic extracts suggests that CLIP may be under neurohumoral control. The nature of the CLIP-releasing factor and the physiological stimuli which regulate its release remain to be established.

Increased pituitary levels and release of CLIP may be associated with other forms of obesity. Intermediate lobe extracts from mice made obese with either a high-fat diet or gold thioglucose treatment also contain elevated concentrations of immunoreactive CLIP (Edwardson and Donaldson, 1979) and the pituitary glands from such animals stimulate insulin release from superfused pancreatic islets, as do pituitaries from young 3 week old *ob/ob* mice (Beloff-Chain *et al.*, personal communication). These results indicate that disturbance of the hypothalamo-hypophysial system and increased release of CLIP could be an important or even essential stimulus for the development of hyperinsulinaemia and consequent in-sulin resistance. Work is required to determine whether plasma levels of CLIP cor-relate with insulin secretion and to elucidate the nature of the physiological stimuli which evoke release of CLIP *in vivo*.

THE HYPOTHALAMO–PITUITARY SYSTEM AND REPRODUCTIVE FUNCTION IN THE *ob/ob* MOUSE

Both male and female *ob/ob* mice demonstrate marked hypogonadism and infer-tility (for reference, see Swerdloff *et al.*, 1976). Experiments by many groups point to the defects of reproductive function being central in origin rather than peripheral. In a detailed study of gonadotrophins and gonadal function in develop-ing male *ob/ob* mice, Swerdloff *et al.* (1976) showed that levels of follicle-stimu-lating hormone (FSH), luteinising hormone (LH) and testosterone were reduced in comparison with lean animals. Since low levels of the gonadal steroids should tend to increase gonadotrophin secretion in the presence of functioning feedback mechanisms at hypothalamo–pituitary level, the *ob/ob* mouse is defective in this respect. The authors point out that this condition resembles that of pre-pubertal animals and suggests persistent immaturity of the hypothalamo–hypophysial axis. Such a conclusion is supported by experiments showing that castrated *ob/ob* males are more sensitive than lean animals to the effects of exogenous testo-sterone on the feedback inhibition of FSH and LH release.

We have measured immunoreactive luteinising hormone-releasing hormone (LH-RH) in extracts of hypothalamus from lean and from *ob/ob* mouse and found no significant difference in content. A similar finding has been made by Batt (personal communication), who has also shown that the LH-RH from *ob/ob* mice has biological activity. In a study of female *ob/ob* mice, the results of which are shown in table 11.1, Batt and Wilson (personal communication) have found that although the pituitary content of LH is reduced, both pituitary and plasma con-centrations are similar to those of lean animals. As exogenous gonadotrophins will readily stimulate gonadal function in these animals, it seems likely that there may

Table 11.1 Immunoreactive luteinising hormone in adult female *ob/ob* and lean (+/+) mice

Group	n	Pituitary weight (mg)	Pituitary LH content (ng)	conc. (ng/mg)	Plasma LH (ng/ml)
ob/ob	6	2.94 ± 0.21	578 ± 47	201 ± 15	0.83 ± 0.14
lean (dioestrus)	8	4.17 ± 0.31*	869 ± 54*	210 ± 8	0.88 ± 0.09
lean (oestrus)	8	4.42 ± 0.26*	909 ± 82*	207 ± 17	0.90 ± 0.15

*$P < 0.01$.
Data supplied by Dr R. Batt and Dr C. Wilson, Royal Veterinary College, London.

be tonic basal release of LH without the marked pre-ovulatory surge of gonado-trophin secretion which characterises the cycle of lean females.

The main features of the hypothalamo–pituitary-gonadal axis in the *ob/ob* mouse are shown in figure 11.5. While there have been few detailed studies in

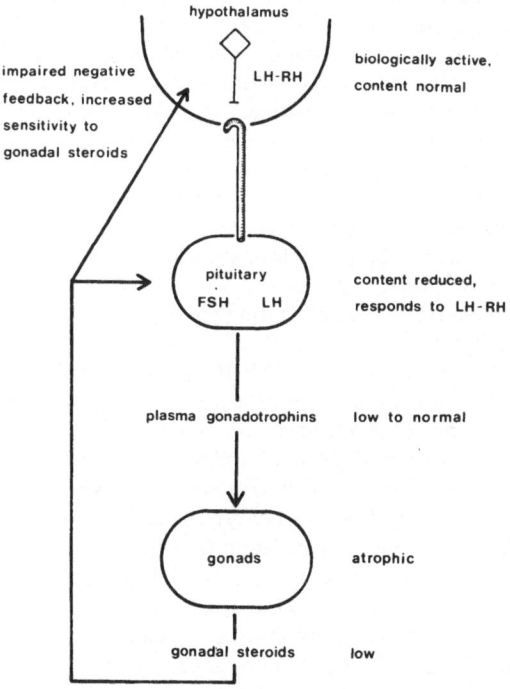

Figure 11.5 Hypothalamo–pituitary–gonadal function in the *ob/ob* mouse.

human obesity, it seems the prevalent view of many clinicians concerned with
obesity that severe adiposity is often marked by disturbances of reproductive
function. Nocturnal profiles of LH secretion in obese patients are low, whereas
prolactin levels are elevated (Kalucy *et al.*, 1976). Reversible gonadotrophin deficiency has been reported in Cushing's disease, which is also characterised by
adiposity (Luton *et al.*, 1977). These workers found that plasma testosterone,
basal and LH-RH stimulated values for plasma LH and FSH were all reduced.
During remission both gonadotrophin and testosterone levels returned to normal
with the supression of hypercortisolism and the authors propose a direct or indirect inhibiting action of adrenal steroids at hypothalamo–pituitary level as being
most consistent with their results. Such a mechanism could also contribute to the
impaired gonadal activity of *ob/ob* mice. It would be of considerable interest to
determine the effects of adrenalectomy and replacement therapy with normal
levels of corticosteroids on reproductive function in these animals.

OTHER PITUITARY HORMONES: GROWTH HORMONE, PROLACTIN
AND THYROID-STIMULATING HORMONE IN *ob/ob* MICE

The most extensive investigation of growth hormone (GH) and prolactin (PRL)
in *ob/ob* mice is that reported by Sinha *et al.* (1975), who showed that, in general,
both pituitary and plasma levels of GH and PRL are reduced in *ob/ob* animals
compared with lean controls. Sex differences in levels of GH and PRL observed
in lean mice were not so marked in *ob/ob* mice, but the most striking observation
was in the response to perphenazine. Lean animals given perphenazine showed a
substantial and sustained increase in plasma PRL, whereas *ob/ob* animals had a
much diminished and more transient response. Since prolactin secretion is under
the dual control of both inhibitory and releasing factors, it is difficult to establish
the mechanisms of this response. However, perphenazine works via a central effect
involving monoaminergic neurons, and the results clearly indicate a difference at
hypothalamic level between *ob/ob* and lean animals.
    Reduced secretion of growth hormone in *ob/ob* mice could involve some change
in the hypothalamic control of somatostatin (SR-IH) release. Patel *et al.* (1977)
have found that the hypothalamus of both *ob/ob* and *db/db* mutants shows a small
but significant increase in the content of SR-IH, whereas levels of this peptide in
the pancreas and stomach are decreased. It is not known whether these differences
reflect increased or decreased secretion in the *ob/ob*. We have also measured SR-IH
in hypothalamic extracts from *ob/ob* mice and were surprised to obtain results
which are completely at variance with those reported above. The content of SR-IH
in lean hypothalamus was 12.15 ± 1.21 ng/hypothalamus ($\pm$ S.E.M.; $n = 12$), compared with only 1.39 ± 0.10 ng/hypothalamus ($n = 6$) in the *ob/ob* mouse. Again,
these levels give no indication of SR-IH release and it is necessary to investigate
this problem using preparations such as the isolated superfused hypothalamus or
incubated hypothalamic synaptosomes in which dynamic measurements can be
made. Such experiments are of crucial importance in view of the inhibitory actions
of SR-IH not only on GH and PRL release, but also on the secretion of thyroid-
stimulating hormone (TSH). There is evidence for impaired thyroid function in the

ob/ob mouse (Joosten and van der Kroon, 1974), and since TSH secretion is also inhibited by SR-IH, the low levels of plasma thyroxine and protein-bound iodide could also be attributable to oversecretion of this hypothalamic peptide.

We have measured the hypothalamic content of immunoreactive thyrotrophin-releasing hormone (TRH) and found a significant reduction in the *ob/ob* mouse. Values for lean animals were 929 ± 76 pg/hypothalamus (n = 6) and for *ob/ob* animals were 472 ± 36 (n = 12). Again, such measurements do not reveal whether TRH release is increased or decreased. Low levels of thyroxine should result in increased TSH output through central feedback control, but the histological appearance of the thyroid gland and its reduced uptake of ^{131}I (Wykes *et al.*, 1958) make this unlikely.

HYPOTHALAMIC NEUROTRANSMITTERS IN THE *ob/ob* MOUSE

The evidence presented above indicates that the secretion of every adenohypophysial hormone is altered to a greater or lesser extent in the *ob/ob* mouse. Since the condition is attributable to a single-gene defect, these abnormalities must reflect a primary disturbance at a more fundamental level. The hypothalamic hormones which regulate secretion from the adenohypophysis are in turn controlled by neurotransmitters, especially the monoamines, which are abundant in this region. An increase in the hypothalamic content of noradrenaline in *ob/ob* mice was reported by Lorden *et al.* (1975), although further studies indicated (Lorden *et al.*, 1976) that this was not accompanied by any change in noradrenaline turnover. Increased levels of noradrenaline were also present in other regions. With a more sensitive enzymatic-isotopic technique this group has been able to measure both noradrenaline and dopamine in hypothalamus and pituitary glands from lean and *ob/ob* mice (Lorden and Oltmans, 1977). This study confirmed the elevated (52 per cent) noradrenaline content in the *ob/ob* hypothalamus and showed that, although there was no significant difference in the dopamine content of this region, there was an increase (76 per cent) in dopamine concentration in the *ob/ob* pituitary gland. However, in a collaborative study with Dr J. S. de Belleroche at Imperial College, we have found, using a similar sensitive method, that the *ob/ob* hypothalamus contains 2–4 times as much dopamine as lean controls. Together these results indicate a widespread disturbance of both dopaminergic and noradrenergic systems in the hypothalamus as well as other brain regions.

DISCUSSION AND CONCLUSIONS

There is direct or strong circumstantial evidence to show that the secretion of all adenohypophysial hormones is altered in the *ob/ob* mouse. The main features of these abnormalities are illustrated in figure 11.6, which indicates that while there is hypersecretion of ACTH, CLIP and α-MSH, three hormones which all originate independently from a common macromolecular precursor, the secretion of the other anterior pituitary trophins is diminished to a variable extent. It is thus un-

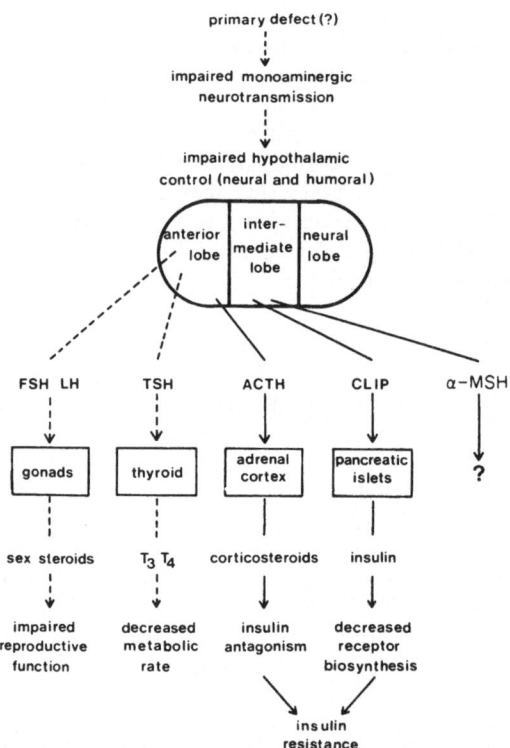

Figure 11.6   Diagram showing proven or likely relationships between the hypo-thalamo–pituitary system and peripheral endocrine and metabolic abnormalities in the *ob/ob* mouse. Solid lines indicate hypersecretion; broken lines indicate hyposecretion.

likely that the primary genetic lesion resides in the hypothalamo–pituitary system. Available evidence does not point clearly to some common element which could account for these changes. Dopamine and noradrenaline are implicated in the in-hibitory control of corticotrophin-related peptides and in the stimulation of FSH, LH, TSH and GH release (for references, see McCann *et al.*, 1974). Thus, some central impairment of catecholaminergic mechanisms could explain the hormonal profile of *ob/ob* mice. However, levels of these neurotransmitters are raised and there is no increase in prolactin secretion, which would be expected, as this hor-mone is strongly under the control of an inhibitory dopaminergic mechanism. If the underlying defect involves hypothalamic catecholamines, it must be expressed in some subtle way such as the derangement of a sub-population of receptors in-volved in hypothalamic control. The concept of a receptor defect has been dis-cussed elsewhere in this volume (Hems, 1979), and an investigation of hypothala-mic receptors for the catecholamines seems warranted.

At least two of the abnormalities of hypothalamo–hypophysial function in *ob/ob* mice could contribute directly to the expression of this syndrome. Elevated secretion of CLIP could contribute to the hyperinsulinaemia of these animals and thus be an indirect cause of the insulin resistance. Kahn (1976) has reviewed the extensive evidence which suggests that hyperinsulinaemia leads directly to a decrease in tissue sensitivity to insulin. Furthermore, the increased output of corticosterone caused by hypersecretion of ACTH would contribute to this condition by its anti-insulin effects.

There is no evidence to suggest that any of the changes in hypothalamo-pituitary function in *ob/ob* mice are adaptive, serving to mitigate the metabolic consequences of obesity. Herbai (1970) reported that following hypophysectomy *ob/ob* mice lose the excess body fat, show increased activity and look grossly healthy. Thus, it seems likely that the hypothalamo–pituitary system makes a significant contribution to the pathophysiology of obesity. Further work is required to determine the nature of the biochemical defects which give rise to these neuroendocrine abnormalities and the extent to which such changes contribute to other forms of obesity.

ACKNOWLEDGEMENTS

Experimental work described in this paper was financed by an MRC Programme Grant to J. A. Edwardson. A. Donaldson is supported by a grant from the Medical Research Committee of St George's Hospital Medical School. We are grateful to Dr R. Batt and Dr C. Wilson, Royal Veterinary College, for allowing us to use unpublished results. Peptides used in this study were kindly supplied by Ciba-Geigy Ltd, Hoechst A.G. and Ayerst Laboratories Ltd. Excellent technical assistance was provided by Mrs J. Pennington.

REFERENCES

Beevor, S., Beloff-Chain, A., Bogdanovic, S., Donaldson, A., Edwardson, J. A. and Hawthorn, J. (1979). Corticotrophin-like peptides, insulin secretion and obesity. Submitted to *Nature*
Beloff-Chain, A. (1979). This volume
Bray, G. A. and York, D. A. (1971). Genetically transmitted obesity in rodents. *Physiol. Rev.*, 51, 598–646
Edwardson, J. A. and Donaldson, A. (1979). Regulation of corticotrophin-related peptides in the intermediate lobe and their possible relation to obesity. In *The Interaction within the Brain-Pituitary-Adrenocortical System* (ed. M. T. Jones, M. F. Dallman and B. Gillham, Academic Press, London (in press)
Edwardson, J. A. and Hough, C. A. M. (1975). The pituitary adrenal system of the genetically obese (*ob/ob*) mouse. *J. Endocrinol.*, 65, 99–107
Garces, L. Y., Kenny, F. M., Drash, A. and Taylor, F. H. (1968). Cortisol secretion rate during fasting of obese adolescent subjects. *J. Clin. Endocrinol. Metab.*, 28, 1843–1847
Hems, D. A. (1979). This volume
Herbai, G. (1970). Weight loss in obese-hyperglycaemic and normal mice following transauricular hypophysectomy by a modified technique. *Acta Endocrinol.*, 65, 712–722
Joosten, H. F. P. and van der Kroon, P. H. W. (1974). Role of the thyroid in the development of the obese-hyperglycaemic syndrome in mice (*ob/ob*). *Metabolism*, 23, 425–436
Kahn, C. R. (1976). Insulin sensitivity and insulin resistance: regulation of insulin receptors in vivo. In *Cell Membrane Receptors for Viruses, Antigens and Antibodies, Polypeptide Hormones and Small Molecules*, Miles International Symposium Number 9 (ed. R. F. Beers and E. G. Bassett), Raven Press, New York, pp. 33–46

Kalucy, R. S., Crisp, A. H., Chard, T., McNeilly, A., Chen, C. N. and Lacey, J. H. (1976). Nocturnal hormonal profiles in massive obesity, anorexia nervosa and normal females. *J. Psychosom. Res.*, **20**, 595–604

Kley, H. K., Herberg, L. and Krueskemper, H. L. (1976). Measurement of urinary free corticosterone as a method for evaluating the adrenal function of small laboratory rodents. *Acta Endocrinol.*, **82**, Suppl. 202, 45–46

Larsson, L. I. (1977). Corticotrophin-like peptides in central nerves and in endocrine cells of gut and pancreas. *Lancet*, **ii**, 1311–1323

Larsson, L. I. (1978). Distribution of ACTH-like immunoreactivity in rat brain and gastrointestinal tract. *Histochemistry*, **55**, 225–228

Lebovitz, H. E. (1973). In *Methods in Investigative and Diagnostic Endocrinology*, Vol. 2A (ed. S. A. Berson and R. S. Yallow), North-Holland, Amsterdam, pp. 349–359

Lorden, J. F. and Oltmans, G. A. (1977). Hypothalamic and pituitary catecholamine levels in genetically obese mice (*ob/ob*). *Brain Res.*, **131**, 162–166

Lorden, J. F., Oltmans, G. A. and Margules, D. L. (1975). Central catecholamine levels in genetically obese mice (*ob/ob* and *db/db*). *Brain Res.*, **96**, 390–394

Lorden, J. F., Oltmans, G. A. and Margules, D. L. (1976). Central catecholamine turnover in genetically obese mice (*ob/ob*). *Brain Res.*, **117**, 357–361

Lowry, P. J. and Chadwick, A. (1970). Purification and amino-acid sequence of melanocyte-stimulating hormone from the dogfish, *Squalus acanthias*. *Biochem. J.*, **118**, 713–718

Luton, J. P., Thieblot, P., Valcke, J.-C., Mahoudeau, J. A. and Bricaire, H. (1977). Reversible gonadotrophin deficiency in male Cushing's disease. *J. Clin. Endocrinol. Metab.*, **45**, 488–495

McCann, S. M., Fawcett, C. P. and Krulich, L. (1974). Hypothalamic hypophysial releasing and inhibiting hormones. In *Endocrine Physiology*, MTP International Review of Science, Physiology Series One, Vol. 5 (ed. S. M. McCann), Butterworths, London, pp. 31–65

Mains, R. E., Eipper, B. A. and Ling, N. (1977). Common precursor to corticotropins and endorphins. *Proc. Nat. Acad. Sci. USA*, **74**, 3014–3018

Migeon, C. J., Green, O. C. and Eckert, J. P. (1963). Study of adrenocortical function in obesity. *Metabolism*, **12**, 718–739

Naeser, P. (1973). Adrenal function in the obese hyperglycaemic syndrome (*ob/ob*) in mice. *Diabetologia*, **9**, 83

Patel, Y. C., Cameron, D. P., Stefan, Y., Malaisse-Lagae, F. and Orci, L. (1977). Somatostatin: widespread abnormality in tissues of spontaneously diabetic mice. *Science*, **198**, 930–931

Scott, A. P., Ratcliffe, J. G., Rees, L. H., Landon, J., Bennett, H. P. J., Lowry, P. J. and McMartin, G. (1973). Pituitary peptide. *Nature New Biol.*, **244**, 65–67

Sinha, Y. N., Salocks, C. B. and Vanderlaan, W. P. (1975). Prolactin and growth hormone secretion in chemically induced and genetically obese mice. *Endocrinology*, **97**, 1386–1393

Solomon, J. and Mayer, J. (1973). The effect of adrenalectomy on the development of the obese hyperglycemic syndrome in *ob/ob* mice. *Endocrinology*, **93**, 510–513

Swerdloff, R. S., Batt, R. A. and Bray, G. A. (1976). Reproductive hormonal function in the genetically obese (*ob/ob*) mouse. *Endocrinology*, **98**, 1359–1364

Tilders, F. J. H. and Smelik, P. G. (1977). Direct neural control of MSH secretion in mammals: the involvement of dopaminergic tubero-hypophysial neurones. *Front. Horm. Res.*, **4**, 80–93

Wykes, A. A., Christian, J. E. and Andrews, F. N. (1958). Radioiodine concentration and thyroid weight in normal, obese and dwarf strains of mice. *Endocrinology*, **62**, 535–538

12

Thermoregulation in genetically obese rodents: the relationship to metabolic efficiency

P. Trayhurn, P. L. Thurlby, C. J. H. Woodward and W. P. T. James (Dunn Nutrition Unit, University of Cambridge and Medical Research Council, Cambridge, UK)

SUMMARY

(1)   The positive energy balance which leads to obesity is due, in several strains of genetically obese rodent, to a combination of hyperphagia and an elevated metabolic efficiency. The significance of differences in efficiency can be illustrated by a pair-feeding experiment with young *ob/ob* and lean mice, where the gross efficiency of the obese individuals was 2.3 times greater than that of the lean.

(2)   The increased gross efficiency of the *ob/ob* mouse is due to a reduced maintenance requirement, and not to any reduction in the energy cost of growth. At normal environmental temperatures (18–25°C) the major components of the maintenance requirement are the basal metabolic rate and the energy cost of maintaining homeothermy—thermoregulatory thermogenesis.

(3)   Adult *ob/ob* mice rapidly die of hypothermia at 4°C, indicating that they have a major thermoregulatory defect. *ob/ob* mice as young as 10 days of age also have an impaired ability to respond to 'cool' environments. A thermoregulatory defect is therefore detectable very early in the development of the *ob/ob* mouse, and can be used as a test for the genotype before obesity is apparent visually.

(4)   Between 25 and 10°C the adult *ob/ob* mouse maintains its body temperature some 2°C below that of lean litter mates. The resting metabolic rate of adult lean and obese *ob/ob* mice is little different at environmental temperatures within, or close to, the thermoneutral zone. However, at temperatures below thermoneutrality the resting metabolic rate of the obese mice *per whole animal* is 20 per cent less than that of the lean.

(5)   At 33°C (thermoneutrality) young lean and obese (*ob/ob*) mice have similar energy requirements for weight maintenance, but at 23°C the maintenance requirement of the obese is 16 per cent less than that of the lean. Pair-feeding young obese mice to the *ad libitum* food intake of their lean litter mates at thermoneutrality eliminates two-thirds of the obese animals' excess energy gain at 23°C.

(6)   The nutritional studies support the metabolic rate measurements in

191

showing that the high gross efficiency of the *ob/ob* mouse is attributable to a reduced energy expenditure on thermoregulatory thermogenesis. We suggest that the primary reason for this reduction is that the *ob/ob* mouse may have a lower hypothalamic 'set-point' for body temperature.

(7) Thermoregulatory differences appear to be a common feature of genetically obese rodents. This suggests that the mechanism by which the *ob/ob* mouse achieves its high efficiency is a general one.

INTRODUCTION

The positive energy balance which leads to obesity can in principle be the result of either hyperphagia or a high efficiency. A high efficiency on a normal energy intake is itself the result of a low level of energy expenditure. In several strains of genetically obese rodent obesity appears to be due to a combination of hyperphagia and a reduced energy expenditure (Bray and York, 1971). Thus, when the food intake of the obese animal is restricted to the level of lean litter mates, obesity still develops—although to a lesser extent than when the animal is allowed to eat *ad libitum*. Several studies have demonstrated the difference in efficiency between lean and obese animals in the obese-hyperglycaemic (*ob/ob*) mouse (Alonso and Maren, 1955; Chlouverakis, 1970; Welton *et al.*, 1973) and in the Zucker or fatty (*fa/fa*) rat (Pullar and Webster, 1974, 1977; Zucker, 1975; Deb *et al.*, 1976). A large difference in efficiency between lean and obese animals has also been shown recently in the diabetic (*db/db*) mouse (Cox and Powley, 1977).

Our main interest in these animals has been in finding an explanation for their high metabolic efficiency. We have concentrated on the question of efficiency, rather than on hyperphagia, for three main reasons. Firstly, by limiting food intake through pair-feeding experiments it is possible to investigate the effects of differences in efficiency without having to take into account the role of hyperphagia; in contrast, studies on the effect of hyperphagia will be complicated by the underlying efficiency component. Secondly, differences in efficiency between individuals may be more important than differences in food intake in explaining the variable propensity to obesity in man (James and Trayhurn, 1976). Thirdly, there is evidence that hyperphagia succeeds rather than precedes the excessive accumulation of fat in *ob/ob* mice, and thus may be a secondary factor (Lin *et al.*, 1977). It has been suggested that, at least in the diabetic mouse, the hyperphagia may be secondary to hyperinsulinaemia (Herberg and Coleman, 1977). Some caution must be exercised, however, in relegating hyperphagia to the position of a secondary factor, because of the difficulty in accurately measuring food (milk) intake during the pre-weaning period.

In this article we have mainly reviewed our own work on metabolic efficiency and thermoregulation in the *ob/ob* mouse. We have tried to show that the major cause of the high efficiency is a reduced energy expenditure on thermoregulatory thermogenesis.

THE IMPORTANCE OF METABOLIC EFFICIENCY

To illustrate the extent to which the *ob/ob* mouse is more efficient than its lean litter mates the results of a recent pair-feeding study have been summarised in table 12.1 (Woodward *et al.*, 1977). In this experiment 4 week old *ob/ob* mice were pair-fed to the *ad libitum* intake of sex-matched lean litter mates. (In all these experiments the *ob/ob* mice were descended from the 'Aston' stock from Imperial College.) The obese animals received their daily ration in either two or three 'meals', in order to minimise the effect of differences in meal pattern. At the end of 6 weeks the animals were killed and their body composition determined.

Table 12.1   Energy gain and gross efficiency of *ob/ob* mice pair-fed to the *ad libitum* food intake of lean litter mates

	Lean	Obese
Digestible energy intake (kJ)	2681 ± 73	2675 ± 72
Energy gain (retention) (kJ)	376 ± 37	850 ± 45
Gross efficiency (%)	13.8 ± 1.0	31.7 ± 1.2

The results are the means ± S.E.M. for 12 pairs of animals (both male and female) pair-fed for 6 weeks beginning at 4 weeks of age.

The body composition of a control group of 4 week old lean and obese mice was also determined, so that the changes during the 6 week experimental period could be estimated. The energy retention and gross efficiency of the obese animals were found to be a substantial 2.3 times greater than in the lean.

Although this experiment shows the magnitude of the difference in efficiency between lean and obese mice, it is important to establish the extent to which the high efficiency and hyperphagia each contribute to the final obesity. In order to do this, the energy gain of obese mice following both *ad libitum* feeding and pair-feeding was compared with the *ad libitum* intake of lean animals. The experiment was conducted for only 10 days with animals aged about 25 days. This procedure was adopted so that early changes could be studied. At the end of the experimental period the carcass energy content was determined and the energy gain calculated by reference to a control group of 25 day old animals. The results of this study are shown in figure 12.1. The energy gain of the pair-fed obese animals was over twice that of the lean, while the gain of the obese animals fed *ad libitum* was three times higher than that of the lean. From this it can be calculated that 61 per cent of the excess gain of the free-living obese mice was due to the enhanced efficiency but only 39 per cent was due to hyperphagia. Efficiency differences are therefore quantitatively more important than differences in food intake in the development of obesity in the young *ob/ob* mouse.

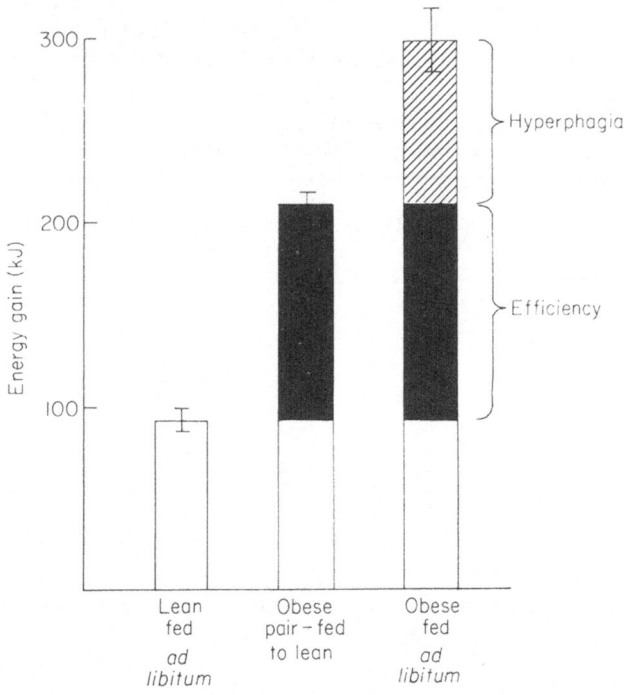

Figure 12.1 Energy gain in *ad libitum* fed obese mice and in obese mice pair-fed to the *ad libitum* intake of lean litter mates. The results are expressed as the means ± S.E.M. (bars) for 8 lean and 8 pair-fed obese animals, and for 6 obese animals fed *ad libitum*. The experiment was performed with males only.

These investigations, like many others on efficiency in genetically obese rodents, were conducted with growing animals. To what extent, therefore, are differences in the energy cost of growth responsible for the differences in gross efficiency? In a recent study the cost of growth of the obese mice was found to be 77.8 kJ/g dry tissue, a value similar to the 78.2 kJ/g dry tissue for lean animals (Woodward *et al.*, 1977). Furthermore, the energy cost of fat deposition was the same for lean and obese animals, at 61.5 and 59.9 kJ/g, respectively. Thus, there is no evidence from this study to support the idea that the greater metabolic efficiency of the *ob/ob* mouse relates to a lower energy requirement for fat or protein deposition.

MAINTENANCE REQUIREMENT

These results indicate that the high efficiency of the *ob/ob* mouse must be due to a low maintenance requirement rather than to a reduced energy cost of growth. Direct measurements have shown that the maintenance requirement *per whole animal* is indeed lower for young *ob/ob* mice than lean (Woodward *et al.*, 1977). At a weight of 30 g, for example, the maintenance requirement of obese mice was found to be lower by 16.8 kJ/day. For adult animals, however, when there is a 2-3-fold difference in body weight, the maintenance requirement *per animal* is higher for the obese than for the lean.

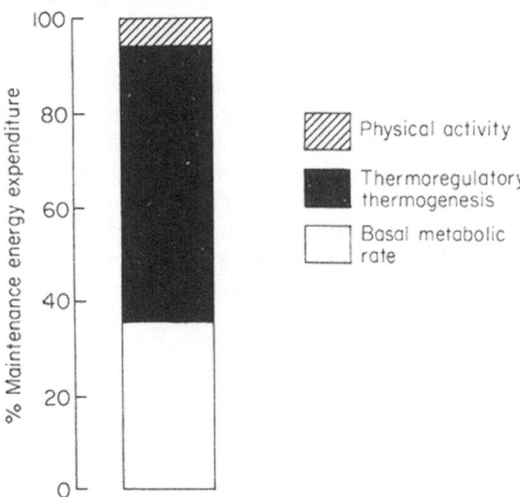

Figure 12.2   An approximate partition of the maintenance energy expenditure of normal mice at 22°C.

The main components of the maintenance requirement of a normal lean mouse at a typical environmental temperature of 22°C are shown in figure 12.2. The two most important factors are the basal metabolic rate (BMR) and the energy cost of maintaining homeothermy—thermoregulatory thermogenesis. The very low energy expenditure attributed to physical activity is based on the calculations of Miller and Mumford (1966). Clearly, because of their quantitative significance, the large efficiency difference between lean and *ob/ob* mice must involve a reduced energy expenditure on the BMR and/or on thermoregulation.

THERMOREGULATION IN *ob/ob* MICE

It has long been recognised that the *ob/ob* mouse has a major thermoregulatory defect, and this is frequently invoked as evidence for hypothalamic dysfunction in these animals. In 1954 Davis and Mayer found that at 3°C *ob/ob* mice very rapidly became hypothermic, whereas lean mice were fully able to maintain a normal body temperature. This observation has been confirmed by several other authors (Yen *et al.*, 1974; Ohtake *et al.*, 1977; Trayhurn *et al.*, 1976; Trayhurn and James, 1978). The effect of cold exposure on obese mice is at first sight surprising, since it is easy to believe that they are very well insulated by the excess body fat. Davis and Mayer (1954) reported, in fact, that the failure of the *ob/ob* mouse to tolerate the cold was due to a failure in heat production—thermogenesis.

The studies showing that the *ob/ob* mouse has a major thermoregulatory defect have all been performed with mature animals. It is therefore impossible to know whether this defect, like many others documented in them, is secondary to the obese state or a primary feature. Davis and Mayer (1954) found that mice made obese with gold thioglucose were able to tolerate the cold, which suggests that obesity *per se* is not important in the defective thermoregulation of the *ob/ob* mouse. In order to investigate this further we have determined the response of pre-weanling *ob/ob* mice to 'cool' environments.

Litters 10–21 days, derived from parents heterozygous for the *ob* gene, were exposed to environmental temperatures of between 4 and 22°C for up to 60 min. The youngest animals were exposed to the highest temperatures. During exposure to the cold rectal temperatures were measured periodically. In general, the individuals from each litter fell into two distinct groups. The larger group showed little or no fall in body temperature during cold exposure, while the smaller group showed a marked fall. The animals in the latter group were ear-

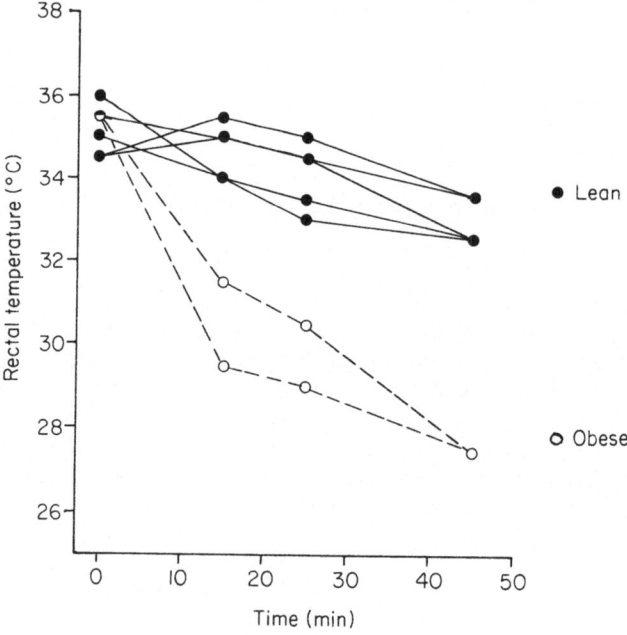

Figure 12.3   Rectal temperature of individual mice, from a single litter aged 12 days, during exposure to an environmental temperature of 15°C. The animals fell into two groups, those subsequently found to be *ob/ob* (○) and those found to be lean (●).

clipped for future identification and the whole litter was returned to the nest. The results obtained for a litter of 12 day old animals are shown in figure 12.3 (Trayhurn *et al.*, 1977). When aged 1 month the animals were examined for obesity. All of the ear-clipped animals were found to be obese, while all the un-marked animals were lean.

These results clearly indicate that the thermoregulatory defect established in adult *ob/ob* mice is not a secondary consequence of the obesity. The defect, al-though not necessarily primary, certainly occurs early in the development of the mutant. It is, in fact, detectable as soon as an appreciable capacity for thermo-regulation has developed in normal mice (at 10–12 days of age), and we have used the defect as a technique for the early identification of animals bearing the *ob/ob* genotype (Trayhurn *et al.*, 1977).

This method of identifying the mutants has been employed to examine early changes in their body composition. *ob/ob* mice have a small excess of body fat, as early as 10–12 days of age, with the 'obesity' developing rapidly after 12 days (Thurlby and Trayhurn, 1978).

BODY TEMPERATURE AT VARIOUS ENVIRONMENTAL TEMPERATURES

A lower body temperature in the adult *ob/ob* mouse has been noted by a number of authors for animals kept at normal environmental temperatures (Joosten and van der Kroon, 1974; Yen *et al.*, 1974; Ohtake, *et al.* 1977). Reduced body tempera-

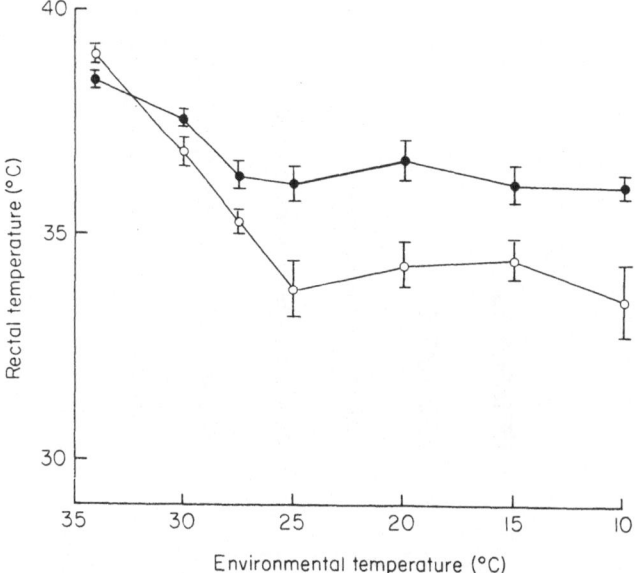

Figure 12.4   Rectal temperatures of adult lean (●——●) and obese (○——○) mice following exposure to a range of environmental temperatures for 1 h. The results are the means ± S.E.M. (bars) for 10 lean and 10 obese animals.

ture have also been found in pre-weanling *ob/ob* mice (Joosten and van der Kroon, 1974; Trayhurn *et al.*, 1977). In an attempt to determine whether the reduced temperature is a result of 'insensitive' thermoregulation, we have investigated the effect of acute exposure to a variety of environmental temperatures on the body temperature of adult lean and obese mice (Trayhurn and James, 1978). In these experiments the animals were exposed to temperatures ranging from 34 to 10°C for 1 h, after which time the rectal temperature was measured. At 34 and 30°C the obese animals had a slightly higher temperature than the lean, while at 27.5°C and below it was consistently lower (figure 12.4). Between 25 and 10°C the obese had body temperatures between 33.5 and 34.5°C, which was a constant 2°C below the range of 36.1–36.7°C found in the lean. Similar differences between lean and obese animals have been reported by Kaplan and Leveille (1974).

These results suggest that the lower body temperature of the obese mouse is not the result of 'inadequate' thermoregulation, but may be the result of a lower 'set-point'. Further support for this idea comes from the observation that the *ob/ob* mouse has a normal diurnal rhythm in body temperature, but at every point during the 24 h cycle its temperature is significantly lower than in lean animals (Joosten and van der Kroon, 1974; Thurlby, unpublished observations).

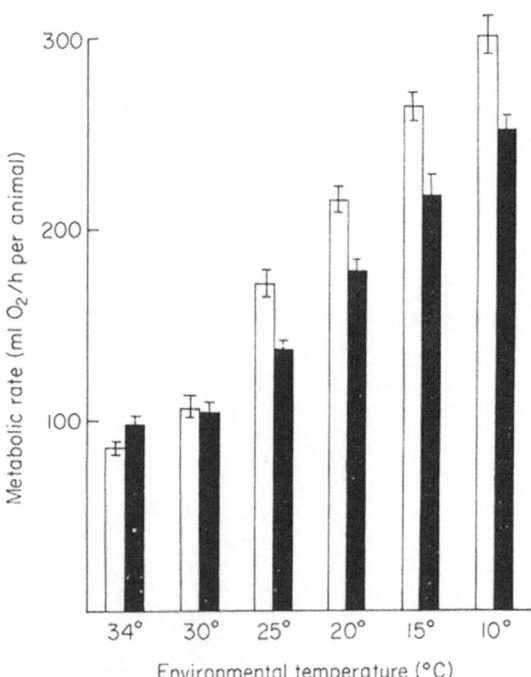

Figure 12.5   Metabolic rate of lean (□) and obese (■) mice during exposure to various environmental temperatures for 40 min. The results are the means ± S.E.M. (bars) for 12 animals in each group.

METABOLIC RATE AT VARIOUS ENVIRONMENTAL TEMPERATURES

The lower body temperature of the obese mice below 30°C implies that the mutants may expend less energy than lean mice on thermoregulation. This need not necessarily be the case, of course, since a lower body temperature is compatible with a high heat flux—through vasodilation or poor vasoconstriction—although this would be energetically wasteful for the obese animal, and reduce its metabolic efficiency.

In order to establish whether the lower body temperature of the obese mouse does indeed reflect a reduced energy expenditure on thermoregulation, the metabolic rate of lean and obese mice was measured at temperatures close to and below the thermoneutral zone (Trayhurn and James, 1978). At 34°C the resting metabolic rate (RMR) of the obese was slightly greater than that of the lean, while at 30°C it was similar. However, at 25°C and below the RMR of the obese, when expressed in absolute units, i.e. *per animal*, was some 20 per cent less than that of the lean. Between 30 and 25°C the increment in the RMR was twice as high in the lean as in the obese animals (figure 12.5). These metabolic rate measurements therefore clearly indicate that the energy expenditure on thermoregulatory thermogenesis is lower in obese mice than in lean.

The RMR of the obese reaches a maximum at about 10°C, while the lean animals show increases down to 0°C. At 4°C the RMR of the lean mice is substantially greater than the obese. The inability of the *ob/ob* mouse to survive low environmental temperatures is therefore clearly due to a reduced capacity for thermoregulatory thermogenesis, as first suggested by Davis and Mayer (1954).

NON-SHIVERING THERMOGENESIS

Thermoregulatory thermogenesis is made up of shivering and non-shivering components. The latter is termed 'regulatory' non-shivering thermogenesis (NST) to distinguish it from the heat produced due to the basal metabolic rate—'obligatory' non-shivering thermogenesis (Jansky, 1973). In adult animals it appears that skeletal muscle and the liver are the main organs involved in the production of heat through regulatory NST, with only a minor contribution coming from brown adipose tissue (Jansky, 1973). There are several methods for measuring the capacity of an animal for regulatory NST. The most commonly used method is to measure the RMR at thermoneutrality, before and after the administration of the catecholamine noradrenaline. This technique has been applied to obese *ob/ob* mice, which were found to have only half the capacity of lean mice for regulatory NST (Trayhurn and James, 1978). It therefore appears that the reduced capacity for thermoregulatory thermogenesis in the *ob/ob* mouse is due primarily to a reduction in the regulatory non-shivering component.

There is evidence which suggests that the reduced NST, as well as a number of the other abnormalities of the *ob/ob* mouse, is due to a defect in the thyroid-induced Na^+-K^+ ATPase (York *et al.*, 1978).

THE EFFECT OF ENVIRONMENTAL TEMPERATURE ON THE
MAINTENANCE REQUIREMENT

We have conducted two types of nutritional study to confirm the view that the *ob/ob* mouse achieves its high efficiency through a reduced energy expenditure on thermoregulatory thermogenesis. In the first study the maintenance requirement of young (24-28 day old) lean and obese mice was assessed at thermoneutrality (33°C) and at a temperature well below the thermoneutral zone (23°C). At the starting age used the body weight of the lean and obese animals was similar. Food intake for weight maintenance was measured for 6 days, and the intake over days 2-6 was averaged. The results obtained are shown in table 12.2 At 33° there was no significant difference in the food intake for maintenance between lean and obese animals. However, at 23°C the lean mice had a maintenance requirement which was 16 per cent greater than that of the obese. Between 33 and 23°C the maintenance requirement of the lean mice increased by 22.4 kJ/day, while that of the obese increased by only 16.6 kJ/day. The requirement for thermoregulatory thermogenesis therefore resulted in a 35 per cent greater increase in the maintenance requirement of lean mice compared with the obese.

Table 12.2   The energy requirement for weight maintenance of lean and obese mice at thermoneutrality and 23°C

	Gross energy intake (kJ/day)	
	33°	23°
Lean	35.0 ± 1.8	58.1 ± 2.2
Obese	32.8 ± 1.1	50.3 ± 2.8
	N. S.	$P < 0.02$

The results are the means ± S.E.M. for 6 pairs of animals at 33°C and 8 pairs at 23°C. The animals weighed approximately 20 g and included both males and females. A paired Student's *t*-test was used for the statistical analysis. N. S. = not significant ($P > 0.05$).

PAIR-FEEDING AT THERMONEUTRALITY AND 23°C

The second nutritional study that was performed to assess the quantitative significance of thermoregulatory thermogenesis in accounting for the high efficiency of the *ob/ob* mouse was to pair-feed young (25 day old) mutants to the *ad libitum* food intake of their lean litter mates, at either 23 or 33°C. The experiment was conducted for 10 days only, in order to cover the period of fast growth while minimising any secondary effects caused by obesity itself.

The results obtained are shown in figure 12.6. They are expressed in terms of the excess energy gain of the obese animal over the lean at the two temperatures. At 23°C the excess gain of the obese over the 10 day experimental period was 116 kJ. However, at 33°C the excess gain was considerably less, at 41 kJ. Thus, by conducting the pair-feeding at thermoneutrality two-thirds of the excess energy gain of the obese mice found at 23°C was abolished (Thurlby *et al.*, 1978).

The two nutritional studies are therefore consistent with the metabolic rate measurements, in showing that the low maintenance requirement and resulting high efficiency of the *ob/ob* mouse are due primarily to a reduced energy expen-

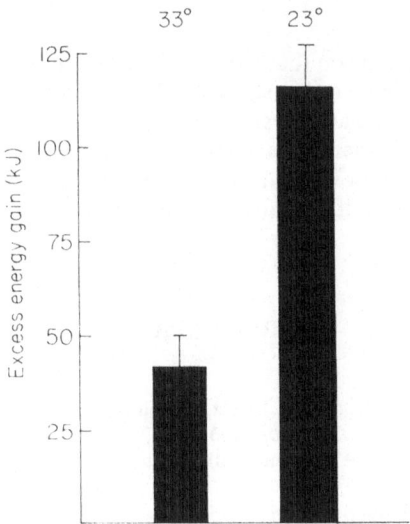

Figure 12.6   Excess energy gain of young obese mice pair-fed to the *ad libitum* food intake of lean litter mates at thermoneutrality and at 23°C. The results are expressed as the means ± S.E.M. (bars) for 8 animals at each temperature. The experiment was performed with males only.

diture on maintaining homeothermy. We would tentatively suggest that the main cause of this reduction in thermoregulatory thermogenesis is a lowered hypothalamic set-point for body temperature. Although there is no definitive evidence for this suggestion—indeed whether or not a 'set-point' as such exists is in dispute—the hypothesis is favoured by the observation that the body temperatures of pre-weanling and adult *ob/ob* mice are lower at normal environmental temperatures, despite the obese animals' spare capacity for thermogenesis, and by the occurrence of a normal diurnal rhythm in body temperature in the mutant.

THERMOREGULATION IN OTHER GENETICALLY OBESE
LABORATORY ANIMALS

The most detailed work on thermoregulation has been conducted with the *ob/ob* mouse. There is evidence, however, of thermoregulatory defects in several other strains of genetically obese rodent. The Zucker rat has been reported to have a lower body temperature than its lean litter mates (York *et al.*, 1972) and to be unable to survive an environmental temperature of 3-4°C (Trayhurn *et al.*, 1976). Similar results have been found with the diabetic mouse (Yen *et al.*, 1974; Trayhurn, unpublished observations). In addition, the yellow obese mouse (see Bray and York, 1971), the spiny mouse (Wise 1977a, b) and the *Ad* mouse (Trayhurn, unpublished observations) all fail to maintain a normal body temperature in the cold.

These observations indicate that thermoregulatory abnormalities are a general feature of genetically obese rodents. It is likely, therefore, that the mechanism by which the *ob/ob* mouse achieves its high metabolic efficiency will also explain the raised efficiency of other obese mutants. Preliminary results with the diabetic mouse (db^{ad}/db^{ad}) have shown that obese animals of this strain do, indeed, expend less energy than their lean litter mates on thermoregulatory thermogenesis.

REFERENCES

Alonso, L. G. and Maren, T. H. (1955). Effect of food restriction on body composition of hereditary obese mice. *Am. J. Physiol.*, **183**, 284–290

Bray, G. A. and York, D. A. (1971). Genetically transmitted obesity in rodents. *Physiol. Rev.*, **51**, 598–646

Chlouverakis, C. (1970). Induction of obesity in obese-hyperglycemic mice on normal food intake. *Experientia*, **26**, 1262–1263

Cox, J. E. and Powley, T. L. (1977). Development of obesity in diabetic mice pair-fed with lean siblings. *J. Comp. Physiol. Psychol.*, **91**, 347–358

Davis, T. R. A. and Mayer, J. (1954). Imperfect homeothermia in the hereditary obese-hyperglycemic syndrome of mice. *Am. J. Physiol.*, **177**, 222–226

Deb, S., Martin, R. J. and Hershberger, T. V. (1976). Maintenance requirement and energetic efficiency of lean and obese Zucker rats. *J. Nutr.*, **106**, 191–197

Herberg, L. and Coleman, D. L. (1977). Laboratory animals exhibiting obesity and diabetes syndromes. *Metabolism*, **26**, 59–99

James, W. P. T. and Trayhurn, P. (1976). An integrated view of the metabolic and genetic basis for obesity. *Lancet*, **ii**, 770–773

Jansky, L. (1973). Non-shivering thermogenesis and its thermoregulatory significance. *Biol. Rev.*, **48**, 85–132

Joosten, H. F. P. and van der Kroon, P. H. W. (1974). Role of the thyroid in the development of the obese-hyperglycemic syndrome in mice *(ob/ob)*. *Metabolism*, **23**, 425–436

Kaplan, M. L. and Leveille, G. A. (1974). Core temperature, O_2 consumption and early detection of *ob/ob* genotype in mice. *Am. J. Physiol.*, **227**, 912–915

Lin, P.-Y., Romsos, D. R. and Leveille, G. A. (1977). Food intake, body weight gain, and body composition of the young obese *(ob/ob)* mouse. *J. Nutr.*, **107**, 1715–1723

Miller, D. S. and Mumford, P. (1966). Obesity: physical activity and nutrition. *Proc. Nutr. Soc.*, **25**, 100–107

Ohtake, M., Bray, G. A. and Azukizawa, M. (1977). Studies on hypothermia and thyroid function in the obese (*ob/ob*) mouse. *Am. J. Physiol.*, **233**, R110–R115

Pullar, J. D. and Webster, A. J. F. (1974). Heat loss and energy retention during growth in congenitally obese and lean rats. *Br. J. Nutr.*, **31**, 377–392

Pullar, J. D. and Webster, A. J. F. (1977). The energy cost of fat and protein deposition in the rat. *Br. J. Nutr.*, **37**, 355–363

Thurlby, P. L. and Trayhurn, P. (1978). The development of obesity in preweanling *ob/ob* mice. *Br. J. Nutr.*, **39**, 391–396

Thurlby, P. L., Trayhurn, P. and James, W. P. T. (1978) An explanation for the elevated efficiency of the genetically obese (*ob/ob*) mouse. *Proc. Nutr. Soc.*, **37**,55A

Trayhurn, P. and James, W. P. T. (1978). Thermoregulation and non-shivering thermogenesis in the genetically obese (*ob/ob*) mouse. *Pflügers Arch. Eur. J. Physiol.*, **373**, 189–193

Trayhurn, P., Thurlby, P. L. and James, W. P. T. (1976). A defective response to cold in the obese (*ob/ob*) mouse and the obese Zucker (*fa/fa*) rat. *Proc. Nutr. Soc.*, **35**, 133A

Trayhurn, P., Thurlby, P. L. and James. W. P. T. (1977). Thermogenic defect in pre-obese *ob/ob* mice. *Nature*, **266**, 60–62

Welton, R. F., Martin, R. J. and Baumgardt, B. R. (1973). Effects of feeding and exercise regimens on adipose tissue glycerokinase activity and body composition of lean and obese mice. *J. Nutr.*, **103**, 1212–1219

Wise, P. H. (1977a). Significance of anomalous thermoregulation in the pre-diabetic spiny mouse (*Acomys cahirinus*): oxygen consumption and temperature regulation. *Aust. J. Exp. Biol. Med. Sci.*, **55**, 463–473

Wise, P. H. (1977b). Significance of anomalous thermoregulation in the pre-diabetic spiny mouse (*Acomys cahirinus*): cold tolerance, blood glucose and food consumption responses to environmental heat. *Aust. J. Exp. Biol. Med. Sci.*, **55**, 475–484

Woodward, C. J. H., Trayhurn, P. and James, W. P. T. (1977). Costs of maintenance and growth in genetically obese (*ob/ob*) mice. *Proc. Nutr. Soc.*, **36**, 115A

Yen, T. T. T., Fuller, R. W. and Pearson, D. V. (1974). The response of 'obese' (*ob/ob*) and 'diabetic' (*db/db*) mice to treatments that influence body temperature. *Comp. Biochem. Physiol.*, **49A**, 377–385

York, D. A., Bray, G. A. and Yukimura, Y. (1978). An enzymatic defect in the obese (*ob/ob*) mouse: loss of the thyroid-induced sodium-potassium dependent adenosine-triphosphatase. *Proc. Nat. Acad. Sci. USA*, **75**, 477–481

York, D. A., Hershman, J. M., Utiger, R. D. and Bray, G. A. (1972). Thyrotropin secretion in genetically obese rats. *Endocrinology*, **90**, 67–72

Zucker, L. M. (1975). Efficiency of energy utilization by the Zucker hereditarily obese rat 'fatty'. *Proc. Soc. Exp. Biol. Med.*, **148**, 498–500

13

Immunity in
genetically obese rodents

C. J. Meade* and J. Sheena (Transplantation Biology Section,
Clinical Research Centre, Harrow, Middlesex, UK)

SUMMARY

(1)   The obese mouse (C57BL/6J-*ob*/*ob*) shows impaired cellular immunity.

(2)   Its T lymphocytes are not irreversibly functionally different from the T
lymphocytes of lean mice; it is just that in the obese mouse these lymphocytes are
working in an environment which depresses their activity.

(3)   Obesity does not depress antibody production, but deposition of material
staining with anti-IgG, anti-IgM and anti-C_3 sera suggests that abnormally high
quantities of autoantibodies are produced.

(4)   Depressed T cell function is not itself a cause of obesity, but consideration
should be given to a possible role for immune and autoimmune processes in the
aetiology of some of the metabolic disturbances associated with the *ob*/*ob* genotype.

(5)   Statements 1 and 2 are also valid for the diabetic mouse (C57BL/KsJ-
db/*db*).

This review will primarily cover work carried out in the Clinical Research Centre
on the genetically obese mouse (C57BL/6J-*ob*/*ob*). It will also refer to American
work on the diabetic mouse (C57BL/KsJ-*db*/*db*). The terms 'obese mouse' and
'diabetic mouse' will be reserved specifically for these mutants. We shall attempt
to show how obese rodents can be used as models to analyse the relationship between obesity, diabetes and immunity (see figure 13.1).

The existence of such relationships in clinical medicine is only now becoming
appreciated. Cellular immunity is impaired in poorly controlled diabetes (Brody
and Merlie, 1970; Bagdade *et al.*, 1974; MacCuish *et al.*, 1974; but see Ragab *et al.*,
1972). This has been suggested as one explanation for the increased prevalence of
infections in patients with diabetes mellitus (Bondy, 1967). Obese subjects are also
subject to a disproportionate number of certain types of infections—e.g. post-operative wound infections (Meares, 1975). Autoimmune processes are of increased
frequency in diabetes, and may have a role in the development of certain forms of
the disease (Irvine *et al.*, 1977). Our special interest has been the area of overlap
between diabetes, obesity and immunity, the heavily shaded area in figure 13.1.

* Present address: Lilly Research Centre Ltd, Earlwood Manor, Windlesham, Surrey, UK

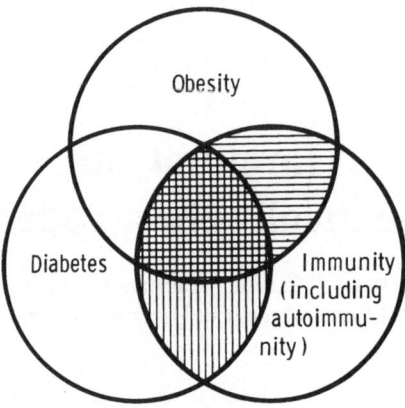

Figure 13.1 Overlap between obesity, diabetes and immunity

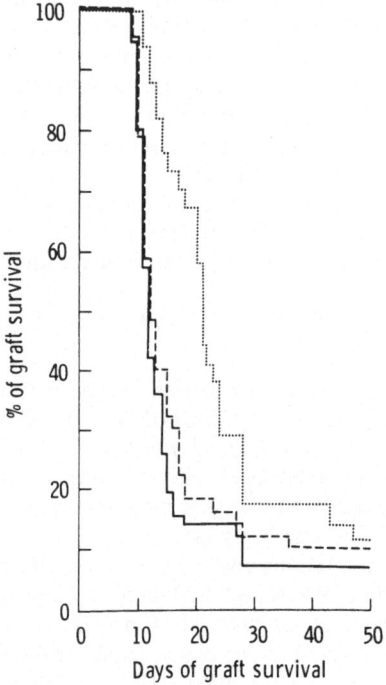

Figure 13.2   Survival of male skin graft on female mice (a weak rejection system).
Mice were primed with a male tail skin graft, then 75 days later challenged with a
second male graft placed contralaterally to the first graft. The graph shows the
survival of the second set grafts. - - - - - Obese mice (*ob/ob*, 36 mice), mean weight
52 g; ——— lean litter mates (+/?, 57 mice), mean weight 24 g; – – – – age- and sex-
matched lean controls (+/+, 50 mice), mean weight 21g. From Sheena and Meade
(1978).

Human obesity and diabetes themselves are such heterogeneous entities that we have turned to the animal model to look for relationships that may be investigated clinically.

Lymphocytes may be divided into two main subclasses—T cells, which require passage through the thymus in order to mature, and B cells, which do not. T cells may be recognised by characteristic proteins on their surface detectable with appropriate antisera—e.g. in mice antisera against the Thy 1.2 antigen. In man T cells may be characterised by their ability to bind sheep erythrocytes. T cells are responsible for cellular immunity—that is, immunity that can be transferred from animal to animal only by transfer of lymphocytes and not by transfer of antibody. B cells, by contrast, produce antibody.

Mice homozygous for the obese mutation have depressed cellular immunity. Thus, the immunological inflammatory reaction to a contact sensitising agent, picryl chloride, is diminished (table 13.1). The ability of female mice to reject a male skin graft is also reduced (figure 13.2; Sheena and Meade, 1978). By contrast, Finger *et al.* (1971) found no difference between obese mice and lean litter mates in their ability to form an antibody against sheep red blood cells, nor was there any difference in the number of spleen cells forming antibody against this antigen. This number was found by overlaying mouse lymphocytes with agar containing sheep red cells and counting the number of 'plaques' (areas of haemolysis) formed round individual antibody-producing cells. Thus, in our study T cell functions were impaired, whereas in the study by Finger and co-workers B cell function was unaffected.

Table 13.1 Response of obese and lean C57BL/6J mice to a contact-sensitising agent

	Number of mice	Body weight (g)	Increase in ear thickness (mm)
Obese mice (*ob/ob*)	20	38 ± 2	0.103 ± 0.008
Lean litter mates (+/*ob* or +/+)	20	24 ± 1	0.185 ± 0.014
Pure breeding age- and sex-matched lean mice (+/+)	9	21 ± 1	0.206 ± 0.009

Comparison of ear swelling in obese and lean 2 month old male mice. Mice were primed by painting with picryl chloride on body and paws, then challenged 7 days later by painting the ears. The increase in ear thickness 24 h after challenge was used as a quantitative measure of the immune response. The difference in ear swelling between lean and obese mice was significant (0.01 > P); the difference between the two types of lean mice (litter mates and pure breeding mice) was not significant (0.05 < P). When unprimed +/+ mice were ear-painted with picryl chloride, the mean increase in ear thickness (7 mice) was only 0.009 mm. Hence, making the initial control reading, which involved applying a mild pressure to the ear, and possible non-immunological irritant effects of the picryl chloride solution, did not cause significant ear swelling. From Sheena and Meade (1978).

Studies on T and B cell functions in diabetic mice have provided very similar results. Mahmoud *et al.* (1976) found a depressed cellular immune reaction to the eggs of *Schistosoma mansoni.* Fernandes *et al.,* (1978) found impaired ability of C57BL/KsJ mice homozygous for the diabetic gene to react against C57BL/6 skin. By contrast, the plaque-forming cell response following *in vivo* challenge with sheep red blood cells was enhanced in diabetic mice in comparison with controls. Skin graft rejection measures predominantly a host versus graft reaction; skin contains very few cells capable of reacting against host tissues. If the organ transplanted contains large numbers of lymphocytes (for example, the spleen), then a graft versus host reaction also occurs. When spleen cells are transplanted from a mouse bearing histocompatibility antigens A to a mouse bearing histocompatibility antigens A and B, then the grafted spleen cells react against host B antigens, but the host cannot react against the spleen cells, since these share A antigens with the host, which is therefore tolerant to such antigens. The F_1 hybrid of strain A and strain B (in our studies C57BL/6J and DBA/2J) therefore provides a suitable host for comparing the ability of cells from obese and lean mice to mount a graft versus host reaction without the complication of a host versus graft reaction. Our experimental design is shown in figure 13.3. Increase in spleen weight is used to quantify immune response.

The experimental design permits spleen cells from lean and obese mice to be compared in the same environment. As table 13.2 shows, in a common environ-

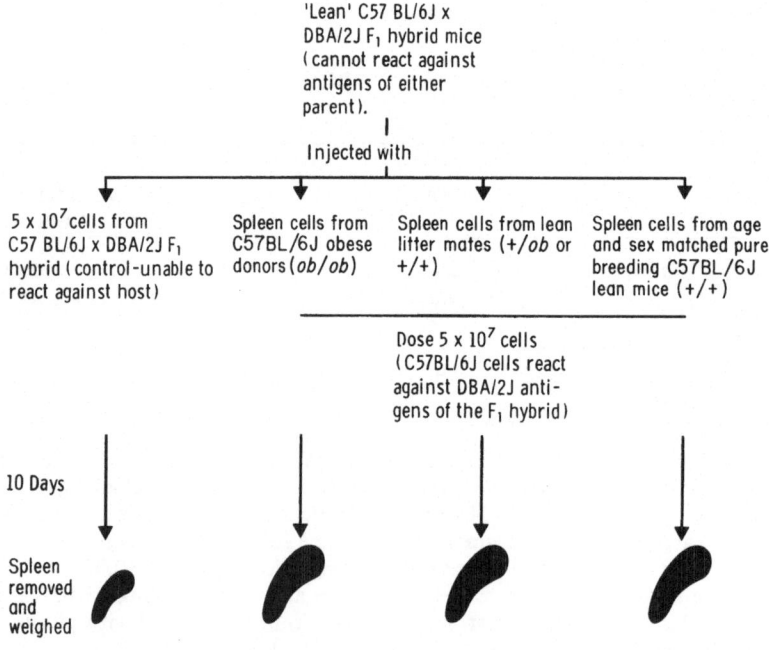

Figure 13.3

Table 13.2 Ability of cells from obese mice and lean controls to mount a graft versus host reaction in lean C57BL/6J × DBA/2J F$_1$ recipients

Donors	Number of C57BL/6J × DBA/2J F$_1$ recipients used		Mean body weight (± S.D.) of spleen cell donors (g)		Weight of C57BL/6J × DBA/2J F$_1$ (recipient) mouse spleens 10 days after injection of cells from lean or obese mice (mg)	
	Experiment 1	Experiment 2	Experiment 1	Experiment 2	Experiment 1	Experiment 2
Obese mice (C57BL/6J ob/ob)	17	8	40 ± 4	45 ± 5	211 (181–245)	181 (139–236)
Lean litter mates (C57BL/6J +/ob or +/+)	18	9	20 ± 1	21 ± 1	247 (223–272)	169 (150–192)
Pure breeding age- and sex-matched lean C57BL/6J mice (+/+)	10	10	18 ± 1	19 ± 1	243 (219–270)	133 (107–164)
B6D2 F$_1$ mice (unstimulated control)	15	10	30 ± 5	31 ± 6	83 (69–100)	78 (76–79)

Mean spleen weight (with 95 per cent confidence limits) after intravenous injection of 5 × 10^7 spleen cells from female mice into female C57BL/6J × DBA/2J F$_1$ recipients. Data have been treated in logarithmic transformation. A two-way analysis of variance and Duncan's multiple range test showed, in both experiments combined, no significant difference in F$_1$ recipient spleen weight between groups injected with cells from either lean or obese C57BL/6J mice. The only significant difference was between the control mice (injected with C57BL/6J × DBA/2J F$_1$ cells) and the three stimulated experimental groups (injected with C57BL/6J cells). For this difference, 0.01 > P. From Sheena and Meade (1978).

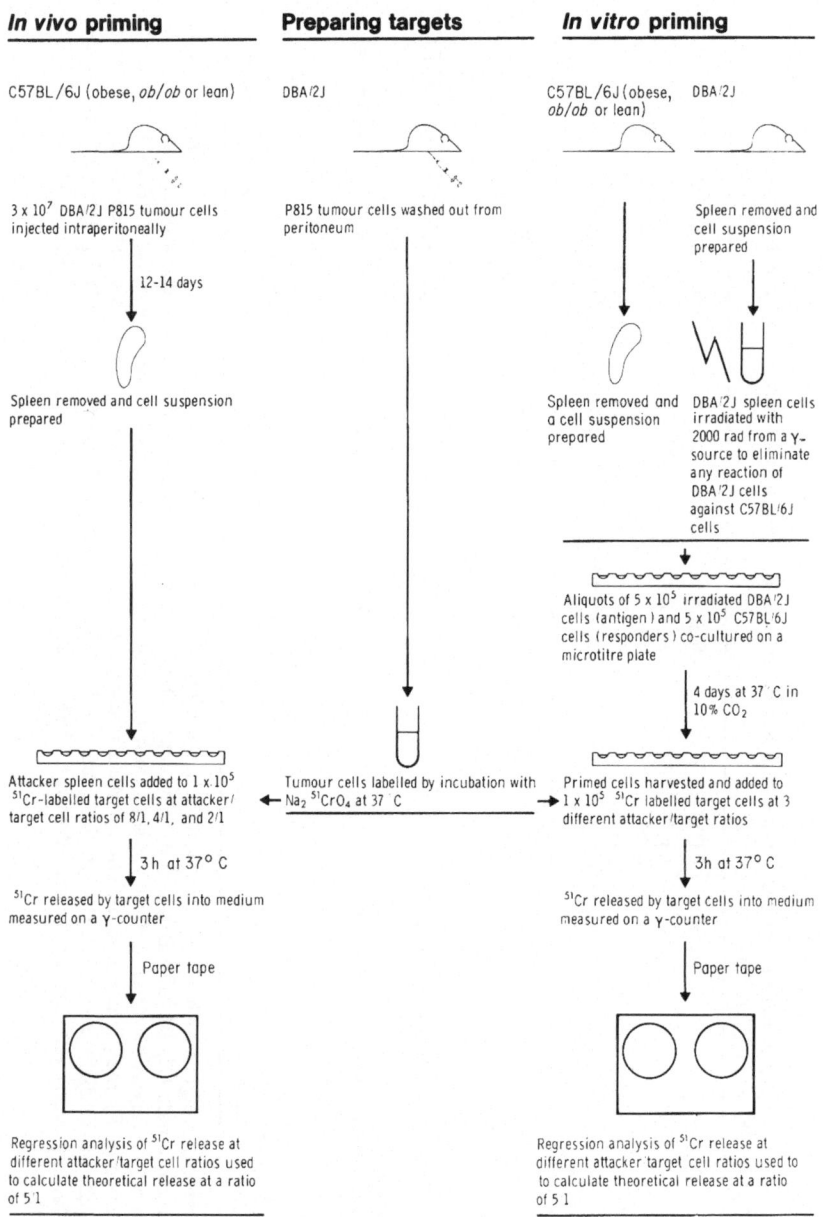

Figure 13.4 Effect of the *ob/ob* genotype on the ability of mice to generate cytotoxic T cells. Experimental design for comparison of *in vitro* and *in vivo* priming.

ment, similar numbers of spleen cells from either lean or obese mice produce a similar response (Sheena and Meade, 1978). There are two probable ways to explain this result. The first explanation is that the depressed cellular immunity of obese mice results from fewer immune cells, but what cells there are in obese mice are, cell for cell, equal in efficiency to those from lean mice. It is certainly true that the spleens of obese mice are slightly smaller than those of lean litter mates, and since the proportion of Thy 1.2-bearing cells is similar, the total number of T cells must be reduced (Meade *et al.*, 1979). The other explanation is that it is factors in the environment (e.g. hormonal environment) of the lymphocytes of obese mice which depress their immune reactivity, and any changes produced are completely reversible in a 'lean' environment.

This second hypothesis may be tested by carrying out the rejection reaction *in vitro*. Figure 13.4 illustrates the experimental design. The experiment consists of two parts. In the first part cells able to kill a tumour graft are generated, in either the different environment of lean or obese host mice (*in vivo* priming) or the same artificial environment provided by tissue culture medium (*in vitro* priming). In the second part the killer cells generated are assayed by their ability to kill tumour cells and make them release an isotope (^{51}Cr) which cannot be released from cells except

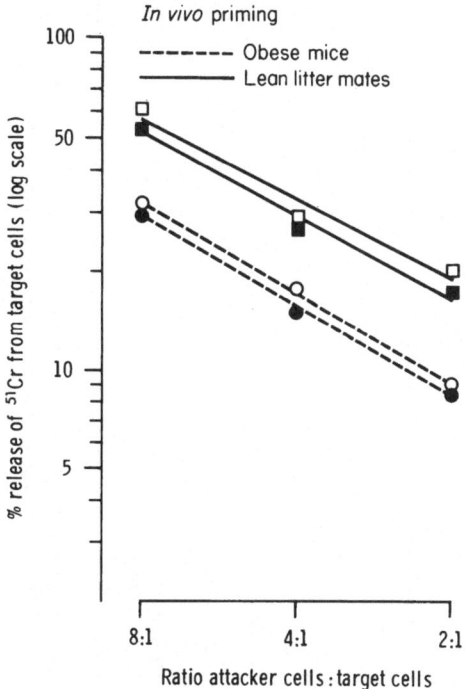

Figure 13.5 Generation of cytotoxic T cells in obese mice and lean litter mates following *in vivo* priming. Results from two obese mice (○ and ●) and two lean litter mates (□ and ■) are shown.

Table 13.3 Comparison of cell-mediated cytotoxicity following *in vivo* or *in vitro* sensitisation

	Percentage corrected lysis at attacker/target ratio 5/1							
	12–14 days after *in vivo* sensitisation with DBA/2 mastocytoma				4 days after *in vitro* sensitisation with irradiated DBA/2 mastocytoma			
	Experiment 1		Experiment 2		Experiment 1		Experiment 2	
Source of cytotoxic attacker cells	Animal I	Animal II	Animal I	Animal II	Animal I	Animal II	Animal I	Animal II
Obese mice (C57BL/6J-*ob/ob*)	7.2 (±2.5)	8.3 (±3.1)	23.2 (±2.0)	21.0 (±3.7)	40.9 (±1.0)	40.5 (±2.4)	38.8 (±0.9)	27.0 (±0.5)
Lean litter mates (+/*ob* or +/+)	21.4 (±4.8)	17.2 (±2.1)	38.7 (±6.8)	43.1 (±9.3)	45.8 (±2.5)	47.0 (±1.1)	36.9 (±3.6)	39.9 (±5.9)
Pure breeding age- and sex-matched lean mice (C57BL/6J-+/+)	40.6 (±2.8)	31.6 (±3.8)	35.7 (±5.3)	53.1 (±7.4)	38.8 (±2.2)	38.8 (±2.5)	39.3 (±4.2)	45.2 (±1.5)

Kill was measured in quadruplicate at three or four attacker/target cell ratios, and a theoretical percentage corrected lysis was determined at an attacker/target ratio of 5/1 from a multiple-fit regression analysis (± one standard error).

$$\text{Percentage corrected lysis} = 100 \times \frac{\text{counts released in presence of primed cells} - \text{counts released in presence of unprimed cells}}{\text{maximum release (obtained after killing all cells with detergent)} - \text{background counts of counting machine}}$$

when they die. As figure 13.5 and table 13.3 show, equal numbers of spleen cells from obese or lean mice, primed *in vivo*, differ in their killing ability. This difference disappears when killer cells are generated in a common, *in vitro* environment (figure 13.6, table 13.3). Thus, reversible environmental effects do contribute to depressed immunity in obese mice (Meade *et al.*, 1979).

Similar results were obtained by Fernandes *et al.* (1978), comparing *in vitro* and *in vivo* priming in diabetic mice. These workers also showed that enhancement by the *db/db* genotype of the plaque-forming cell response to sheep red cells was no longer found when priming was *in vitro* rather than *in vivo*.

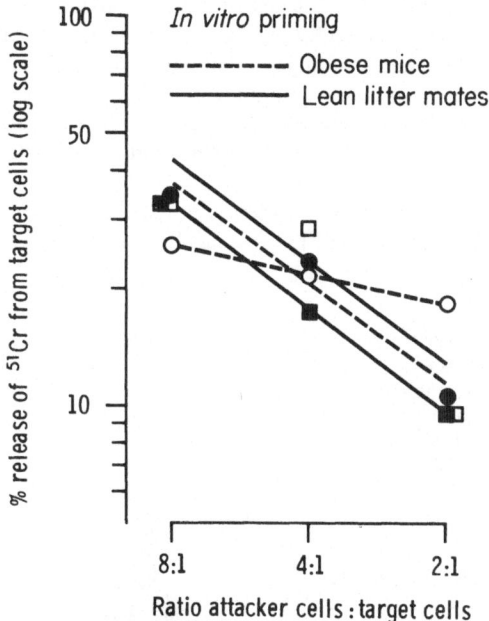

Figure 13.6   Generation of cytotoxic T cells in obese mice and lean litter mates following *in vitro* priming. Symbols have the same meaning as in figure 13.5.

The nature of the environmental factor depressing immune function in obese mice is unclear. Adrenalectomy has some effect on the ability of the *ob/ob* genotype to depress spleen cell numbers (Meade *et al.*, 1978), but the hypercorticosteroidism of obese mice is unlikely to be the sole factor involved

Immunological changes observed following treatment of massive obesity by small intestinal shunts suggest that changes in the immune system similar to those seen in obese rodents may also occur in human obesity. Hallberg *et al.* (1976) found increased reactivity to purified protein derivative of tuberculin (PPD) following treatment of obesity. A discrepancy between alterations in *in vitro* and *in vivo* immune parameters was reported reminiscent of the discrepancy between the *in vitro* and *in vivo* immune depressive effects of the *ob/ob* or *db/db* genotypes.

Thus, an *in vivo* reaction in which T cells are important, the delayed skin reaction to PPD, was markedly enhanced after treatment of obesity, although the ability of PPD *in vitro* to stimulate lymphocyte [^{14}C]-thymidine uptake was hardly affected. Nor was there any change in the number of T cells in the blood (as assayed by counting cells able to form rosettes of bound sheep erythrocytes). *In vitro* measurement of reactivity to phytohaemagglutinin (a T cell stimulant) showed only a slight post-operative increase. Serum concentrations of three major classes of immunoglobulin, IgA, IgG and IgM, were unaffected and 12 months post-operatively there was a small decrease in the titre of antistapholysin antibodies. Other evidence suggesting that immunological abnormalities may be a feature of obesity in general and not just of the particular expression of the *ob/ob* or *db/db* genotypes was provided by Newberne (1966) in his study of the effects of over-nutrition on resistance of dogs to distemper virus. This study unfortunately included no assays for cellular immune reactivity, but spleen follicle size was more depressed by infection in overfed dogs, and the incidence of paralytic encephalitis was higher. Overfed animals had normal or high titres of antiviral antibodies.

There is a frequent association between depressed T cell function and auto-immune disease (Allison *et al.,* 1971). This has been explained either by ascribing

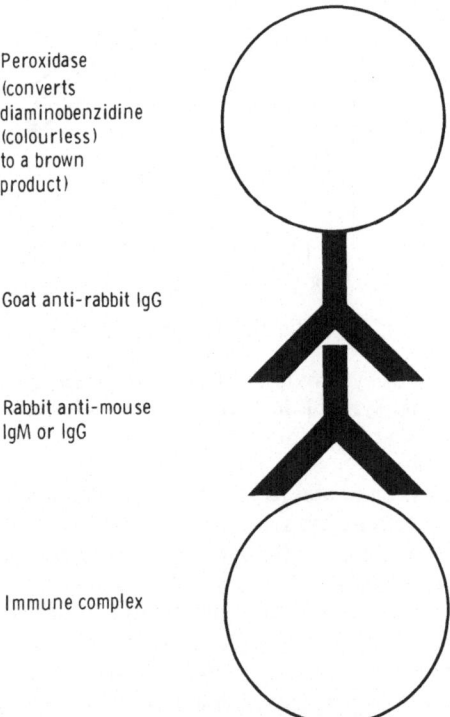

Peroxidase
(converts
diaminobenzidine
(colourless)
to a brown
product)

Goat anti-rabbit IgG

Rabbit anti-mouse
IgM or IgG

Immune complex

Figure 13.7   Immunoglobulin in immune complexes can be demonstrated histochemically by attaching peroxidase through an antibody 'sandwich'. Peroxidase is revealed by flooding the section with diaminobenzidine.

to a subpopulation of T cells a specific function suppressing the activity of lymphocytes directed against self-antigens (Allison *et al.*, 1971) or by granting T cells a role in the removal of foreign material and effete self-antigens (Bellanti and Green, 1971).

Immune complexes formed by reaction of autoantibodies against circulating self-antigens will be deposited in the kidney, where they may be detected using appropriate anti-immunoglobulin antisera attached either to a fluorescent label or to peroxidase which can convert diaminobenzidine from a colourless to a brown product (figure 13.7). The glomeruli of 12 week old obese mice showed significantly increased quantities of deposited IgG or IgM in comparison with lean litter mates. The difference between lean and obese mice was particularly marked when anti-IgM antisera were used; obese mice were all strongly positive whilst lean litter mates showed virtually no deposits (figures 13.8 and 13.9). C_3 binds immune complexes and deposits also reacted with anti-C_3 sera (Bourne 1975; Meade, Sowter and Bourne, 1978). Damage resulting from immune complex deposition could explain the glomerular lesions reported in older obese mice (Nathorst-Windahl and Hellman, 1964; Bergstrand *et al.*, 1968).

The glomeruli of obese mice also reacted with anti-insulin antisera. This reaction was blocked by added excess insulin. This suggests that insulin is one of the self-antigens with which autoantibodies are complexed in the glomeruli. The possible contribution of insulin-anti-insulin complexes to insulin resistance in obese mice remains to be explored. Since insulin has several antigenic determinants to any of which antibody may be bound, it is possible that the usual radio-immunoassay procedures for serum insulin detect insulin to which antibody is already bound and which may or may not retain its biological activity. It is not known to what extent insulin–anti-insulin complexes are able to occupy insulin receptors and, once these receptors are occupied, trigger a response to this hormone.

Anti-insulin antibodies were described in a study including human obese adult onset diabetics by Rao *et al.* (1974). However, since the patients studied received exogenous insulin, it is not clear to what extent autoantibodies were caused by insulin injection. Farrant and Shedden (1965) have demonstrated fluorescein-labelled insulin in renal complexes from patients never treated ('immunised') with exogenous insulin. Possibly the obese mouse provides a model for understanding the origin of 'spontaneous' anti-insulin autoantibodies.

Another way in which impaired T cell function might predispose to diabetes is by predisposing to infection with diabetogenic viruses. Cellular immunity plays an important protective role in many viral infections. Webb *et al.* (1976) reported that mice homozygous for the diabetic gene (*db/db*) were more susceptible to Coxsackie virus B4 than normal (+/+) mice. Although the heterozygote (+/*db*) is, in many respects, indistinguishable from the lean homozygote (+/+), it shows a greater mortality after Coxsackie infection in comparison with the lean homozygote, yet less susceptibility than the *db/db* mouse. Unlike infected +/+ variants, there was no inflammatory cell infiltrate in the pancreases of any of the infected *db/db* or +/*db* mice examined.

It is clear that the obesity of the *ob/ob* mouse is not caused by the T cell defect, since mice homozygous for the nude gene (*nu/nu*) as well as the obese gene (*ob/ob*) are still considerably more obese than age- and sex- matched *nu/nu* +/+ mice, although they lack functional circulating T cells. Lean nude mice develop early mani-

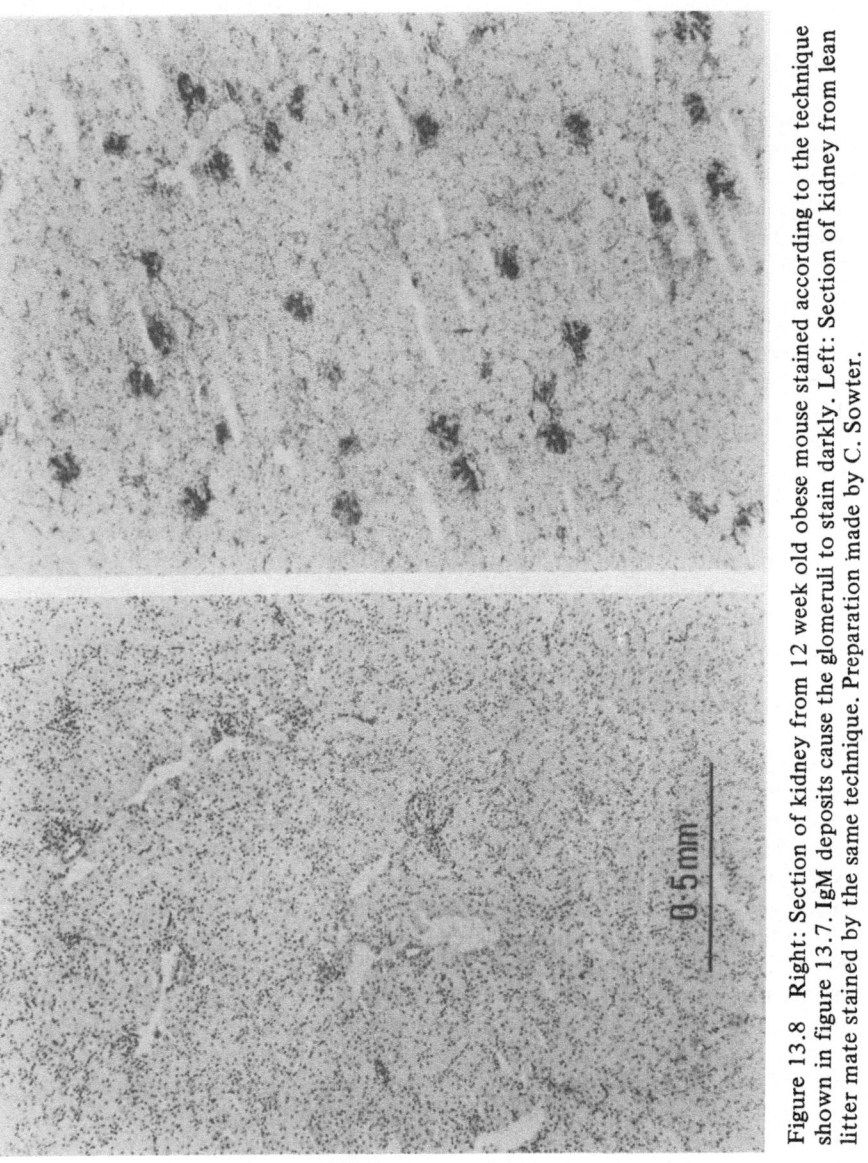

Figure 13.8 Right: Section of kidney from 12 week old obese mouse stained according to the technique shown in figure 13.7. IgM deposits cause the glomeruli to stain darkly. Left: Section of kidney from lean litter mate stained by the same technique. Preparation made by C. Sowter.

0·5 mm

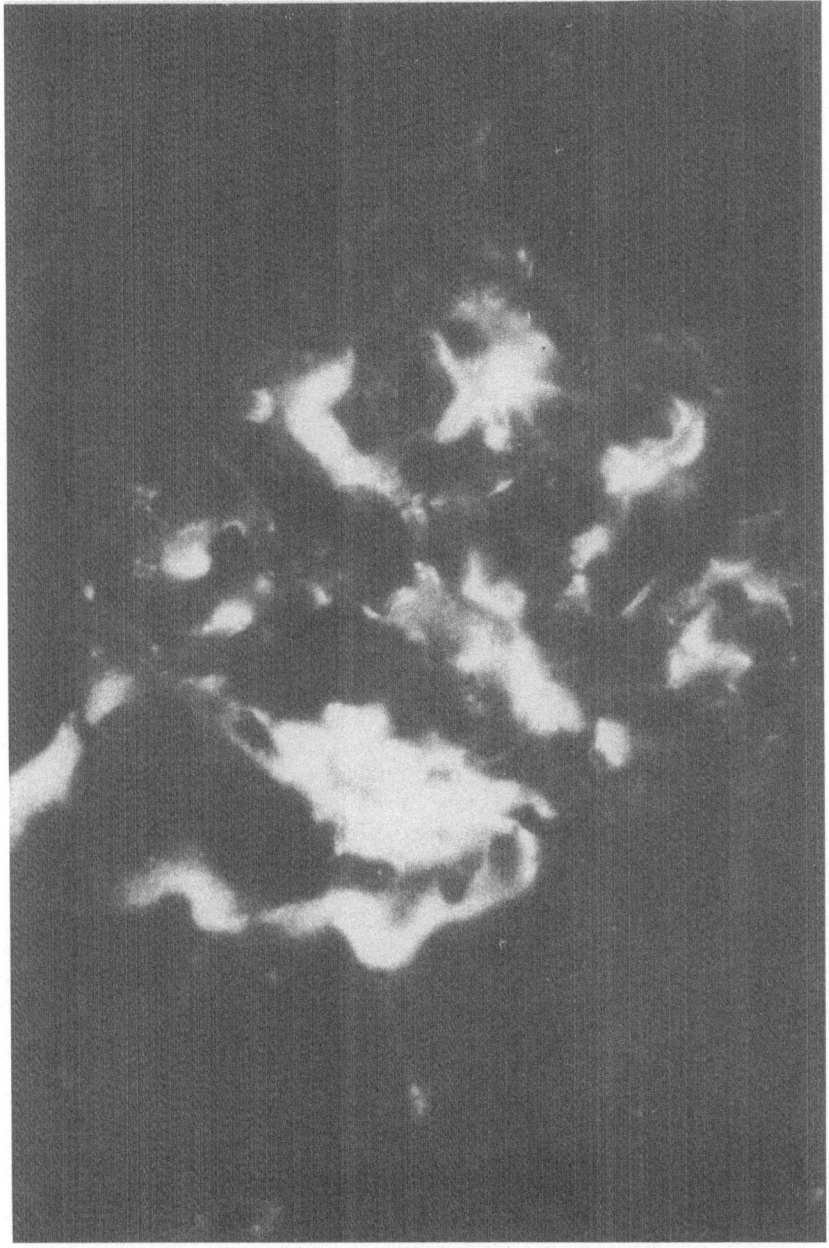

Figure 13.9 High-power photomicrograph of glomerulus from 12 week old obese mouse stained with anti-mouse IgM to which a fluorescent label was attached. Preparation made by C. Sowter.

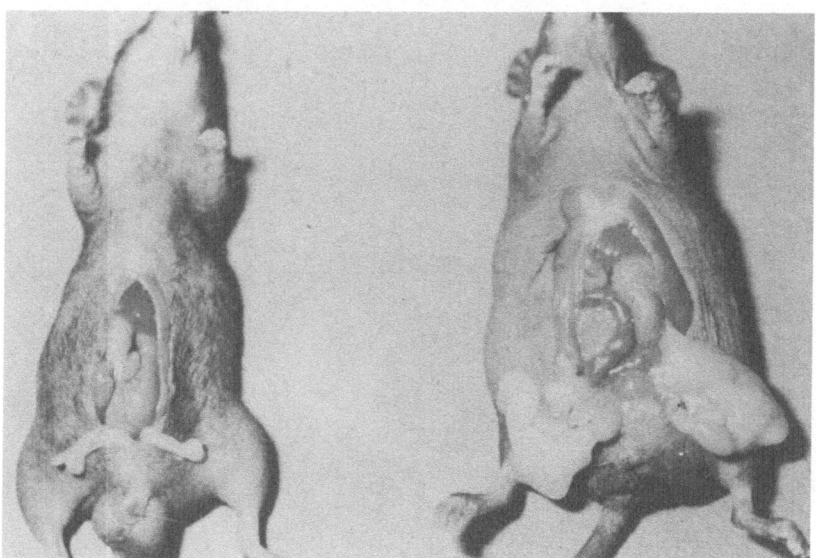

Figure 13.10   Right: Mouse of genotype *nu/nu ob/ob*. Left: Lean *nu/nu* litter mate. Bred by David Tucker, MRC Laboratories, Carshalton.

festations of spontaneous autoimmunisation, including immunoglobulin deposits in glomeruli of the kidney (Pelletier *et al.,* 1975). Thus obesity is also not a consequence of a generalised autoimmune reaction. However, a role for autoimmune phenomena in the development of diabetes and other metabolic disturbances is not excluded. Lean *nu/nu* mice themselves suffer from a number of endocrine disturbances (Shire and Pantelouris, 1974; Ruitenberg and Berkvens, 1977).

One other example of an association between autoimmune disease and obesity has been reported—the obese chicken, which suffers from autoimmune thyroiditis (Wick, 1970). In view of the other studies reported here, a final question must be asked, a question which repeatedly appears in studying the relationship of diabetes, obesity and immunity. Obesity and autoimmune thyroiditis: which came first— the chicken or the egg?

REFERENCES

Allison, A. C., Denman, A. M. and Barnes, R. D., (1971). Cooperating and controlling functions of thymus-derived lymphocytes in relation to autoimmunity. *Lancet,* ii, 135–140

Bagdade, J. D., Root, R. K. and Bulger, R. J. (1974). Impaired leukocyte function in patients with poorly controlled diabetes. *Diabetes,* **23,** 9–15

Bellanti, J. A. and Green R. E. (1971). Immunological reactivity. Expression of efficiency in elimination of foreignness. *Lancet,* ii, 526–529

Bergstrand, A., Nathorst-Windahl, G. and Hellman, B. (1968). The electron microscopic appearance of the glomerular lesions in obese-hyperglycaemic mice. *Acta Pathol. Microbiol. Scand.*, 74, 161–168

Bondy, P. K. (1967). In *Textbook of Medicine*, 13th edn. (ed. P. B. Beeson and W. McDermott), Saunders, Philadelphia, p. 1654

Bourne, R. C. (1975). Insulin resistance in maturity-onset diabetes: studies of the obese hyperglycaemic mouse (*ob/ob*) as a model for evaluating the aetiology of the human disease. M. I. Biol. Thesis, Ewell County Technical College

Brody, J. I. and Merlie, K. (1970). Metabolic and biosynthetic features of lymphocytes from patients with diabetes mellitus: similarities to lymphocytes in chronic lymphatic leukemia. *Br. J. Haematol.*, 19, 193–201

Farrant, P. C. and Shedden, W. I. (1965). Observations on the uptake of insulin conjugated with fluorescein isothiocyanate by diabetic kidney tissue. *Diabetes*, 14, 274–278

Fernandes, G., Handwerger, B. S., Yunis, E. J. and Brown, D. M. (1978). Immune response in the mutant diabetic C57BL/Ks-db + mouse: discrepancies between *in vitro* and *in vivo* immunological assays. *J. Clin. Invest.*, 61, 243–250

Finger, H., Rakow, L. and Beneke, G. (1971). Fettsucht und Antikörperbildungsvermögen. *Pathol. Microbiol.*, 37, 449–458

Hallberg, D., Nilsson, B. S. and Backman, L. (1976). Immunological function in patients operated on with small intestinal shunts for morbid obesity. *Scand. J. Gastroenterol.*, 11, 41–48

Irvine, W. J., McCallum, C. J., Gray, R. S., Campbell, C. J., Duncan, L. J. P., Farquhar, J. W., Vaughan, H. and Morris, P. J. (1977). Pancreatic islet-cell antibodies in diabetes mellitus correlated with the duration and type of diabetes, coexistent autoimmune disease and HLA type. *Diabetes*, 26, 138–147

MacCuish, A. C., Urbaniak, S. J., Campbell, C. J., Duncan, L. J. P. and Irvine, W. J. (1974). Phytohaemagglutinin transformation and circulating lymphocyte subpopulations in insulin-dependent diabetic patients. *Diabetes*, 23, 708–712

Mahmoud, A. A., Rodman, H. M., Mandel, M. A. and Warren, K. S. (1976). Induced and spontaneous diabetes mellitus and suppression of cell-mediated immunologic responses. Granuloma formation, delayed dermal reactivity, and allograft rejection. *J. Clin. Invest.* 57, 362–367

Meade, C. J., Sheena, J. and Mertin, J. (1979). Effects of the obese (*ob/ob*) genotype on spleen cell immune function. *Int. Arch. Allergy Appl. Immunol.* (in press)

Meade, C. J. and Sowter, C. (1979). Autoantibody production associated with the obese (*ob/ob*) genotype (in preparation)

Meares, E. M. (1975). Factors that influence surgical wound infections. *Urology*, 6, 535–545

Nathorst-Windahl, G. and Hellman, B. (1964). Lipohyalin glomerular lesions in ageing obese-hyperglycemic mice. *Med. Exp.*, 10, 67–71

Newberne, P. M. (1966). Overnutrition on resistance of dogs to distemper virus. *Fed. Proc.*, 25, 1701–1710

Pelletier, M., Hinglais, N. and Bach, J. F. (1975). Characteristic immunohistochemical and ultrastructural glomerular lesions in nude mice. *Lab. Invest.*, 32, 388–396

Ragab, A. H., Hazlett, B. and Cowan, D. H. (1972). Response of peripheral lymphocytes from patients with diabetes mellitus to phytohaemagglutinin and *Candida albicans* antigen. *Diabetes*, 21, 906–907

Rao, K. J., Page Faulk, W., Karam, J. H., Grodsky, G. M. and Forsham, P. H. (1974). Evidence in support of the concept of immune complex disease in insulin-treated diabetics. In *Immunity and Autoimmunity in Diabetes Mellitus* (ed. P. A. Bastenie and W. Gepts), Excerpta Medica, Amsterdam, and American Elsevier, New York, pp. 255–263

Ruitenberg, E. J. and Berkvens, J. M. (1977). The morphology of the endocrine system in cogenitally athymic (nude) mice. *J. Pathol. Bacteriol.*, 121, 225–231

Sheena, J. and Meade, C. J. (1978). Mice bearing the *ob/ob* mutation have impaired immunity. *Int. Arch. Allergy Appl. Immunol.* 57, 263–268

Shire, J. G. M. and Pantelouris, E. M. (1974). Comparison of endocrine function in normal and genetically athymic mice. *Comp. Biochem. Physiol.,* 47A, 93–100

Webb, S. R., Loria, R. M., Madge, G. E. and Kibrick, S. (1976). Susceptibility to mice to group B Coxsacki virus is influenced by the diabetic gene. *J. Exp. Med.,* 143, 1239–1247

Wick, G. (1970). The effect of bursectomy, thymectomy and X-irradiation on the incidence of precipitating liver and kidney autoantibodies in chickens of the obese strain (OS). *Clin. Exp. Immunol.,* 7, 187–199

14

Comparison of
genetic models
of obesity in animals
with obesity in man

W. P. T. James, M. J. Dauncey, R. T. Jung, P. S. Shetty and P. Trayhurn
(Dunn Clinical Nutrition Centre, Addenbrooke's Hospital, Cambridge, UK)

SUMMARY

There are many parallels between the genetic models and the obese patient, but
care is needed to distinguish between those features which appear as a consequence of the obese state—factors which the species are likely to share—and
the features primarily responsible for the development of obesity in the different
species. The problem in man is more complex than in the genetically obese rodents,
since social factors play a much more important part. Although there is strong evidence for a familial pattern in human obesity and studies on twins emphasise the
similarity of body weight even when twins are raised apart, there is increasing evidence that social factors play an important role in determining the prevalence of
obesity, particularly in affluent societies. Both animal and human studies show
that individuals who are prone to obesity have a subnormal rate of energy expenditure, but there is no evidence for a low BMR. In the *ob/ob* mouse the metabolic
defect is largely attributable to a subnormal rate of non-shivering thermogenesis,
whereas in man the lower energy requirement in obesity-prone individuals is
probably associated with subnormal thermogenic responses to food and other
stimuli—e.g. caffeine. In man the reduced energy output can be balanced by an
increase in the RMR which rises on overfeeding as the lean body mass increases
in conjunction with the accumulation of body fat. The hyperphagia of the genetically obese animals is also found in some, but not all, obese patients.

These considerations of energy balance do not seem to relate to changes in the
total number of fat cells in the study of obese animals or patients. We conclude
that much of the emphasis on hyperplastic obesity has been misplaced, since the
true number of total adipocytes in the body is markedly underestimated in normal
individuals if reliance is placed for determining fat cell number on subcutaneous
fat samples only. Since internal fat depots contain small adipocytes, it is wrong to
assume that a subcutaneous sample is representative of all body fat. In addition,
if individuals of normal body weight are studied, it becomes clear that the normal
variation in total fat cell number is so large that few, if any, obese patients have
true hyperplasia.

Continued effort to identify the primary factors determining the hyperphagia and metabolic efficiency of genetically obese rodents is a prerequisite for improving our understanding of the clinical condition in man, but it must be recognised that the causes of obesity in man are many and that social factors may well be more important than either genetics or acquired metabolic conditions in determining the development of obesity in an individual.

INTRODUCTION

It would be surprising if human obesity did not include as one of its principal factors a genetic predisposition to greater metabolic efficiency and perhaps to hyperphagia. A wide range of diseases is now being found to occur more frequently in genetic subgroups of populations and it is probably only a matter of time before relationships are found between genetic markers and obesity in man. Animal husbandry has for years used genetic selection to improve the fattening qualities of pigs and cattle, and this successful breeding policy has emphasised the importance of genetic factors in determining the size of an animal and the proportion of body fat laid down during growth. Given the heterogeneity of man and the recognised wide range of values for such nutritional indices as the body's requirement for protein, it seems foolhardy to presume that the requirement for energy and the predilection to obesity do not show a similar variability. Perhaps with these views in mind Astwood (1962), some years ago, regarded obesity as a hereditable disease, which was so strongly determined by genetic factors as to preclude the usefulness of attempts at treatment.

Nevertheless, some nutritionists and physicians regard the role of genetics in human obesity as irrelevant. Obesity is considered to result from an eating disorder (Bruch, 1973), or from a reduced amount of physical activity. This decrease in energy expenditure may occur in middle age and be compounded by the gradual decline with age in the metabolic cost of maintaining essential body functions. When food intake is not reduced to match the falling energy expenditure, then obesity will result. Cultural factors, including the ready availability of appetising food and a steady reduction in the physical exercise which many occupations now require, also increase the pressure on the physiological mechanisms for maintaining body weight.

These widely differing views on the factors involved in the development of obesity can be reconciled if it is accepted that most human obesity differs from the more extreme genetic models of obesity in rodents in having several interacting causes, including social factors. The problem for the geneticist then becomes one of quantifying the contribution which genetics make to the development of obesity. If the prevalence of obesity increases within a population, then it is likely that social factors are primarily responsible for these changes. Individuals becoming obese within such an environment include those with a greater propensity for fat deposition, but any genetic predisposition may not be apparent on a population basis, because some subjects will become obese because of overriding social pressures even though they are not unduly metabolically efficient. Thus, the *increase* in the prevalence of obesity in the USA and Europe (Hejda, 1978) probably results from social pressures rather than any genetic process such as outbreeding and hybrid vigour. In other parts of the world, however, there may be continuing selection pressure in communities with high infant mortality rates: in these areas the metabolically more efficient children may survive the rigours of semi-starvation and infection to become obese adults as soon as food becomes readily available (James and Trayhurn, 1976).

Analyses of the relative importance of environmental and genetic influences have depended on characterising family patterns of obesity and relating the weights of adopted or foster children to those of their families. A few studies of twins have also compared the weights of identical and non-identical twins with other siblings, and occasionally it has been possible to obtain weights of twins brought up in different environments. The results of several of these studies have been summarised by Hartz *et al.* (1977), who also set out some of the considerable statistical problems involved in evaluating the contribution of genetics to obesity.

The familial nature of obesity has been documented in several countries. In Boston, USA, the prevalence of obesity rises from 9 per cent in children of non-obese parents to 40 per cent when one parent is obese and to 80 per cent when both parents are obese (Johnson *et al.*, 1956). Similar figures have been obtained in other studies (Mayer, 1965). More recent studies from the USA (Garn and Clark, 1976) have confirmed the strong familial nature of obesity and the similarity of both weights and skinfold thicknesses of siblings. However, these latei studies also showed that obesity in one parent was often associated with obesity in the partner. This could reflect a concentration of genetic traits within families if individuals of similar and genetically determined weights tended to marry each other. However, social factors could be of equal or of greater importance.

Early twin studies in the USA (Newman *et al.*, 1937) showed that the weight difference between identical twins reared in the same environment was much less than that between non-identical twins, who, in turn, had similar weight differences to those of siblings measured at equivalent ages. Thus, genetic factors did seem to relate to weight differences. The strongest evidence for the role of genetics in obesity comes from analyses of the weights of twins reared in separate environments. Verschuer reported that weights of identical twins differed by 1.4 per cent when they were reared together and this weight difference increased to only 3.6 per cent when they were brought up separately (Verschuer, 1927). On this basis one might expect to find that fostered or adopted children would maintain weights appropriate to their genetic background rather than the new family environment. Information on fostered or adopted children is far more readily obtainable than that on twins, but unfortunately there is rarely any information relating to the children's biological parents or family, and reliance has to be placed on the degree to which the adopted children differ in weight from other biologically related siblings in the family.

Garn *et al.* (1976) have recently examined the correlation between the skinfold thickness of 429 adopted children, aged from 1 to 18 years, and the skinfold thickness of their adopted siblings or parents. Unrelated siblings had skinfold thicknesses which were not more disparate than those of biologically related siblings. The triceps skinfold measurements of the adopted children were also as well correlated with those of their adopted parents as the correlations of skinfold thickness between parents and their offspring. Garn *et al.* (1976) therefore concluded that the similarities of genetically related individuals living within a family environment did not necessarily indicate the working of genetic factors. Garn and Clark's (1976) earlier analysis of skinfold measurements obtained on black and white families of different social classes in the Ten-State Study had shown different results for males and females as their occupation and social class changed. Thus, females became fatter in adult life when poor, but thinner in adult life when in a higher income group; in contrast, the adult males became fatter in the higher income groups. This, again, has been interpreted as reflecting the dominance of social pressures. Given the high prevalence of obesity in the USA, this interpretation may be correct, with the social factors overriding any genetic contribution to obesity.

Hartz *et al.* (1977) carried out a much larger study on 5573 adopted, fostered or step-children and on nearly 20 times this number of biologically related children of women attending commercially run slimming groups. The data were based on the weight–height relationships reported in a questionnaire survey of what was probably a genetically selected group within the population. The family environment accounted for 32 per cent of the variation in the degree of obesity recorded in all the children and seemed to be more important than heredity, which contributed about 11 per cent to the total variability in the weight–height relationship. Thus, the more recent studies from the USA suggest that in affluent societies social factors and child-rearing practices exert a dominating influence on the prevalence of obesity. The studies do not exclude, however, a genetic and metabolic 'predisposition' to obesity.

The most striking indication of a possible metabolic component in familial obesity comes from the recent work of Griffiths and Payne (1976), who found that 3–4 year old non-obese children of obese parents were expending 22 per cent less energy at rest than children of equivalent weight and age whose parents were non-obese. The children of obese parents were in approximate energy balance because they were also eating 22 per cent less than the children from non-obese families. These results are unlikely to reflect the effects of parental constraints on eating in the obese family, with a secondary reduction in energy expenditure, and it would be more reasonable to conclude that these children were of a genetic stock with an energy requirement appreciably below normal. It was also inferred that the amount of physical activity taken by the children of obese families was less, but whether these differences in energy expenditure were metabolically or behaviourally determined remains unclear. Nevertheless, any social pressure on the children of the obese families to eat even an average amount of food would lead to energy imbalance and the risk of obesity developing.

METABOLIC RATES

Traditionally, investigators have favoured measurement of the basal or resting metabolic rate in an attempt to identify a low metabolic rate both in the animal models of obesity and in human obesity. We have shown in Chapter 12 the fallacy of this approach in studies on the *ob/ob* mouse. In the animal studies we conclude that the BMR (i.e. the metabolic rate measured under thermoneutral conditions) is 'normal' in the obese mice. If, in the obese state, the metabolic rate is calculated per 100 g body weight or in relation to some power of weight (e.g. weight to the power of 0.75) then this form of expression may lead to the erroneous conclusion that the metabolic rate is 'subnormal'.

Similar considerations apply to our results on obese adults. The conditions for measuring the resting metabolic rates (RMR) differed from those required for monitoring the BMR, since studies were conducted at 20–22°C rather than at thermoneutrality. However, the patients wore ample clothing, and separate studies on volunteers did not show any systematic difference between the RMR measured in this way and the BMR measured under thermoneutral conditions at 28°C in a whole-body calorimeter (James *et al.*, 1978). Thus, we consider the RMR values to be close to those which would have been obtained with subjects at thermoneutrality. The measurements on obese patients were conducted after an overnight fast and under conditions of weight stability, so that the effects of a state of energy imbalance would be minimised: dieting is known to reduce the metabolic rate.

When the RMR was expressed in absolute units (e.g. as kJ/min), we found that the obese men and women had a high, not a low, metabolic rate. This conclusion and its implications have been obscured for years, because the results have usually been expressed in terms of the subject's surface area. The latter expression follows traditional nutritional and physiological practice but it detracts from an understanding of energy balance in the individual patient. In addition, we have compared the metabolic rate of an obese patient with the RMR values for healthy subjects of approximately equivalent height, age and sex, but of normal weight. Table 14.1 shows that the obese person has an RMR in excess of that for the normal weight individual. Figure 14.1 plots the individual RMR values obtained on the obese women. The majority of the women had an RMR which was above the mean RMR expected for normal weight women aged 20 years. The results for normal weight individuals are constructed from the data on normal British subjects which were painstakingly collected by Robertson and Reid (1952). Since the average age of our patients was 43 years, this graph tends to minimise the degree to which the values in the obese women exceed the expected RMR.

One cause for the elevated RMR in obesity is apparent from Table 14.1. Not only is there a marked excess of body fat in the obese patient, but also there is an associated increase of 32–36 per cent in lean body mass (LBM). The hypertrophy of skeletal muscle may be expected as obesity develops, since the physical load imposed by the need to carry the excess weight will induce muscular hypertrophy. Since muscle mass normally comprises about 50 per cent of LBM, muscular tissue would have to hypertrophy by 60–70 per cent in order to account for the observed rise in LBM. In addition, skeletal muscle is considered normally to make only a small contribution to the BMR (Holliday *et al.*, 1967), and therefore muscular hypertrophy alone seems an inadequate explanation for the elevated BMR. It should be noted that this increase in muscle mass in human obesity is quite different from that observed in the genetically obese rodents. In these animals skeletal muscle is often poorly developed, and in the *ob/ob* mouse this defect appears to be an intrinsic part of the syndrome (Bergen *et al.*, 1975). Whether 'pre-obese' humans have a relatively small muscle mass is unknown.

Another organ which will have increased its fat-free mass (FFM) is the adipose

Table 14.1 The body composition and metabolic rates of obese and non-obese adults in energy balance (mean ± S.D.)

	Height (m)	Weight (kg)	Body fat (kg)	Lean body mass (kg)	Resting metabolic rate (kJ/min)
Female					
Controls (15)	1.62 ± 0.06	55.7 ± 6.8	16.0 ± 3.3	39.7 ± 4.3	3.93 ± 0.75
Obese (61)	1.63 ± 0.07	96.0 ± 18.4	43.4 ± 14.2	52.6 ± 7.7	4.55 ± 0.42
Male					
Controls (11)	1.76 ± 0.07	61.8 ± 8.0	14.8 ± 5.0	52.0 ± 6.6	4.17 ± 0.09
Obese (11)	1.75 ± 0.08	114.7 ± 22.1	44.5 ± 20.2	70.3 ± 9.8	6.01 ± 0.06

The resting metabolic rate significantly correlated with the lean body mass ($r = 0.829$). The contribution of body fat to the variability in RMR was only 0.2 per cent once the co-correlation between the lean body mass and body fat had been taken into account. Data taken from James et al. (1978).

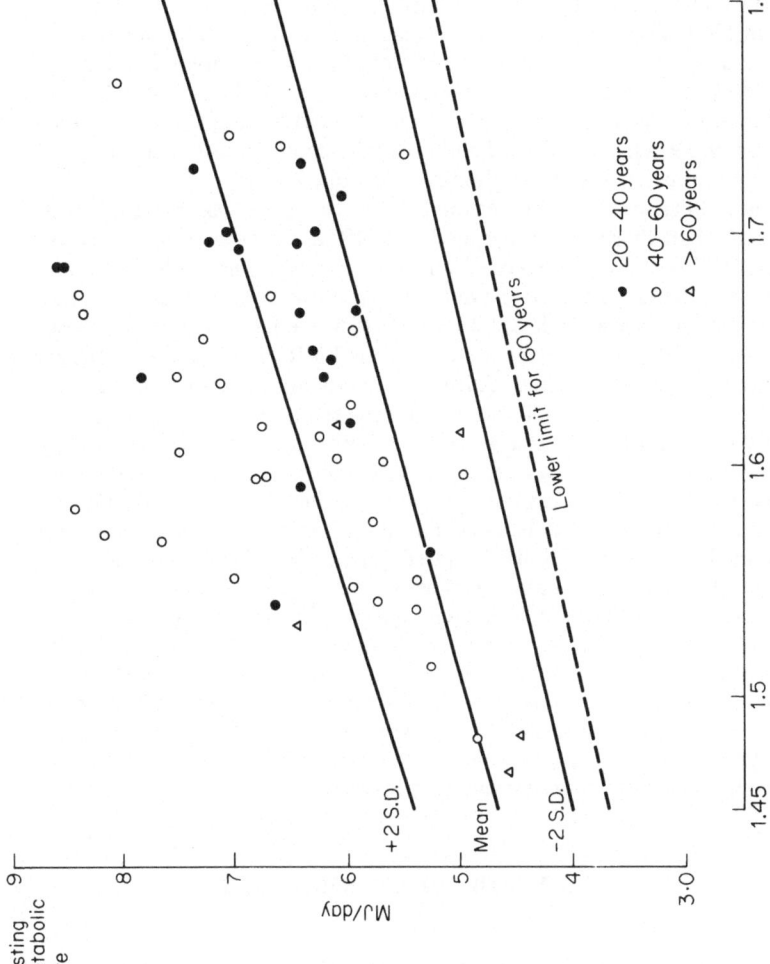

Figure 14.1 The resting metabolic rates of obese women compared with the range of metabolic rates expected for normal weight women of equivalent height (regression lines).

tissue itself. If the concentration of triglyceride in the adipose tissue of lean and obese individuals is taken as 78 and 83 per cent, respectively (Garrow, 1974), then the calculated increase in FFM associated with the excess 27.4 kg of fat in the obese women (table 14.1) will amount to 3.8 kg—i.e. 30 per cent of the increase in LBM. This increase in fat-free mass reflects not only an increase in adipose tissue water, but also an increase in adipose tissue protein. Protein maintains its concentration on a weight basis despite the marked increase in both the absolute mass of triglyceride and the concentration of lipid in the adipose tissue itself (Garrow, 1974). The enlarged adipose organ could therefore contribute appreciably to the increase in metabolism in the obese state.

The question arises as to whether the BMR of the 'pre-obese' patient is 'subnormal' and might thus account for the low RMR data presented by Griffiths and Payne (1976) in non-fasting children from obese families. With the values of the obese patients lying in or above the upper normal range (figure 14.1) it seems unlikely that a low BMR can be the single prime factor in the pre-obese state, because on this basis one would expect the BMR to increase with obesity until the energy expenditure had returned once more to a 'normal', not a high, value, Thus, the high BMR in obesity implies either that true hyperphagia is commonly a primary factor—a suggestion not borne out by the work of Griffiths and Payne—or that some other component of energy expenditure is 'defective' in the obese and that the increase in the BMR can then be considered to be a 'compensatory' phenomenon. This generalisation does not exclude the possibility that in individual cases a low BMR is indeed a major factor contributing to the constitutional tendency to obesity. If these subjects then consumed an 'average' amount of energy for their sex, age, height and weight, they would slowly become obese until the BMR was brought up to the average level. At this stage they would be obese with a normal BMR and a normal energy turnover. We have observed, in cross-sectional studies, that the RMR increases by 380 kJ/24 h for each 10 kg increase in body weight above the desirable weight. Thus an appreciable state of obesity is needed to compensate for an initial imbalance of only 1 MJ if other components of energy expenditure do not also adjust.

DIETARY-INDUCED THERMOGENESIS

The recent emphasis on diminished thermogenesis as the metabolic basis for obesity owes much to the work of Miller and his colleagues (Miller, 1975). Miller emphasises the thermogenic response to diet as a mechanism for adjusting energy expenditure when energy intakes change. This adjustment will then buffer any tendency to change the size of the body's energy stores. Miller favours the postprandial phase as the important time for the regulation of energy expenditure, since overfeeding six subjects for 3 weeks had no measurable effect on the BMR. However, studies by Goldman and his colleagues (1973) showed that three of four volunteers fed a mean daily energy excess of 3.6 MJ as fat for 83 days had an increase in BMR of 6, 12 and 29 per cent, the BMR of the fourth subject falling by 5 per cent. Body fat increased by 11, 13, 18 and 19 per cent. Similarly, four subjects overfed a mean of 7.6 MJ daily as extra carbohydrate for 18 days increased their BMR by a mean of 12 per cent and body weight by 5 per cent. Measurements by Apfelbaum *et al.* (1971) with a respiratory chamber also documented that the increase in metabolic rate which occurred after eight subjects consumed 6.3 MJ extra energy for 15 days persisted throughout the 24 h period. These studies therefore imply not only that the thermogenic response reflects the inflow of substrate for storage or immediate metabolism, but also that overfeeding leads to secondary metabolic changes which may take more than a few days to develop. However, accurate measurements of the metabolic rate in man over a 24 h period are needed to establish the time course of the response in the BMR to overfeeding and will require the use of large human calorimeters. The evidence that overfeeding can increase the BMR accords with the well-documented reduction in BMR with underfeeding (Apfelbaum *et al.*, 1971) and means that this component of energy expenditure must be considered when the different routes whereby energy expenditure changes are assessed. Our own cross-sectional studies of obese patients with elevated RMRs were conducted under conditions of energy balance and the increase in RMR seemed to be accounted for by the increase in LBM. The adaptation in the metabolic rate seen in both overfeeding and starvation studies, however, exceeds that expected from the change in LBM and is an additional adjustment limiting changes in energy balance.

Thermogenic responses to a meal are also known to adjust to changes in energy balance, and postprandial thermogenesis increases (although not linearly) as the energy content of a meal increases. An additional small increment is evident if food and exercise are combined (Miller, 1975), so that food and exercise seem to have a small synergistic effect on energy expenditure. Whether or not these changes are sufficient to exert effects of quantitative importance in the control of energy balance is unknown, although changes of this magnitude are sufficient on a cumulative basis to account for the slow changes in weight actually observed in adult man (Gordon and Kannel, 1973). The increase in metabolic rate after a meal usually amounts to 10–20 per cent of the BMR. If this response is reduced by only 5 per cent in an individual, then this reduction will put the subject into positive energy balance unless there is a corresponding fall in energy intake.

The recognition of the importance of small changes in energy balance has led to several studies seeking to establish whether the obese person has a lower thermogenic response to food. The conflicting evidence on this point has been summarised by Garrow (1974) and Miller and Mumford (1973). Since then a carefully conducted study by Pittet and co-workers has shown a difference in the response of lean and obese women to a standard 50 g dose of glucose (Pittet *et al.*, 1976). These results have been recalculated to illustrate the possible net thermogenic difference between the lean and the obese subjects and are worked out on a daily basis (table 14.2). The thermogenic responses are assumed to be complete within $2\frac{1}{2}$ h following the meal, with the subjects normally consuming 8 MJ per day in three meals. Miller and Mumford (1973) found that the thermogenic response to energy intakes is linear over the energy range being considered, so we may assume a proportional effect when calculating from the 800 kJ of glucose ingested in the experiment to an assumed 8000 kJ in the hypothetical daily intake (Miller, 1975). In this particular example the energy difference in dietary thermogenesis amounts to 537 kJ per day, which, if unmatched by any increase in other components of energy output or a fall in energy intake, would lead to a net positive balance of

Table 14.2 Basal metabolic rate and thermogenic responses to food in normal weight
and obese women

				Dietary thermogenic response	
	Height (m)	Weight (kg)	BMR (kJ per 24 h)	Observed 800 kJ meal as 50 g glucose	Calculated for 8000 kJ per day
Normal	1.69	56.6	6956	94.2	942
Obese	1.64	83.7	7478	40.5	405
Difference: lean − obese			−523	+53.7	+537

Recalculated from Pittet *et al.* (1976).

196 MJ or perhaps an increase of 6 kg body weight per year. Energy expenditure
seemed to be approximately equal in both groups because of the higher BMR in
the obese group. Although the increment in BMR in Pittet's obese women is less
than we have observed, the rise does counteract the reduced dietary thermogenesis of the obese and demonstrates how the elevated BMR in the obese can be considered as a compensatory phenomenon.

COLD-INDUCED THERMOGENESIS

We have given in Chapter 12 our reasons for regarding reduced non-shivering
thermogenesis as a quantitatively important cause of the enhanced metabolic
efficiency of the *ob/ob* mouse. Studies on dietary-induced thermogenesis (DIT)
in genetically obese rodents have not been undertaken at thermoneutrality—a
prerequisite for the demonstration of DIT, since DIT and cold-induced thermogenesis interact. Thermoregulatory thermogenesis may be reduced as heat becomes
available from the dietary-induced increase in metabolism and these additional
adjustments can mask and reduce the apparent thermic effect of food.

   We have outlined elsewhere (James and Trayhurn, 1976) the analogies which
can be drawn from studies of thermoregulation in obese rodents and obese man.
It must be recognised, however, that acute cold stress is an unusual event in everyday life and that a failure of obese adults to respond to cooling may reflect either
the benefits of the additional insulation from the excess subcutaneous fat or, in
cases where a fall in body temperature does occur in obese adults, an acquired
defect rather than a primary abnormality. It must also be recognised that the proportion of energy turnover needed for the maintenance of body temperature declines in the larger species from the level found in small animals, so that man may
normally expend only a small fraction of his energy intake on maintaining body
temperature. In addition, man adapts behaviourly to changes in environmental
temperature not only by creating an artificial temperature in the home, but also

by adjusting his micro-environment through changes in the clothing worn. Never-theless, we have recently observed that ten women of normal weight maintained in identical clothing in a calorimeter for periods in excess of 24 h adjusted their 24 h heat production by about 6 per cent when the calorimeter temperature was lowered from 28°C to only 22°C (unpublished observations). Thus, the component of cold-induced thermogenesis in man may not be as insignificant as was once thought.

If dietary-induced and cold-induced non-shivering thermogenesis are biochemi-cally and hormonally related, then similar mechanisms may apply in both man and experimental animals. In man the dietary manifestation may predominate, with the thermoregulatory defect being the more important component in rodents.

HORMONAL CHANGES

There are many hormonal changes in the *ob/ob* mouse which are similar to those observed in obese men and women. Nevertheless. many of these changes—e.g. the hyperinsulinemia with insulin resistance, the changes in glucagon secretion and its responsiveness, the increase in adrenal cortical activity and the alterations in anterior pituitary function—are probably secondary consequences of the obese state rather than primary defects which determine the difference in metabolic efficiency between lean and obese subjects (James, 1976). Indeed, if one's con-cern is to identify the primary factors determining either the hyperphagia and/or the metabolic efficiency in the obese animal or patient, it could be argued that most of the studies conducted hitherto demonstrate epiphenomena, because they have neglected to consider either the impact of the immediate diet, the state of energy balance or the secondary effects of the obesity on hormonal responses.

Our demonstration of the importance of non-shivering thermogenesis in obese mice has focused our attention on the two hormonal systems which are known to be involved—i.e. the thyroidal and sympathetic catecholaminergic systems (LeBlanc, 1975). Both thyroid hormones and catecholamines are known to be thermogenic and the role of thyroid hormones in obesity has been studied for many years. In *ob/ob* mice thyroidal function appears to be depressed, with the obese animal having a smaller thyroid and lower circulating levels of thyroid hormones (Joosten and van der Kroon, 1974). Administering thyroid hormones helps in overcoming the reduced thermogenic response to cold, but whether this reversal signifies that hypothyroidism is the primary lesion in the *ob/ob* mouse rather than a response secondary to an hypothalamic mechanism for re-ducing thermogenesis seems doubtful (Ohtake *et al.*, 1977). The picture in obese man is different, since hypothyroidism is rarely found as a cause for obesity, and circulating T_4 and T_3 levels tend to be high rather than low (Bray *et al.*, 1976).

Nothing is known as yet about peripheral concentrations of catecholamines and catecholamine turnover in genetically obese animals, but our preliminary information on man suggests that venous noradrenaline levels are low in some obese subjects (table 14.3) and that this difference persists despite a variety of stimuli, including postural changes, the cooling of the extremities and the administration of oral caffeine—a known catecholamine stimulant. Whether these changes prove to be yet another secondary consequence of obesity or factors associated with a

Table 14.3   The fasting plasma catecholamine levels of a group of obese
            and thin women

Subjects	Number	Noradrenaline (μg/ml)	Adrenaline (μg/ml)
Obese	18	0.15 ± 0.02	0.05 ± 0.01
Thin	13	0.31 ± 0.04	0.03 ± 0.01

Both groups studied while in energy balance and on an intake of at least
50 mmol sodium daily. Mean ± S.E.M.

reduced thermogenic response depends on showing that these differences persist
when obese patients return to an appropriate weight for their height and then re-
adjust their intake to approximately 'normal' levels. Nevertheless, we have found
that there are differences in the metabolic response of lean and obese women to a
standard dose of oral caffeine (table 14.4). Thus, in practice an everyday stimulus
to thermogenesis is more potent in the lean than in the obese individual, whether
or not this difference is constitutive. The reduction in response in the obese person
therefore imposes an additional burden unless energy intake is correspondingly
reduced.

Table 14.4   The thermogenic response to oral caffeine in obese and thin women

Subjects	RMR (kJ/min)	10–30 min	% increase RMR 45–60 min	105–120 min
Obese (6)	4.52 ± 0.20	7.3 ± 2.3	8.9 ± 2.0	6.7 ± 1.7
Thin (6)	3.27 ± 0.09	13.5 ± 2.3	13.3 ± 1.7	13.9 ± 2.2

Oral caffeine was given in water at a dose of 3 mg per kg of desirable weight for
height, taking the desirable values from those appropriate for the midpoint of
the medium frame size of the Metropolitan Life Insurance tables. Mean ± S.E.M.

FAT CELLS

The last ten years have seen a renewed interest in the fat cell in obesity, with
particular attention being paid to the possibility that animals and obese adults
are prone to obesity if they develop an excess number of fat cells. There is an increase in the number of triglyceride-filled adipocytes in the *ob/ob* mouse, but this
does not mean that the increase is an important factor promoting obesity—the
excess number of adipocytes may merely reflect the need to recruit additional
cells to accommodate the enlarging depot of triglyceride. The fat transplantation

Table 14.5 The effect of neglecting the smaller omental adipocytes when calculating total fat cell number from subcutaneous samples

Patients	Total body fat derived from skinfold measurements (kg)	Fat cell size (μg triglyceride)		'Apparent' number based on subcutaneous sites only	Fat cell number × 10^{10} 'True' number		
		Subcutaneous	Omental		Subcutaneous	Omental	Total
Obese							
Female (14)	30.54	0.729	0.293	4.19	2.81	3.44	6.25
Male (5)	26.00	0.602	0.406	4.32	2.89	2.11	5.00
Non-obese							
Female (16)	17.84	0.458	0.134	3.90	2.61	4.39	7.00
Male (9)	14.60	0.408	0.189	3.58	2.40	2.55	4.95

Data taken from Jung et al. (1978).

studies of Ashwell *et al.* (1977) have clearly established that in several different strains of genetically obese rodents the intrinsic properties of the fat cells are not a primary factor in the development of obesity. In man it has been suggested that obese adults may have an hyperplastic adipose tissue, particularly if the obesity was of childhood origin. This excess of adipocytes has been considered to reflect either a genetic abnormality or the result of overfeeding during a critical period in infancy when the final number of adipocytes may be determined.

We have re-examined this problem in 125 patients and find little to support the current ideas on adipose tissue hyperplasia. Not only is the calculated number of adipocytes normal in most obese patients, but also there is no relationship between adipocyte number and the age of onset of the obesity. These conclusions are based on the same methods for sampling subcutaneous adipose tissue and for calculating fat cell number as those used by others. In addition, we find that adipocytes from intra-abdominal and intrathoracic fat are smaller than subcutaneous fat cells (table 14.5). The usual methods for calculating the body's content of adipocytes therefore seriously underestimate the true number of fat cells. With the expansion of internal fat cells with obesity, the size of internal and external fat cells may become equal. This equivalence will then, for the first time, give a true indication of the unchanged but large number of adipocytes in the body. If subcutaneous samples alone are obtained, then there will appear to have been an increase in fat cell number in the very obese individual. Thus, spurious changes in fat cell number can be obtained and there may be no need to invoke the idea of recruiting stem cells as new adipocytes.

Since the majority of our obese patients were less than 200 per cent of the ideal weight, their internal adipocytes were still smaller than the subcutaneous cells and no definite case of hyperplastic obesity was found. We conclude, therefore, that the importance of hyperplastic obesity has been overemphasised. If one's chief interest is in determining the causes of obesity, then there seems little point in pursuing research in this field if the methodological problems of obtaining valid figures for the true number of fat cells cannot be resolved.

ACKNOWLEDGEMENTS

The results presented in this chapter were obtained with the skilled assistance of P. Murgatroyd, H. Davies, T. Crisp, R. Hawkins and R. Spires. We also thank Drs M. A. Barrand and B. A. Callingham for the catecholamine assays.

REFERENCES

Apfelbaum, M., Bostsarron, J. and Locatis, D. (1971). Effect of caloric restriction and excessive caloric intake on energy expenditure. *Am. J. Clin. Nutr.*, **24**, 1405–1409

Ashwell, M., Meade, C. J., Medawar, P. and Sowter, C. (1977). Adipose tissue: Contributions of nature and nuture to the obesity of an obese mutant mouse (*ob/ob*). *Proc. Roy. Soc. London*, **B195**, 343–353

Astwood, E. B. (1962). The heritage of corpulence. *Endocrinology*, **71**, 337–341

Bergen, W. G., Kaplan, M. L., Merkel, R. A. and Leveille, G. A. (1975). Growth of adipose and lean tissue mass in hindlimbs of genetically obese mice during pre-obese and obese phases of development. *Am. J. Clin. Nutr.*, 28, 157–161

Bray, G. A., Fisher, D. A. and Chopra, I. T. (1976). Relation of thyroid hormones to body-weight. *Lancet*, i, 1206–1208

Bruch, H. (1973). *Eating Disorders – Obesity and Anorexia Nervosa and the Person Within*, Routledge and Kegan Paul, London

Garn, S. M., Bailes, S. M. and Higgins, I. T. T. (1976). Fatness similarities in adopted pairs. *Am. J. Clin. Nutr.*, 29, 1067–1068

Garn, S. M. and Clark, D. C. (1976). Trends in fatness and the origins of obesity. *Pediatrics*, 57, 443–456

Garrow, J. S. (1974). *Energy Balance and Obesity in Man*, North-Holland, Amsterdam

Goldman, R. F., Haisman, M. F., Bynum, G., Horton, E. S. and Sims, E. A. H. (1973). Experimental obesity in man: metabolic rate in relation to dietary intake. In *Obesity in Perspective*, Vol. 2, Pt. 2, Fogarty International Center Series on Preventive Medicine, pp. 165–186

Gordon, T. and Kannel, W. B. (1973). The effects of overweight on cardiovascular disease. *Geriatrics*, 28, 80–88

Griffiths, M. and Payne, P. R. (1976). Energy expenditure in small children of obese and non-obese parents. *Nature*, 260, 698–700

Hartz, A., Giefer, E. and Rimm, A. A. (1977). Relative importance of the effect of family environment and heredity on obesity. *Ann. Human. Genet.*, 41, 185–193

Hejda, S. (1978). Problems of obesity in Czechoslovakia and comparable countries. Supplement Nr. 15 till *Naringsforskning*: Argang 22, pp. 46–55

Holliday, M. A., Potter, D., Jarrah, A. and Bearg, S. (1967). The relation of metabolic rate to body weight and organ size. *Pediat. Res.*, 1, 185–195

James, W. P. T. (1976). *Research on Obesity*, DHSS/MRC Group Report, HMSO, pp. 53–56

James, W. P. T., Davies, H. L., Bailes, J. and Dauncey, M. J. (1978). Elevated metabolic rates in obesity. *Lancet* i, 1122–1125

James, W. P. T. and Trayhurn, P. (1976). An integrated view of the metabolic and genetic basis for obesity. *Lancet*, ii, 770–773

Johnson, M. L., Burke, B. S. and Mayer, J. (1956). Incidence and prevalence of obesity in a section of schoolchildren in the Boston area. *Am. J. Clin. Nutr.*, 4, 231–238

Joosten, H. F. P. and van der Kroon, P. H. W. (1974). Role of the thyroid in the development of the obese-hyperglycaemic syndrome in mice (*ob/ob*). *Metabolism*, 23, 425–436

Jung, R. T., Gurr, M. I., Robinson, M. P. and James, W. P. T. (1978). Does adipocyte hyper-cellularity in obesity exist? *Br. Med. J.* ii, 319–321

LeBlanc, J. (1975). *Man in the Cold*, Charles C. Thomas, Springfield, Illinois

Mayer, J. (1965). Genetic factors in human obesity. *Ann. N. Y. Acad. Sci.*, 131, 412–421

Miller, D. S. (1975). Thermogenesis in everyday life. In *Second International Congress on Energy Balance* (ed. E. Jequier), Editions Médecine et Hygiene, Geneva, pp. 198–208

Miller, D. S. and Mumford, P. M. (1973). Luxuskonsumption. In *Energy Balance in Man* (ed. M. Apfelbaum), Masson, Paris, pp. 195–207

Newman, H. H., Freeman, F. N. and Holzinger, K. J. (1937). *Twins: A Study of Heredity and Environment*, University of Chicago Press, Chicago, pp. 335–349

Ohtake, M., Bray, G. A. and Azukizawa, M. (1977). Studies on hypothermia and thyroid function in the obese (*ob/ob*) mouse. *Am. J. Physiol.*, 233, R110–R115

Pittet, Ph., Chappius, Ph., Acheson, K., de Techtermann, F. and Jéquier, E. (1976). Thermic effect of glucose in obese subjects studied by direct and indirect calorimetry. *Br. J. Nutr.*, 35, 281–292

Robertson, J. D. and Reid, D. D. (1952). Standards for the basal metabolism of normal people in Britain. *Lancet*, i, 940–943

Verschuer, O. V. (1927). Die vererbungsbiologische Zwillingsforschung. *Ergebn. Med. Kinderheilk.*, 31, 35–120

15

The clinician's approach

J. S. Garrow (Clinical Research Centre, Harrow, Middlesex, UK)

SUMMARY

About one-third of the adult population of this country is overweight, and a
DHSS/MRC study group recently said that obesity 'is . . . common enough to constitute one of the most important medical and public health hazards of our time,
whether we judge importance by shorter expectation of life, increased morbidity
or cost to the community in terms of both money and anxiety.' We do not know
how many obese people treat themselves effectively without asking medical help,
but those cases who present to a doctor are difficult to treat effectively. The
clinician therefore hopes that those working with animal models will point the
way to better methods of treatment or prevention. However, it is not easy to
establish fruitful cooperation along these lines, for reasons which are discussed in
this paper. Having worked in close association with both clinicians and laboratory
workers for the last 25 years, I am sure that it is important that each recognises
the limitations of the other group if animal experiments are to be used to best
advantage.

It is obvious that human obesity is a heterogeneous condition: it is hard to believe that a defect in a particular metabolic pathway is a main cause of most human obesity, although it may be the sole cause of obesity in a particular strain of
rodent. Since my prejudices are those of a clinician, I believe that the onus is on
the student of animal models to suggest a way in which the defect he has identified in the animal could be found in human subjects. This is often a very difficult
task. The clinician cannot ethically submit his patients to investigations which make
no contribution to therapy, but some progress may be made by investigating human
volunteers, who are preferably the investigators themselves.

THE CLINICAL PROBLEM

The clinician's approach to obesity is different from that of a laboratory worker
because his starting point is totally different. An effort should be made to bridge

237

the chasm between the two types of work, otherwise opportunities for progress will be lost.

The problem confronting a clinician interested in obesity is illustrated in figure 15.1. The data are taken from a paper by Binnie (1977), who is a general practitioner in the north of England with about 1200 patients, of which 43 have been under treatment for obesity, and have been followed for 10 years. Of course, this does not represent the entire obese population in such a practice, since only about half the people who are trying to lose weight ever ask for medical advice (Ashwell 1973), some overweight people attend slimming clubs rather than general practitioners, and some make no significant attempt to do anything about losing weight. In figure 15.1 the excess weight (in pounds above the life insurance 'desirable' weight for height) for each patient is plotted on the vertical axis, and the weight change over the 10 year follow-up period is plotted horizontally. Both the clinician and his patients need help. Ideally, all the points in the figure should be clustered round the broken line, which represents 'cure', since on this line weight loss over 10 years equals excess weight at the beginning of the study. Few patients are near the broken line, many are grossly overweight, and 19 of the 43 patients

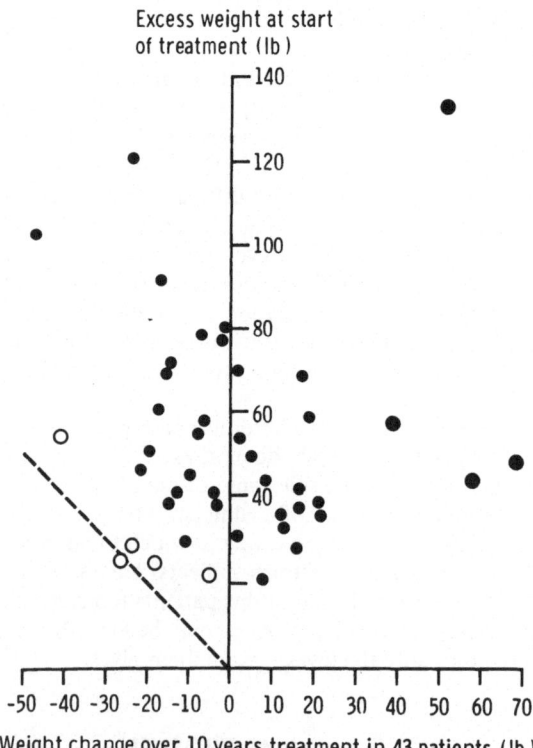

Figure 15.1 Excess weight at the start of treatment, and weight change over 10 years, in obese patients in a typical general practice (data of Binnie, 1977). 'Cured' individuals should cluster around the dashed line (open circles).

have actually increased their weight excess rather than losing weight. Not many general practitioners have published results which permit such an objective analysis of the course of human obesity over a long time span, but there is certainly no reason to suppose that Dr Binnie's results are worse than average. This is the challenge to those who work with obese laboratory animals: what do your studies tell us which is of help to the clinician and his patient?

Some laboratory workers reject this question. If various models of obesity are developed in experimental animals, and the causes of the obesity in these models are elucidated by laboratory workers, then it is up to the clinician to pick up the relevant clues and apply them to the problems of human obesity. This is true. I have listened to the papers over the last two days of this symposium and this is what I have been trying to do. However, the chasm between clinical practice and laboratory models is real and deep: it is not just a matter of social attitudes, although mutual hostility between basic scientists and clinicians sometimes makes matters even worse than they need be. For more than 20 years I have been fortunate to work in the field of human nutrition with both medical and non-medical colleagues, and progress can only be made if each takes the trouble to find out what the other lot are good at, and are trying to do, and the constraints within which they work.

Since this paper is on behalf of the clinical side of the house, let me indicate the points where the clinician hopes to receive help from basic scientists. By far the greatest contribution from basic science to clinical work is that we are sure that the laws of thermodynamics apply also to man. The patients in figure 15.1 who gained weight must have taken in more energy from their food than they expended, and, no doubt, if they had been imprisoned and fed a sufficiently restricted diet, they could all have been brought within the desirable range of weight for height. However, these are patients, not laboratory animals, so it is little help to say: 'Since the immediate causes of obesity are overeating and underexercising, the remedies are available to all, but many patients require much help in using them.' (Davidson and Passmore, 1969.) *Why* do they need so much help, and why do some patients respond so much better than others with the same degree of help? On this important question clinicians offer widely differing explanations, which is often the case when we have no reliable data to guide us. It was once fashionable to attribute any failure in treatment to inexcusable frailty in the patient. Thus, Gray and Kallenbach (1939) discharged with dishonour any patient who had failed to lose weight during the first month of treatment: 'The man or woman with obesity who just will not try a treatment for one month had better eat on and be happy though fat.' Jollife and Alpert (1951) were more scientific but equally unsympathetic. They devised a formula by which they believed they could predict the amount of weight a given patient should lose on the prescribed diet. 'It is on this basis that we categorically accuse each subject of dietary errors whenever they fail to equal at least 85% of prediction. Whenever they do better than 115% of prediction we tell them that they are losing excess fluid.' This bigoted attitude, reminiscent of mediaeval theology, is no longer openly expressed but pessimism is still common: 'Gross obesity may well be incurable and death from a late complication is almost inescapable.' (Kemp 1972), and 'One might well question whether refractory obese patients should be treated at all.' (Goldrick *et al.,* 1973).

Goldrick's question 'whether refractory obese patients should be treated at all' cannot be answered unless we really understand what is meant by the term 'refractory obese patients'. If it means patients who have failed to respond to treatment which was inappropriate, then obviously the solution is not to give up treatment, but to give the correct treatment. However, if obesity is sometimes truly incurable, we need to know how to identify such patients. Ideally, we would like to know *why* obesity in a particular patient was incurable.

ASTWOOD'S HYPOTHESIS

An interesting theory was put forward by Astwood (1962): 'I wish to propose that obesity is an inherited disorder and due to a genetically determined defect in an enzyme: in other words that people who are fat are born fat, and nothing much can be done about it.' Lean people can be made fat by overfeeding them (Sims *et al.*, 1968), and even the most grossly obese can be made thin by sufficiently prolonged food restriction (Bortz, 1969), but it is also true that there are some people who remain within the normal range of weight for many years without conscious effort (Fox, 1973), while for others this can only be achieved by constant vigilance and self-denial. Astwood's theory is plausible and provocative, but difficult to prove or disprove.

In man the *distribution* of body fat is genetically influenced. This is shown in figure 15.2, which is taken from the data of Borjeson (1976) on 101 pairs of twins. He measured the thickness of subcutaneous fat at three sites on the body, triceps, subscapular and abdominal, and showed that in monozygotic twins the thickness of skinfold at these sites differed less than in dizygotic twin pairs. His was not a random study of twin pairs: on the Swedish school register there were

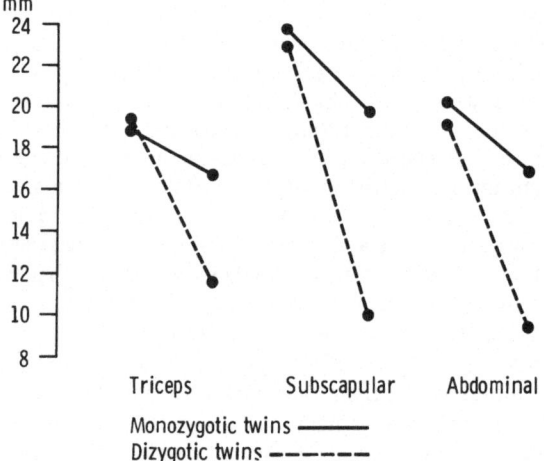

Figure 15.2   Average skinfold thickness at three sites in 40 pairs of monozygotic twins and 61 dizygotic twins of the same sex (data of Borjeson, 1976).

5008 pairs of twins at school. Of these, 160 pairs were of the same sex, with one or both of the twins more than two standard deviations above the normal weight for height. It was from the subsample that Borjeson's subjects were taken, so it is difficult to conclude from his results to what extent genetic factors determine the total amount of fat in the body.

If human obesity is strongly influenced by genetic factors, with a fairly simple mode of inheritance, it should be easy to show parent–child or sibling–sibling similarity in obesity. Data from the Ten State Nutrition Survey in America are shown in figure 15.3. Here the thickness of the triceps skinfold has been used to define obesity (Garn and Clark, 1976), and by this criterion two lean parents tended to have leaner children than parents who were both obese, but there is considerable overlap. Similarly, fat children were more likely to have fat siblings,

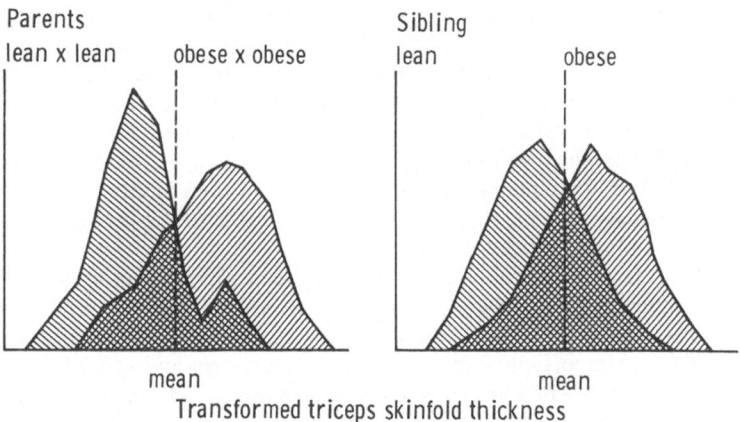

Figure 15.3 Triceps skinfold thickness in children with parents who were both lean or both obese, and in children whose siblings were either lean or obese (data of Ten State Nutrition Survey: after Garn and Clark, 1976).

and lean children to have lean siblings, but there is no clear division between the two populations. It is very difficult to extract information about genetic influences from data of this sort, since environmental influences are hopelessly intermingled (Garrow, 1976). Genetic factors may indeed be involved, but many other factors must modify the expression of these factors. Obesity in children has no characteristic time of onset, which is a striking factor in most of the genetic obese syndromes in rodents. A survey of all the obese children aged 10 years in Newcastle showed that roughly half were already obese by the age of 5, and in the remainder they became obese between 5 and 10 years (Wilkinson *et al.*, 1977). Many studies have shown that there is a statistically significant association between obesity in childhood and in later life, but the majority of obese adults were not obese children, and the majority of obese children do not become obese adults (Garrow, 1976). When all the data on growth in the first year of life are analysed, there is no formula which will predict with useful accurary which children will be obese at school age (Mellbin and Vuille, 1973).

This does not disprove Astwood's thesis: perhaps there is at least a genetic pre-disposition to a metabolic defect which renders some people far more liable to obesity than others. This is where the clinician looks to the animal experimentalist for help. There are many 'models' of obesity which have been described in this symposium, and to varying degrees the biochemical explanation for the obesity in these animals is known.

Given free access to food, all the varieties of genetically obese rodents reviewed by Bray and York (1971) ate more than their controls, but in some strains the difference in food intake was quite small, while in the *db/db* mouse the obese ani-mal ate twice as much as its lean counterpart. Of course, if obese patients admit that they eat far more than normal, this is not evidence of any metabolic defect, hereditary or otherwise. Astwood (1962) suggested that, if in obese patients the ability to mobilise fat is impaired, they would be deprived of the normal energy-producing substrates in the blood, and this is the situation which in normal people leads to hunger: '. . . it is easy to imagine that a minor derangement could be responsible for a ravenous appetite.' However, the majority of obese patients do not have a ravenous appetite, and dietary surveys repeatedly fail to show that obese people on average eat more than lean ones. This may reflect the fallibility of dietary surveys, but when we look elsewhere for support for Astwood's hypothesis, it is hard to find. If obese people had an impairment of their ability to mobilise fuel from their energy stores, they should have low concentrations of glucose and fatty acid in the blood, but this is not so.

It is difficult to draw useful parallels between hunger in rats and in men. We infer that a rat is hungry if it eats the food which is available to it, but man would not eat rat chow anyway, and we do not know what the rat thinks of it (Lane-Petter, 1970). It is important to distinguish between hunger and appetite (Yudkin, 1963): man often eats when he is not hungry if the available food is particularly palatable, and this is also true of some rats. It is easy to make the Osborne–Mendel rat obese with a high-fat diet which it evidently finds highly palatable (Faust *et al.*, 1976). Human subjects are unable to detect covert changes in the energy den-sity of food with much accuracy (Wooley *et al.*, 1972; Pudel and Oetting, 1977; Porikos *et al.*, 1977), although they will usually make a guess in the correct direc-tion when the energy content of the food is altered by a factor of 2 (Durrant and Mann, 1977).

MODELS OF 'METABOLIC' OBESITY

Obesity in a man or rodent is not necessarily associated with increased food intake. If the utilisation of food is more efficient, the obese animal can save energy, and store it as fat, when pair-fed with a lean counterpart. A constantly recurring sug-gestion is that obese individuals may absorb energy from ingested food with greater efficiency, but this implies that the engery lost in faeces would be reduced in the obese subject, and, so far as I know, this has never been shown in human or animal models.

Much more plausible, and more interesting, is the idea that the absorbed energy is used more economically by using metabolic paths with a higher yield of end-product for a given input of energy. In this field the clinician is better equipped to work alongside the animal experimentalist, because, even if he cannot understand the biochemistry of 'futile cycles' which waste energy, he can measure the energy expenditure of his patients. At the end of the day, whatever happens in the black box of human metabolism, the intake of energy must be accounted for either by heat produced or by a change in the energy stores of the body. Recently there have been great improvements in techniques with which to measure human body composition, and by direct calorimetry the heat output of patients over 24 h can be measured with an error of only about 1 per cent (Garrow, 1978). It is easy to show that some obese patients have a much lower resting metabolic rate than others of the same sex and weight. While almost everyone shows a decrease in metabolic rate when food intake is reduced, this decrease is much more marked in some patients than in others. Figure 15.4 is taken from the data of Garrow and Warwick (1978). Among 27 obese women admitted to our research ward, there was a highly significant correlation between body weight and resting metabolic rate measured by indirect calorimetry. However, there is also a large scatter about the regression line: if we consider only the open symbols, which show weight and metabolic rate on admission to the ward, there is a range of about 50 ml O_2 per minute in resting metabolic rate among patients of the same weight. This is equivalent to about 350 kcal (1.5 MJ) daily resting energy expenditure.

It is reasonable to suppose that those obese patients with a low metabolic rate for their weight were likely to be those in whom the energy-wasting 'futile cycles'

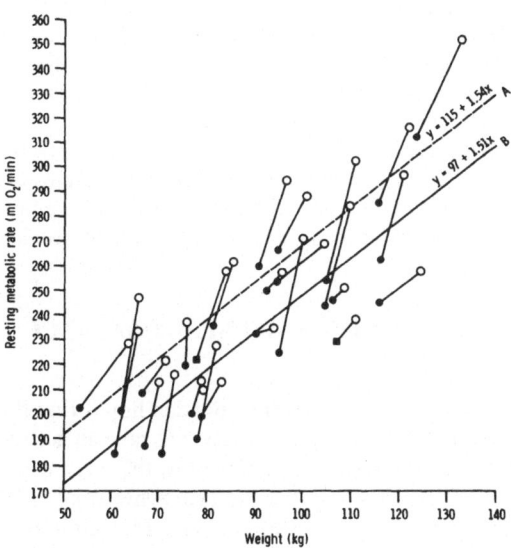

Figure 15.4   Resting metabolic rate and body weight in obese women before (open symbols) and after (filled symbols) a period of 3 weeks on a diet supplying 800 kcal (3.4 MJ) daily (data of Garrow and Warwick, 1978).

were absent or suppressed, and that these patients would not show a further decrease in metabolic rate when they were kept strictly to a diet supplying only 800 kcal (3.4 MJ) in the research ward. In fact, as figure 15.4 shows, there is some suggestion that the patients with initially low metabolic rates show a slightly smaller decrease than those with initially high metabolic rates, but there is no very clear relationship between these two characteristics. Certainly the decrease in metabolic rate is not related to the original weight, since the regression lines for values on admission and discharge values are almost exactly parallel, indicating that there is on average a decrease of about 18 ml O_2 per minute in these patients, and that this decrease is no greater in the very heavy patients than in those who are only slightly overweight.

IMPAIRED THERMOGENESIS AS A CAUSE OF OBESITY

Some obese rodents fail to raise their metabolic rate normally in response to cold exposure, and it is tempting to suggest that impaired cold-induced thermogenesis is a factor in human obesity also. When lightly clothed subjects are exposed to severe cold, lean subjects react more than fat ones (Buskirk *et al.*, 1963; Quaade, 1963; Wyndham *et al.*, 1968), but in experiments which stop short of heroism there is no difference in thermogenesis, although differences in insulation can be shown between fat and thin subjects (Jéquier *et al.*, 1974). There have been conflicting reports about the thermogenic response to food in fat and lean human subjects, but at least one report (Pittet *et al.*, 1976) found that obese subjects showed a smaller increase in metabolic rate after food than lean controls. The differences are small, but in the regulation of energy balance small differences may be very important over a long period of time. It seems that the metabolic response which increases energy expenditure in response to cold is not the same as that which is induced by food, since men who are both fed and chilled show an increase in metabolic rate which is equal to the sum of the response to each stimulus separately (Buskirk *et al.*, 1960). However, it is not clear exactly what energy-consuming reactions are responsible for each of these thermogenic responses. This is a question to which animal experimentalists may be able to provide an answer.

INACTIVITY AND ENERGY BALANCE

The main message of this paper is that the clinician hopes that those who are expert with various animal 'models' of obesity will take an active interest in applying their models to the clinical problem. However, there is a danger that evangelical fervour may lead research workers to find excessive similarities between the laboratory and clinical situation. Figures 15.5 and 15.6 show an example of this danger.

Figure 15.5 shows a most interesting effect observed by Mayer *et al.* (1954). They found that normal adult female rats not only gained more weight, but also actually ate more food ($P < 0.05$) if they were given no exercise than if they were

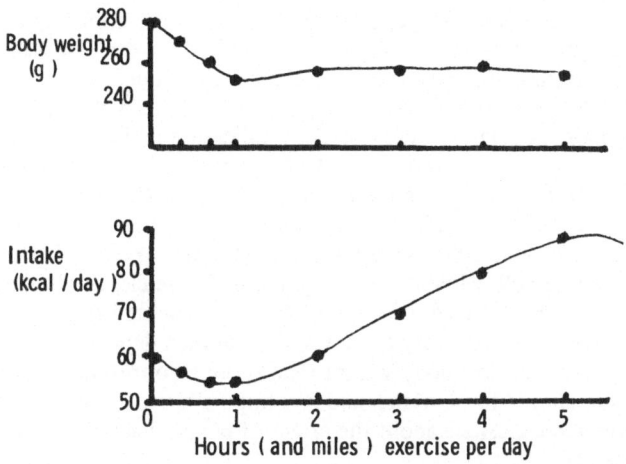

Figure 15.5 Body weight and energy intake in normal adult rats exercised on a treadmill at 1 mile per hour for 0–5 h per day (data of Mayer *et al.*, 1954).

made to run on a treadmill at one mile per hour for 1 h/day. If the duration of exercise was increased beyond 1 h, the food intake of the rats increased to a maximum with 5 h exercise, and thereafter declined as the rats became exhausted. They therefore concluded that at very low levels of activity, which they termed the 'sedentary range', intake was inversely related to activity, while at higher levels (the 'normal activity' range) it was directly related to activity, and set out to test this hypothesis on workers in a jute mill in West Bengal (Mayer *et al.*, 1956). The results of the second study are shown in figure 15.6, which is gratifyingly similar to

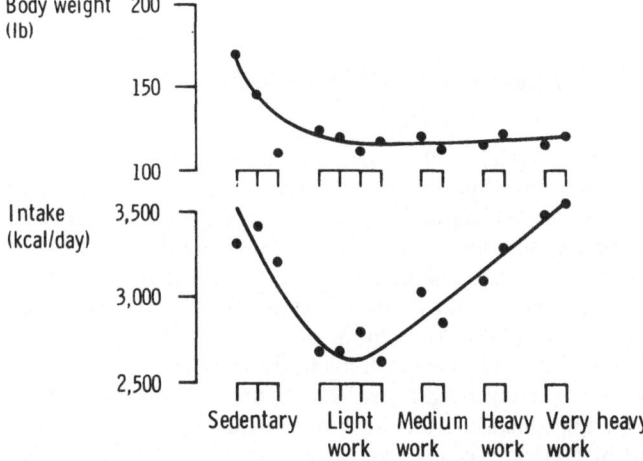

Figure 15.6 Body weight and energy intake of workers in a jute mill in West Bengal (data of Mayer *et al.*, 1956).

figure 15.5. They conclude (Mayer *et al.*, 1956) that the relationship between activity and intake in the 'normal activity' and 'sedentary zone' in man is similar to that previously found in experimental animals. This conclusion is still frequently quoted, but it does not withstand close examination: indeed, the story is too good to be true. A more detailed criticism of this publication is given elsewhere (Garrow, 1978) but the flaw can be seen in figure 15.6. A group of workers classed as sedentary were assessed to have a dietary intake equal to those engaged in heavy work, and about 700 kcal (3 MJ) per day greater than those in light work. The 'sedentary' workers were either supervisors, stall-holders or clerks living in the premises of the jute mill, while those in 'light work' were clerks who had to come in to work from some distance. It is upon the difference between these two groups that the evidence for a 'sedentary zone' response in man depends. The clerks living on the premises had a *lower* body weight than those who travelled to work, despite their higher alleged food intake, and it is impossible to reconcile this fact with the authors' conclusions about the effect of physical activity on food intake or energy balance.

This study has been discussed in some detail, and critically, because it would be very useful to know whether the authors' conclusions are correct or not. The publications are now more than 20 years old, and it is time they were repeated. Do *all* rats eat more when they are sedentary? Is this effect found in both lean and obese strains, and with all types of food? Can the effect be replicated, since the effect shown in figure 15.5 is small and only just statistically significant? If it is replicable in rodents, what is the mechanism responsible for the phenomenon, and is it practicable to look for the effect in man? I know of only one carefully controlled study (Warnold and Lenner, 1977) of human subjects on measured *ad libitum* food intake, under both sedentary and light work conditions, and this does not show the 'sedentary zone' effect postulated by Mayer *et al.* (1956), but it was not specifically designed to investigate this point. We have very little reliable information about the role of physical activity in the regulation of energy balance in man.

HUMAN MODELS OF HUMAN OBESITY

Spontaneous human obesity, such as that illustrated in figure 15.1, is obviously a heterogeneous condition. It is unrealistic to hope that any animal model will do more than explain part of the problem, but any assistance at all would be very welcome. I firmly believe that when we come to understand the mechanism controlling energy balance in man, it will be revealed as a many-tiered hierarchy. The weight curves of laboratory animals are relatively smooth, whereas those in man are curiously notched and distorted. In the population of Framingham, who were followed over a period of 18 years with medical examinations every 2 years, the average fluctuation of body weight was 10 kg (Gordon and Kannel, 1973), and it is hard to escape the conclusion that in some individuals body weight is kept fairly constant by conscious effort when change in weight reaches unacceptable proportions. This makes it very difficult to test simple hypotheses about the control system, since the only person who can know if change in body weight is spontaneous or deliberately induced, is the individual himself.

Figure 15.7 is taken from a paper by Forbes and Reina (1970) which was primarily concerned with changes in lean body mass with increasing age. It is at least as interesting to observe the fluctuations in adipose tissue mass in these two individuals. In subject A.B. adipose tissue increases at the rate of about 2 kg per year,

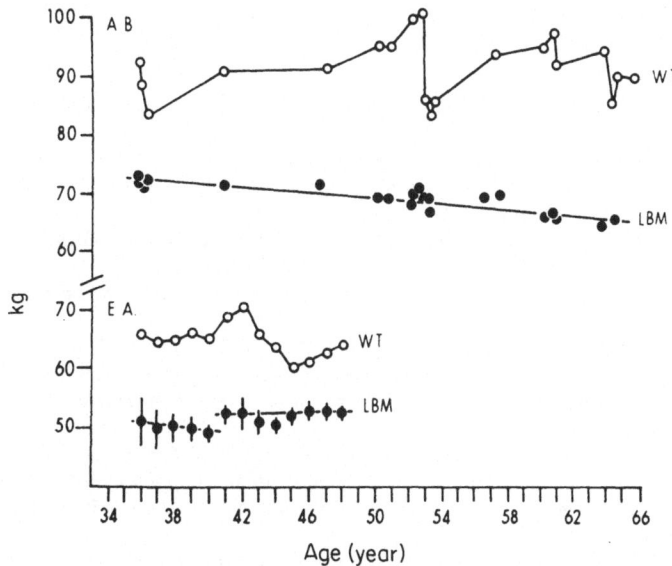

Figure 15.7 Body weight (open symbols) and lean body mass (filled symbols) in two individuals over many years (data of Forbes and Reina, 1970). Reproduced with permission.

and long-term regulation is achieved by short periods of rapid weight loss. In subject E.A. fluctuations of similar amplitude are seen, but the rate of decease is much less rapid than in A.B. Is there any animal model which would behave in this way? If not, it must be important, when trying to fit together animal models and the clinical problem of human obesity, to recognise factors in human obesity which may be genetically or metabolically determined, and those which are the effects of social or cultural factors which cannot apply in the rodent.

To end this apologia of a clinician's approach, may I suggest that it is profitable to combine a study of animal models of human obesity with a study of human models also. Why does body weight in man fluctuate by about 10 kg, and what circumstances predispose to increasing or decreasing weight? These are questions which it is virtually impossible for a clinician to answer by observing his patients, or for a laboratory worker to answer by observing laboratory rodents. However, if every person engaged in research on patients, or animal models of obesity, also observed the factors which influenced his own body weight, I believe we would come closer to answering this important question. The clinician's prejudice was well expressed by Alexander Pope:

Know then thyself, presume not God to scan,
The proper study of mankind is man.

REFERENCES

Ashwell, M. (1973). A survey of patients' views on doctors' treatment of obesity. *Practitioner*, **221**, 653–658

Astwood, E. B. (1962). The heritage of corpulence. *Endocrinology*, **71**, 337–341

Binnie, C. C. (1977). Obesity in general practice. Ten year follow-up of obesity. *J. R. College General Pract.* **27**, 492–495

Borjeson, M. (1976). The aetiology of obesity in children. *Acta Paediatr. Scand.*, **65**, 279–287

Bortz, W. M. (1969). A 500 pound weight loss. *Am. J. Med.*, **47**, 325–331

Bray, G. A. and York, D. A. (1971). Genetically transmitted obesity in rodents. *Physiol. Rev.*, **51**, 598–646

Buskirk, E. R., Thompson, R. H., Moore, R. and Whedon, G. D. (1960). Human energy expenditure studies in the National Institute of Arthritis and Metabolic Diseases Metabolic Chamber. 1. Interaction of cold environment and specific dynamic action. 2. Sleep. *Am. J. Clin. Nutr.*, **8**, 602–613

Buskirk, E. R., Thompson, R. H. and Whedon, G. D. (1963). Metabolic response to cold air in men and women in relation to total body fat content. *J. Appl. Physiol.*, **18**, 603–612

Davidson, S. and Passmore, R. (1969). *Human Nutrition and Dietetics*, 4th ed., Livingstone, Edinburgh, p. 385

Durrant, M. and Mann, S. (1977). Investigations into patient responses to feeding low- and high-energy foods. *Proc. Nutr. Soc.*, **36**, 113A

Faust, I. M., Johnson, P. R. and Hirsch, J. (1976). Noncompensation of adipose mass in partially lipectomized mice and rats. *Am. J. Physiol.*, **231**, 538–544

Forbes, G. B. and Reina, J. C. (1970). Adult lean body mass declines with age: some longitudinal observations. *Metabolism*, **19**, 653–663

Fox, F. W. (1973). The enigma of obesity. *Lancet*, **ii**, 1487–1488.

Garn, S. M. and Clark, D. C. (1976). Trends in fatness and the origins of obesity. *Pediatrics*, **57**, 443–456

Garrow, J. S. (1976). Upbringing, appetite and adult obesity. In *Early Nutrition and Later Development* (ed. A. W. Wilkinson), Pitman Medical., London, pp. 219–228

Garrow, J. S. (1978). *Energy Balance and Obesity in Man*, 2nd edn., North-Holland, Amsterdam

Garrow, J. S. and Warwick, P. (1978). Diet and obesity. In *Diet of Man: Needs and Wants* (ed. J. Yudkin), Applied Science, Barking, pp. 127–144.

Goldrick, R. B., Havenstein, N. and Whyte, H. M. (1973). Effects of caloric restriction and fenfluramine on weight loss and personality profiles of patients with long-standing obesity. *Aust. New Z. J. Med.*, **3**, 131–141

Gordon, T. and Kannel, W. B. (1973). The effects of overweight on cardiovascular disease. *Geriatrics*, **28**, 80–88

Gray, H. and Kallenbach, D. C. (1939). Obesity treatment: results in 212 outpatients. *J. Am. Dietetic Assoc.*, **15**, 239–245

Jéquier, E., Gygax, P.-H., Pittet, P. and Vanotti, A. (1974). Increased thermal body insulation: relationship to the development of obesity. *J. Appl. Physiol.*, **36**, 674–678

Jolliffe, N. and Alpert, E. (1951). The 'performance index' as a method for estimating effectiveness of reducing regimens. *Postgrad. Med.* **9**, 106–115

Kemp, R. (1972). The overall picture of obesity. *Practitioner*, **209**, 654–660

Lane-Petter, W. (1970). Do laboratory animals like eating? *Proc. Nutr. Soc.*, **29**, 335–338

Mayer, J., Marshall, N. B., Vitale, J. J., Cristensen, J. H., Mashayekhi, M. B. and Stare, 1 . J. (1954). Exercise, food intake and body weight in normal rats and genetically obese adult mice. *Am. J. Physiol.* **177**, 544–548

Mayer, J., Roy, P. and Mitra, K. P. (1956). Relation between caloric intake, body weight and physical work: studies in an industrial male population in West Bengal. *Am. J. Clin. Nutr.*, **4**, 169–174

Mellbin, T. and Vuille, J.-C. (1973). Physical development at 7 years of age in relation to velocity of weight gain in infancy with special reference to incidence of overweight. *Br. J. Prevent. Sociol. Med.*, **27**, 223–235

Pittet, Ph, Chappuis, Ph., Acheson, K., de Techtermann, F. and Jéquier, E. (1976). Thermic effect of glucose in obese subjects studied by direct and indirect calorimetry. *Br. J. Nutr.*, **35**, 281–292

Porikos, K. P., Booth, G. and van Itallie, T. B. (1977). Effect of correct nutritive dilution on the spontaneous food intake of obese individuals: a pilot study. *Am. J. Clin. Nutr.*, **30**, 1638–1644

Pudel, V. E. and Oetting, M. (1977). Eating in the laboratory: behavioural aspects of the positive energy balance. *Int. J. Obesity*, **1**, 369–386

Quaade, F. (1963). Insulation in leanness and obesity. *Lancet*, **ii**, 429–432

Sims, E. A. H., Goldman, R. F., Gluck, C. M., Horton, E. S., Kelleher, P. C. and Rowe, D. W. (1968). Experimental obesity in man. *Trans. Assoc. Am. Physicians*, **81**, 153–170

Warnold, I. and Lenner, R. A. (1977). Evaluation of the heart rate method to determine the daily energy expenditure in disease. A study on juvenile diabetics. *Am. J. Clin. Nutr.*, **30**, 304–315

Wilkinson, P. W., Parkin, J. M., Pearlson, J., Philips, P. R. and Sykes, P. (1977). Obesity in childhood: a community study in Newcastle upon Tyne. *Lancet*, **i**, 350–352

Wooley, O. W., Wooley, S. C. and Dunham, R. B. (1972). Can calories be perceived, and do they affect hunger in obese and non-obese humans? *J. Comp. Physiol. Psychol.*, **80** 250–258

Wyndham, C. H., Williams, C. G. and Loots, H. (1968). Reactions to cold. *J. Appl. Physiol.*, **24**, 282–287

Yudkin, J. (1963). Nutrition and palatability with special reference to obesity, myocardial infarction and other diseases of civilization. *Lancet*, **i**, 1335–1338

Name Index*

*First author only given. Reference list page numbers in italic.

250

Subject Index